## Parasitology Research Monographs

Volume 19

**Series Editor**
Heinz Mehlhorn, Department of Parasitology, Heinrich Heine University, Düsseldorf, Germany

This book series "Parasitology Research Monographs" presents carefully refereed volumes on selected parasitological topics. Parasites have an increasing impact on animal and human health in the present times of globalization and global warming. Parasites may be agents of diseases and- often at the same time- vectors of other agents of disease such as viruses, bacteria, fungi, protozoa and/or worms. The growth in knowledge of parasitic physiology, cell structure, biotechnological and genetic approaches, ecology, therapeutic capabilities, vaccination, immunology, diagnosis, transmission pathways and many other aspects of parasitology is increasing dramatically, even in the face of the breakthroughs that have already been made. Reflecting these most recent achievements and the importance of parasites as a threat to human and animal health, the series' broad scope concentrates on particularly hot topics emerging from the scientific community. Chapters offer compact but intense insights into the ongoing research and into the methods and technologies used to control parasites. The volumes in the series build on these topics, and the volume editors are well-known experts in their respective fields. Each volume offers 10 to 20 comprehensive reviews covering all relevant aspects of the topic in focus.

Jian Li • Wei Wang • Heinz Mehlhorn
Editors

# Echinococcus: Control and Elimination of *Echinococcosis* with a Focus on China and Europe

Springer

*Editors*
Jian Li
Hubei University of Medicine
Shiyan, China

Wei Wang
Jiangsu Institute of Parasitic Diseases
Wuxi, China

Heinz Mehlhorn
Heinrich-Heine-University
Düsseldorf, Germany

ISSN 2192-3671 ISSN 2192-368X (electronic)
Parasitology Research Monographs
ISBN 978-3-031-54478-1 ISBN 978-3-031-54479-8 (eBook)
https://doi.org/10.1007/978-3-031-54479-8



This Springer imprint is published by the registered company Springer Nature Switzerland AG
The registered company address is: Gewerbestrasse 11, 6330 Cham, Switzerland

# Preface

Humans and their animals live since thousands of years close together and thus both groups are endangered by life-threatening diseases being transmitted vice versa. Echinococcosis is based on infections by the so-called tapeworms of the genus *Echinococcus*. The present book offers insider knowledge on the situation in China and Europe, since common standards of diagnosis and treatment have been developed and become constantly ameliorated.

| | |
|---|---|
| Shiyan, China | Jian Li |
| Wuxi, China | Wei Wang |
| Düsseldorf, Germany | Heinz Mehlhorn |

# Contents

# About the Editors

**Jian Li** PhD, is Professor and Master supervisor. His research interests encompass the epidemiology, drug resistance, multi-omics (genomics, transcriptomics, proteomics, immunomics, metabolomics, and the microbiome) of parasites, as well as methods of rapid diagnosis using high-throughput immunoscreening platforms and PCR-based tools and test strips.

**Heinz Mehlhorn** has retired from Düsseldorf University, was editor of the Springer journal Parasitology Research for 25 years and was President of the German and World Societies of Parasitology and Protozoology. He has published 50 parasitology books, more than 100 scientific articles and is currently chairman of a German company, which is developing safe products against parasites.

**Wei Wang** is Associate Professor in the Jiangsu Institute of Parasitic Diseases, China. He is an expert in the field of infectious diseases and tropical diseases. His research interests focus on emerging and reemerging infectious diseases, neglected tropical diseases, and drug resistance. He has published more than 150 papers in international and national peer-reviewed journals, including The New England Journal of Medicine, The Lancet Infectious Diseases, Journal of Travel Medicine, Stem Cell Research & Therapy, The Journal of Infectious Diseases, and has published more than 10 book chapters.

# Brief Recognition History of *Echinococcus* Tapeworm

# 1

Xu Wang, Shuai Han, and Jianping Cao

**Abstract**

The cognitive history of *Echinococcus* began with the description of the diseases it caused by doctors more than 2000 years ago, but it was not until the seventeenth century that scholars began to discover the pathogens of the parasite, and in the next 200–300 years, its classification system was continuously debated and revised through morphology and molecular methods. After entering the twenty-first century, through the phylogenetic researches, nine existing valid natural species of *Echinococcus* were ultimately identified and preserved.

**Keywords**

*Echinococcus* · Echinococcosis · Cognitive history · Hydatids · Species

The larva of *Echinococcus*, a tapeworm of the phylum Platyphyllum, parasitizes animals and can cause liver, lung, brain, and other organ lesions, called echinococcosis, which can lead to death in severe cases (Eckert and Thompson 2017). It is a zoonotic parasitic disease that has attracted worldwide attention. Among the various

X. Wang · S. Han · J. Cao (✉)
National Institute of Parasitic Diseases, Chinese Center for Disease Control and Prevention; Chinese Center for Tropical Diseases Research; National Key Laboratory of Intelligent Tracking and Forecasting for Infectious Diseases; Key Laboratory on Parasite and Vector Biology, Ministry of Health; WHO Centre for Tropical Diseases; National Center for International Research on Tropical Diseases, Ministry of Science and Technology, Shanghai, China
e-mail: caojp@chinacdc.cn



J. Li et al. (eds.), *Echinococcus: Control and Elimination of Echinococcosis with a Focus on China and Europe*, Parasitology Research Monographs 19,
https://doi.org/10.1007/978-3-031-54479-8_1

types of echinococcosis, the most common is the cystic echinococcosis (CE) caused by the *Echinococcus granulosus*. Its appearance is characterized by a number of translucent fluid-filled cysts of varying sizes, which appear as space-occupying swelling in the lesion of humans or other animals. Therefore, echinococcosis is also collectively referred to as hydatids.

## 1.1 Some Vague and Trivial Records of Cystic Lesions

Early physicians did not have a clear understanding of echinococcosis. So, we can only correlate them from some of the probable pathologic descriptions they left behind. For example, in ancient Greece, Hippocrates (~460–377 BC), the founder of Western medicine, wrote in his classic *Aphorisms* (VII, 55): "In those who water stuffed lives open into the //omentum, the bell is filled with water, and they die" (Eckert and Thompson 2017). Almost at the same time in ancient China, the *Miraculous Pivot* (475–221 BC), the foundation literature for the formation of the Chinese theoretical system of traditional medicine, described several cases of abdominal swelling, such as (IX, 57): "when you press the abdomen of a patient with your hand, the depressed part will rise with your hand, as if pressed on a bag containing water," and "at the beginning, the abdominal mass is as large as an egg, and with the development of the disease, the mass also gradually increases, and when it is fully formed, the abdominal bulge is like pregnancy, and the course of the disease is long, which can go through several years. When pressed by hand, the lump is hard and can move when pushed" (Jia 2015). Due to the limitations of early observation techniques and methods, the ancient doctors' cognition of echinococcosis was often limited to the description of appearance characteristics, and did not give an essential definition of this disease, so that we cannot conclude that these diseases are what modern scientists call echinococcosis from the perspective of later generations. This dilemma has undergone some changes with the rise and development of anatomy. During the Roman Empire, Aretaeus (50 AD) from Cappadocia described in his work of *De causis et signa acutorum morborum* as "there were many small fluid filled blisters in the patient's abdomen, and some fluid flowed out during needle puncture." In the seventeenth century, scientists began to discover similar pathological features in other animals. For example, Bartholin first observed transparent egg-like lesions in pigs in a slaughterhouse in 1653. Wille recorded in his monograph the presence of grape clusters on the surface of the liver and lungs of livestock in 1675. These descriptions are very similar to those of cystic echinococcosis (CE). However, due to the fact that microscopic observation technology and systematic parasitology have not yet been developed, there have been various opinions on the cause and true nature of echinococcosis. Some ancient scholars even believed that this disease came from the pathological changes of patients' own tissues, possibly from the accumulation of mucus between layer cells, or from the differentiation of blood vessels and their contents (Eckert and Thompson 2017).

## 1.2 An Infectious Disease Caused by Parasites

In 1684, Francesco Redi (1626–1697) discovered the mobile cysticerci in patients in Florence, proposing the hypothesis that this disease may be caused by *Taenia* spp (Grove 1990). tapeworm. Facts have proven that this is a correct direction. The following year (1685), Professor Philip Jacob Hartmann (1648–1707) once again demonstrated in a research at the University of Konigsberg that this cysticerci has the same scolex as a tapeworm, with its tail connected to a spherical bladder, which is believed to be the metacestode of *Taenia hydatigena* and referred to as *Cysticercus tenuicollis*, which is the first report on the morphology of *Echinococcus*. Soon, in 1687, Edward Tyson (1650–1708), a physician at Oxford University, described the characteristics of the motor behavior of *Cysticercus tenuicollis* in his report, called it "lumbrici hydropici," which means that the fluid-filled hydatids in animals are actually worms with independent life forms, providing a new way for continuing to study the origin of echinococcosis (Enigk 1986). In 1760, Peter Simon Pallas (1741–1811) of the University of Leiden described in his medical paper that the hydatids were remarkably similar in sick people and animals, with some small brood capsules attached to the inner walls, redefining the parasites as bladder worms (Enigk 1986). In 1782, Priest John August Ephraim Goeze (1731–1793) of Germany studied the structure of the hydatids through the microscope and discovered a large number of protoscoleces, as well as tapeworm-like scoleces in the inner capsules, confirming their association with the *Taenia* family, and defined it as *Taenia visceral socialis granulosa (*Enigk 1986*)*. In 1786, professor Batsch summarized the previous research results and named the parasite *Hydatigena granulosa*, which causes cystic lesions in internal organs in humans and animals, and usually completes its life cycle with domestic animals and dogs as intermediate and definitive hosts (Romig et al. 2015). In 1800, Zedder described the cysts in sheep and classified them into genus *Polycephalus*. Finally, Dr. Carl Asmund Rudolphi (1771–1832) classified this species as a new genus *Echinococcus* in 1801, formally included in the zoology classification, and has been used up to now (Rudolphi 1801).

## 1.3 Multiplex Echinococcosis and *Echinococcus*

### 1.3.1 Dualism of CE and AE

According to previous research, cysts formed by the larvae of *E. granulosus* are generally unilocular hydatid, but in some patients, doctors have discovered some abnormalities. In 1852, Ludwig Buhl (1816–1880), a pathologist in Germany, found a rare tumor in the human liver, initially described it as "alveolarcolloid," and speculated that it might be caused by tissue canceration (Eckert and Thompson 2017). In 1855, Rudolf Ludwig Virchow (1821–1902), a professor of pathology at the University of Würzburg, described in detail the infiltrating and growing multilocular hydatid, which grew like a tumor. Unlike the unilocular hydatid previously seen in humans and livestock, he believes that this lesion is caused by parasite invasion of surrounding tissue, rather than malignant hyperplasia of the patient's tissue itself (Virchow 1856). In 1863, German scientist Rudolf Leuckart proposed to refer to this parasite within multilocular hydatid as *Echinococcus multilocularis* (Leuckart 1862). However, due to the limitations of parasitic morphological techniques at the time, there was a heated debate at that time over whether the previously discovered unilocular hydatid and the later reported multilocular hydatid were caused by the same parasite or by different species. Many scholars, including R. Virchow, initially believed that both hydatids were caused by *E. granulosus*, and that the differences in pathological features were dependent on patient's own specificity. In contrast, professor A. Morin of the University of Bern, Switzerland, is a dualist in his paper, arguing that multilocular hydatid is caused by a new pathogen. His view was supported by Adolf Poselt (1867–1936), a professor at the University of Innsbruck in Austria later, and confirmed by the morphological, pathological, histological, and clinical manifestations of infection in dogs conducted by Posselt in 1901 (Eckert and Thompson 2017). The cause of multilocular hydatid was identified as a new pathogenic species, which was the worm of *E. multilocularis* mentioned earlier. In the 1950s, Robert L. Rausch and E.L. Schiller found a similar hydatid lesion in the lungs of Inuit people in Alaska, and related it to the alveolar metacestodes from local voles and the *Echinococcus* eggs in Arctic fox. The parasite was named *Echinococcus sibiricensis* (Rausch and Schiller 1951; Rausch and Schiller 1954; Rausch and Schiller 1956). However, in 1955, Johannes Vogel (1900–1980), a German helminthologist, found there is no significant difference between *E. sibiriensis* and *E. multilocularis* after conducting morphological analysis of echinococcosis infections in voles, foxes, dogs in Europe and Alaska. Therefore, the combination of the two kinds of *Echinococcus* was defined as the same species, officially named *E. multilocular*is Leuckart, 1863, could cause the alveolar echinococcosis (AE) of human but with a completed life cycle by small rodents (intermediate hosts) and foxes (definitive hosts) in the wild (Vogel 1955; Vogel 1957a). Because the tapeworm has a wide variety of morphological and biological characteristics, some scholars even proposed to separate it into a new genus, *Alveococcus* Abuladze, 1959, but this proposal was not recognized by the academic community (Lukashenko and Zorikhina 1961). So far, this parasite still belongs to *Echinococcus*. Thus, the pathogen discovery and species dispute of the major echinococcoses (AE and CE) and *Echinococcus* (*E. multilocularis* and *E. granulosus*) in Eurasia came to an end.

### 1.3.2 Echinococcus in the New World (America)

At the same time, in the New World, i.e., the American continent, the process of understanding *Echinococcus* is also taking place simultaneously. With the exception of *E. granulosus* and *E. multilocularis*, which are found in North America and common to Eurasia, the earliest records of endemic species can be traced back to 1836 (Eckert and Thompson 2017). Austrian scientist Johann Naterer (1781–1843) collected some parasites in a cougar (*Puma concolor*) in Brazil (Tappe et al. 2008); In 1863, Karl Moritz Diesing (1800–1867) classified these parasites as a new species,

*Taenia oligarthra*, preserved in the Hofmuseum in Vienna (Luhe 1910). In 1909, Dr. Maximilian F.L. Luhe (1870–1916) of the University of Konigsberg carefully reviewed previous materials and believed that this parasite was closely related to the genus *Echinococcus*. In 1914, Brumpt (1877–1951) and Joyeux (1881–1966), French parasitologists, discovered some multilocular metacesteodes from agoutis in Brazil, temporarily named it as *Echinococcus cruzi* and speculated to be related to the previous *T. oligarthra* (Brumpt and Joyeux 1924). In 1926, professor Thomas W.M. Cameron (1894–1980) formally classified this parasite as *Echinococcus* based on the autopsy of a jaguarundi (*Felis yagouaroundi*) in London Zoo, namely *E. oligarthrus* (Cameron 1926). Eventually, it was confirmed that the species was transmitted and propagated with cats (cougar, jaguar, jaguarundi, etc.) and rodents (opossum agoutis, pacas, opossums, etc.) from South America as the definitive host and intermediate host. At present, sporadic cases of human infection have been reported, and the characteristics of the single cyst caused by *E. oligarthrus* are also known as unicystic echinococcosis (UE) (Eckert et al. 2001).

### 1.3.3 Echinococcus in Africa

On the other hand, in the African continent, the earliest record of *Echinococcus* is from Gough's note in 1908, which described *E. granulosus* obtained from silver-backed jackal (*Canis mesomelas*) in South Africa and successfully infecting sheep (Hüttner and Romig 2009). The native species of Africa was discovered in 1926 by Thomas W. M. Cameron in the African wild dog (*Lycaon pictus*) in South Africa and named it as *Echinococcus logimanubrius* because of its unique small hook with long-handle morphology (in the same publication, Cameron also mentioned an *Echinococcus minimus* found in *Canis lupus* in Macedonia) (Cameron 1926). In 1934, Ortlepp also discovered another species with a different large hook shape, *Echinocccus lycaontis*, from *L. pictus* (in that article, Ortlepp renamed a tapeworm collected from foxes in the United Kingdom as a new species of *Echinococcus cameroni*) (Ortlepp 1934). Today, however, the population of African wild dog populations are in rapid decline and threatened with extinction, making further research difficult (Hüttner and Romig 2009). A major discovery was in 1937 when Ortlepp continued to discover a new species of *Echinococcus* tapeworm in the intestines of South African lions (*Panthera leo*), and recognized as a separate species based on the morphology of obvious rough folds on the rostellar hooks and characteristics of the feline as the nature definitive host, it was confirmed as an independent species, named *Echinococcus felidis* (Ortlepp 1937). Subsequently, scientists successively found signs of infection of *E. felidis* in wild animals such as zebras (*Equus quagga*, *Equus zebra*), warthogs (*Phacochorus* sp.), and bushpig (*Potamochoerus larvatus*, *Potamochoerus porcus*) (Hüttner and Romig 2009). It has been identified as an *Echinococcus* species with independent transmission ecological chain among wildlife in African. Interestingly, in 1943, Lopez Neyra and Soler Planas first described the presence of *E. felidis* and *E. lycaontis* in domestic dogs in southern Spain, as well as other two new species, *E. ortleppi* and *E. intermedius* (Lopez-Neyra and Soler Planas 1943).

## 1.4 A Large and Complex Genus Group

### 1.4.1 Morphological Method

With the recognition of the "taxonomic status" of the genus *Echinococcus* Rudoiphi, 1801, and the common characteristic of the dual host life-cycle, scientists around the world have successively reported various newly discovered species, but most of them are identified based on the description of morphological structure (Rausch 1953). However, some scholars have proposed that these morphological differences (such as the size or length of the rostellar hooks) may be within the range of normal individual differences, and it is necessary to reevaluate these features as a basis for species differentiation (Hüttner and Romig 2009). In 1953, Robert Rausch conducted morphological analysis based on stored sample materials and believed that the hook size of adults with mature eggs in the uterus was the only reference for species classification. In addition, the shape of the hook, the number of testes, and the species specificity of the host were also important reference criteria. Therefore, seven species of *Echinococcus* (*E. granulosus* Batsch, 1786; *E. oligarthrus* Dieing, 1863; *E. longimanubrius* Cameron, 1926; *E. minimus* Cameron, 1926; *E. cameroni* Ortlepp, 1934; *E. lycaontis* Ortlepp, 1934; *E. felidis* Ortlepp, 1937) were reclassified for biological classification, and proposed *E. longimanubrius* and *E. minimus* are the species inquirendae to be studied, and it is believed that they may be *E. granulosus*; *E. cameroni* is also considered to be conspecific with *E. granulosus*; but there was no response to *E. intermedius* and *E. ortleppi* for evaluation. Combining with previously identified *E. multilocularis*, a total of six species are considered valid (Rausch 1953). In 1963, Rausch and Nelson reviewed their previous classification studies and concluded that there were some omissions in their early conclusions, and collected *E. granulosus* isolates from all over the world (Britain, Germany, and Yugoslavia in Europe, Alaska in North America, Siberia in Asia, Australia in Oceania, and Kenya in Africa), as well as some *Echinococcus* previously reported, such as *E. cameroni*, *E lycaontis*, *E. intermedius*, and *E. ortleppi* (Rausch and Nelson 1963). According to the systemic morphological characteristics, it was concluded that *E. cameroni*, *E. intermedius*, *E. longimanubrius*, *E. lycaontis*, *E. minimus*, and *E. ortleppi* were of the same species as *E. granulosus*, replacing them as independent valid species. And ultimately, only five nature species were left, including three species that can be clearly distinguished by morphology (*E. granulosus*, *E. multilocularis*, and *E. oligarthrus*), and two species to be confirmed: one is *E. felidis* because of the uniqueness of felines as terminal hosts rather than morphological differences (Rausch and Nelson 1963); another one is *Echinococcus patagonicus*, which was reported by Szidat in 1960 and found in the intestine of the foxes in Argentine (Szldat 1960). Due to the lack of sufficient information, *E.*

*patagonicus* needs to be further identified, but is presumed to be closely related to *E. granulosus* (Rausch and Nelson 1963).

### 1.4.2 Naming of Subspecies

In 1960, according to the high incidence of echinococcosis among Indians in northwestern Canada, Cameron successfully identified a species of *Echinococcus* that only found cysts in the lungs of various deer (wolves are the natural host for the adults, but dogs are also susceptible to infection), and the anatomical structures of both the adults and the larvae were different from those previously found. The *Echinococcus* was found mainly in the high latitudes of North America and Eurasia, but is still considered to be a variant of *E. granulosus*, so Cameron named it as *Echinococcus granulosus* var. *canadensis* (or *Echinococcus granulosus canadensis*) (Cameron 1960). Thus, Rausch and Nelson realized that some morphological characteristics of the adult stage may be due to collection and preservation methods, while differences in the larval stage may be influenced by different hosts, and that these so-called species may actually be only subspecies or synonymous species of valid natural species (Rausch 1967). In this period, some scholars have classified the isolates of *E. granulosus* found from native wolves, moose, and other deer in North America as *E. granulosus orealis* Sweatman, 1963 (Sweatman and Williams 1963); isolates obtained from silverback jackals infected by the larvae inside sheep in African were described as *E. granulosus africanus* Verster, 1965 (Verstr 1965); isolates collected from livestock in New Zealand was defined as *E. granulosus newzealandensis* Verster, 1965 (Verstr 1965); isolates found in horses in the United Kingdom was identified as *E. granulosus equinus* Williams and Sweatman, 1963 (Williams and Sweatman 1963); the *E. sibiricensis* Rausch and Schiller, 1954, previously discovered in North America, redefined as *E. multilocularis sibiricensis* Vogel, 1957 (Vogel 1957b); and isolates of *E. multilocularis* found in sheep in Kazakhstan was identified as *E. multilocularis Kazakhensis* Shul'ts, 1961 (may actually be *E. granulosus*) (Shul'ts 1962). In addition, Lothar Szidat (1892–1973) described two new species of *Echinococcus* in South America, *Echinococcus cepanzoi* and *Echinococcus pampeanus* (Szidat 1971). However, according to Verster's revised classification standard of *Echinococcus* in 1965, number of adult segments, arrangement of mature segments, number and distribution of testis, and host preference were also used as the basis for division. Finally, only *E. granulosus* and *E. multiloculus* retained their independent classification status as natural species. *E. oligarthra* is considered to be the species inquirendae, and all other species are considered to be congeners or subspecies of these three species (Verstr 1965). According to this approach, *E. patagonicus* and *E. cepanzoi* were identified as *E. granulosus*, while *E. pampeanus* was same with *E. oligarthrus* (Eckert and Thompson 2017).

It is worth mentioning that in 1970, another new species of *Echinococcus* with distinct morphological characteristics from native America was isolated by Rausch and Bernstein from bush dog (*Speothos venaticus*) captured from Ecuador and kept at the Los Angeles Zoo. It is named after Professor Johannes Vogel as *Echinococcus vogeli* in honor of his remarkable contribution to the study of *Echinococcus* (Rausch and Bernstein 1972). Researches have shown that *E. vogeli* completes its life cycle primarily through food chain relationships between bush dogs (terminal hosts) and paca (intermediate hosts), which can cause a predominance of polycystic echinococcosis (PE) in humans (Rodrigues-Silva et al. 2002).

### 1.4.3 Concept of Strains (Genotypes)

The proposal of subspecies was of great help to the classification of *Echinococcus* with certain scientific significance, but it also causes some problems. Due to the deviation of sampling, preservation and measurement methods of observers, scholars proposed a lot of subspecies. According to statistics, there were more than 85 species or subspecies of *Echinococcus* with binomial or trinomial Latin names at that time, which caused some confusion in the taxonomy of *Echinococcus* (Romig et al. 2015). In 1967, according to the definition of subspecies (part of the population of a species inhabits a local geographic area within its distribution range is affected by the living environment in the area leading to specific changes in morphological structure or physiological function, which is called a subspecies) given by Ernst Walter Mayr, a famous evolutionary biologist (Mayr 1963), Rauch equated different host combinations of *Echinococcus* with different geographical habitat divisions of free living animals. He believed that different tapeworms will form a certain isolation in the stable relationship of food chains between hosts, and then evolve into different subspecies. In this view, he refuted the previous concept of various subspecies of *Echinococcus*, because many of the so-called subspecies coexist within the same biological link of a clear predator (final host) – prey (intermediate host), and do not form significant reproductive isolation (Rausch 1967). However, due to the limitations of research technology at the time, it could not provide more proposals to revise taxonomy of *Echinococcus*. Therefore, some scholars have proposed to temporarily classify all these taxa in the form of strains within the *E. granulosus* complex, called *E. granulosus* sensu lato (Bowles et al. 1992). In the 1990s, Bowles used the emerging DNA sequencing technology to test the sequence of some mitochondrial gene fragments (*cox*1 and *nad*1) of *Echinococcus* from different host sources and summarized the genotypes of eight strains of *E. granulosus*: the G1–G8 genotypes represent the common sheep strain, Tasmania sheep strain, India buffalo strain (buffalo strain), UK horse strain (horse strain), Holland bovine strain (cattle strain), African camel strain (camel strain), Poland pig strain (pig strain), and USA cervid strain, respectively; there are also two strains (genotypes) of *E. multilocularis*: the North American strain of the M1 genotype and the European strain of the M2 genotype (Bowles et al. 1992; Bowles and McManus 1993; Bowles et al. 1995). In 1997, Scott identified a new genotype in human echinococcosis cyst by molecular method and named it G9 genotype or Poland human strain (human strain) (Scott et al. 1997), but this strain was not accepted and was later considered as a microvariant of G7, so it was also called pig variant strain (or human–pig strain) (Kedra et al. 1999). Until the end of the twentieth century, it was recognized that there were four valid species of *Echinococcus*, namely *E. granulosus* with nine genotypes/strains (including the lion strain, also known as *E. felidis*), *E. multilocularis* with two genotypes/strains, *E. oligarthrus*, and *E. vogeli (*Eckert et al. 2001*)*.

## 1.5 Increasingly Defined Population Relationships

### 1.5.1 New Genotypes, New Species

In 2003, Lavikainen collected five isolates of *E. granulosus* from four reindeer and a moose in northeastern Finland. Through DNA sequencing, it was found that the genotype of the isolates was closely related to the G5 and G7 genotypes, but also had some similarities with G8. This isolate is considered to be a unique and previously undocumented new genotype, named G10 or Fennoscandian cervid strain (Lavikainen et al. 2003). So far, a total of 11 genotypes/strains have been reported in the *E. granulosus* system (Romig et al. 2015):

G1/common sheep strain;
G2/Tasmanian sheep strain;
G3/buffalo strain;
G4/horse strain;
G5/cattle strain;
G6/camel strain;
G7/pig strain;
G8/American cervid strain;
G9/variant pig strain;
G10/Fennoscandian cervid strain;
lion strain/*E. felidis*.

In 2005, Xiao discovered and named a native species in the Qinghai-Tibet Plateau of China in East Asia. By combining more abundant morphological, molecular, ecology, geographical distribution, and other methods, he found that it has interspecific differences, and named it *Echinococcus shiquicus* after Shiqu County where it was first found. *E. shiquicus* is phylogenetically closely related to *E. multilocularis* and has overlapping ecological niches with the latter, so it has been previously mistaken for *E. multilocularis* or one of its subspecies (Xiao et al. 2006). *E. shiquicus* mainly circulates between its terminal host, the Tibetan fox (*Vulpes ferrilata*), and its intermediate host, the pika (*Ochotona curzoniae*), but it is not clear whether it can cause disease in humans (Xiao et al. 2006). Subsequently, in 2007, Tang discovered a kind of new tapeworm from *Vulpes corsac* in the Hulunbuer Grassland in northeast China. She believed that its morphology, especially the shape of uterus, was different from *E. multilocularis*, and named it *Echinococcus russicensis* (Tang et al. 2007). However, this discovery still lacks sufficient evidence to establish the status of independent species. In 2009, Nakao analyzed the mitochondrial genes and found that there was a significant separation of isolates of *E. multilocularis* from different regions, and divided the existing of *E. multilocularis* into four genotypes, namely Asian type, European type, North American type, and Mongolian type. It replaces a variety of previously described species or subspecies related to *E. multilocularis*, including *E. sibiricensis* Rausch and Schiller 1954 and *E. russicensis* Tang 2007 (Nakao et al. 2009).

### 1.5.2 Phylogenetic Species

After more than 20 years of research and accumulation on the epidemiology, biology, and geographical distribution of *E. granulosus*, with more nuclear and mitochondrial gene loci have been sequenced and analyzed, the previous method of describing *Echinococcus* using genotypes or strains has great limitations and some contradictions (Romig et al. 2015). On the one hand, the genetic distance and relationship between genotypes are not balanced, and some genotypes (e.g., G1 and G3) are so similar that they can only be distinguished at the information level of the mitochondrial genome, but should only be treated as single genotype in taxonomy, and also shown more distant relationships with other genotypes (e.g., G4 or G5), while genotype such as G2 have been considered invalid (Kinkar et al. 2017). On the other hand, naming strains (genotypes) after animals can often cause some confusion. For example, the common sheep strain is actually the most mainstream strain found in the world, spread almost all over the world, and can infect a wide range of herbivores or livestock such as cattle, sheep, pigs, camels, and kangaroos as its intermediate host, while the horse strain parasitizes almost exclusively in horses (Rahman et al. 2014). In 2013, Nakao introduced the concept of phylogenetic species, combined with the phylogenetic analysis of nuclear and mitochondrial genes, and revised all the recognized species and genotypes (strains) of *Echinococcus* into nine naturally valid species, namely *E. granulosus* (combining G1 and G3 of the previous *E. granulosus*), *E. equinus* (the previous G4), *E. ortleppi* (the previous G5), and *E. canadensis* (combined with the previous G6, G7, G8, and G10), *E. felidis* (the previous lion strain), *E. multilocularis*, *E. shiquicus*, *E. oligarthrus*, and *E. vogeli* (Nakao et al. 2013). This classification version was accepted by most of the later scholars and is still in use today. However, in 2014, Lymbery applied the concept of evolutionary species (a single group of organisms with a shared evolutionary trajectory) and proposed reinstating the use of *E. intermedius*, *E. canadensis*, and *E. borealis* to refer to the G6/G7, G10, and G8 as separate species, respectively (Lymbery et al. 2015). This classification is consistent with mitochondrial data analysis and may become an acceptable nomenclature in the future, given the relationship and transmission chain in hosts of the three. In order to accomplish this work, it is first necessary to prove that the genetic characteristics can be stable among the three without being affected by interspecific hybridization. Perhaps obtaining more nuclear loci data from the isolates can help solve this issue (Romig et al. 2015).

## 1.6 Summary

According to a 2015 estimate by the WHO Foodborne Disease Burden Epidemiology Reference Group (FERG), echinococcosis leads to 19,300 deaths and approximately 871,000 disability-adjusted life years (DALYs) globally each year. The annual global burden of AE and CE are approximately 666,000 DALYs with 18,200 cases and 184,000 DALYs with 188,000 new cases, respectively. Moreover, echinococcosis caused up to US$ 2 billion in economic losses to the global livestock industry every year (Fu et al. 2021; Cardona and Carmena 2013). Echinococcosis has been a highly harmful but neglected infectious disease, and now we may need concerted efforts and increased cooperation on a global scale to control. With the passage of time, the scientific and physicians community's understanding of *Echinococcus* and echinococcosis is gradually deepening, but there are still many questions that need to be further explored. The taxonomic study is only a small part of *Echinococcus* researches, but it carries a high degree of summary and induction of the biological law of *Echinococcus*, and lays a foundation for more far-reaching ecology, epidemiology, diagnosis and treatment, prevention, and control. It took more than 2000 years to go from discovering the *Echinococcus* and artificially expanding it into a large and cumbersome complex to radically simplifying its taxon, finally to using molecular methods to separate the independent species with a never-ending explore. Finally, we would like to state that through the process of literature search, we have found that there are very few relevant materials in China. So, this article will also be rewritten into a review in China and published in Chinese journals for easy access by native Chinese readers.

## References


Bowles J, Blair D, McManus DP (1992) Genetic variants within the genus *Echinococcus* identified by mitochondrial DNA sequencing. Mol Biochem Parasitol 54(2):165–173

Bowles J, Blair D, McManus DP (1995) A molecular phylogeny of the genus *Echinococcus*. Parasitology 110(Pt3):317–328

Bowles J, McManus DP (1993) NADH dehydrogenase 1 gene sequences compared for species and strains of the genus *Echinococcus*. Int J Parasitol 23(7):969–972

Brumpt E, Joyeux C (1924) Description d'un nouvel echinococcque: *Echinococcus cruzi* n.sp. Ann Parasitol 2(3):226–231

Cameron TWM (1926) Observations on the genus *Echinococcus* Rudolphi 1801. J Helminthol 4:13–22

Cameron TWM (1960) The incidence and diagnosis of hydatid cyst in Canada. *Echinococcus granulosus var. canadensis*. Parassitologia 2(3):381–390

Cardona GA, Carmena D (2013) A review of the global prevalence, molecular epidemiology and economics of cystic echinococcosis in production animals. Vet Parasitol 192(1–3):10–32

Eckert J, Gemmell MA, Meslin F, Pawłowski ZS (2001) WHO/OIE manual on echinococcosis in humans and animals: a public health problem of global concern. World Health Organization, Paris

Eckert J, Thompson RC (2017) Historical aspects of echinococcosis. Adv Parasitol 95:1–64

Enigk K (1986) Geschichte der Helminthologie im deutschsprachigen Raum. New York, Stuttgart

Fu MH, Wang X, Han S, Guan YY, Bergquist R, Wu WP (2021) Advances in research on echinococcoses epidemiology in China. Acta Trop 219:105921

Grove DI (1990) A history of human helminthology. CAB International, Wallingford

Hüttner M, Romig T (2009) *Echinococcus* species in African wildlife. Parasitology 136(10):1089–1095

Jia W (2015) Echinococcosis. China Agriculture Press, Beijing

Kedra AH, Swiderski Z, Tkach VV, Dubinsky P, Pawlowski Z, Stefaniak J, Pawlowski J (1999) Genetic analysis of *Echinococcus granulosus* from humans and pigs in Poland, Slovakia and Ukraine. A multicenter study. Acta Parasitol 44:248–254

Kinkar L, Laurimäe T, Sharbatkhori M, Mirhendi H, Kia EB, Ponce-Gordo F, Andresiuk V, Simsek S, Lavikainen A, Irshadullah M, Umhang G, Oudni-M'rad M, Acosta-Jamett G, Rehbein S, Saarma U (2017) New mitogenome and nuclear evidence on the phylogeny and taxonomy of the highly zoonotic tapeworm *Echinococcus granulosus* sensu stricto. Infect Genet Evol 52:52–58

Lavikainen A, Lehtinen MJ, Meri T, Hirvelä-Koski V, Meri S (2003) Molecular genetic characterization of the Fennoscandian cervid strain, a new genotypic group (G10) of *Echinococcus granulosus*. Parasitology 127:207–215

Leuckart R (1862) Die menschlichen Parasiten und die von ihnen herruhrenden Krankheiten. Winter, Leipzig

Lopez-Neyra CR, Soler Planas MA (1943) Revision del genero *Echinococcus* Rud. y description de una especie nuéva Parasita intestinal del porro en Almeria. Revista Ibérica de Parasitologia 142:169–201

Luhe M (1910) Cystotanien sudamerikanischer Feliden. Zoologische Jahrbücher 12:687–710

Lukashenko NP, Zorikhina VL (1961) Epidemiology of alveolar echinococcosis in central areas of the Barabinsk forest-steppe in the Novosibirsk region. Meditsinskaya Parazitologiya i Parazitarnye Bolezni 30(2):159–168

Lymbery AJ, Jenkins EJ, Schurer JM, Thompson RC (2015) *Echinococcus canadensis, E. borealis*, and *E. intermedius*. What's in a name? Trends Parasitol 31(1):23–29

Mayr E (1963) Animal species and evolution. Harvard University Press, Cambridge

Nakao M, Lavikainen A, Yanagida T, Ito A (2013) Phylogenetic systematics of the genus *Echinococcus* (Cestoda: Taeniidae). Int J Parasitol 43(12–13):1017–1029

Nakao M, Xiao N, Okamoto M, Yanagida T, Sako Y, Ito A (2009) Geographic pattern of genetic variation in the fox tapeworm *Echinococcus multilocularis*. Parasitol Int 58(4):384–349

Ortlepp JR (1934) *Echinococcus* in dogs from Pretoria and vicinity. Onderstepoort J Vet Sci Anim Ind 3:97–108

Ortlepp JR (1937) South African helminths. Onderstepoort J Vet Sci Anim Ind 9:311–336

Rahman WA, Elmajdoub LE, Mohd Nor SA, Wajidi MF (2014) A revisit to the infamous zoonotic echinococcosis: a molecular review. J Veterinar Sci Technol 5(5):208

Rausch RL (1953) The taxonomic value and variability of certain structures in the cestode genus *Echinococcus* (Rud., 1801) and a review of recognized species. Faculty Publications from the Harold W. Manter Laboratory of Parasitology, Lincoln

Rausch RL (1967) A consideration of infraspecific categories in the genus *Echinococcus* Rudolphi, 1801 (Cestoda: Taeniidae). J Parasitol 53(3):484–491

Rausch RL, Bernstein JJ (1972) *Echinococcus vogeli sp. n.* (Cestoda: Taeniidae) from the bush dog, *Speothos venaticus* (Lund). Z Tropenmed Parasitol 23(1):25–34

Rausch RL, Nelson GS (1963) A review of the genus *Echinococcus* Rudolphi, 1801. Ann Trop Med Parasitol 57:127–135

Rausch RL, Schiller EL (1951) Hydatid disease (echinococcosis) in Alaska and the importance of rodent intermediate hosts. Science 113:57–58

Rausch RL, Schiller EL (1954) Studies on the helminth fauna of Alaska. XXIV. *Echinococcus sibiricensis n. sp.*, from St. Lawrence Island. J Parasitol 40:659–662

Rausch RL, Schiller EL (1956) Studies on the helminth fauna of Alaska. XXV. The ecology and public health significance of *Echinococcus sibiricensis* Rausch & Schiller, 1954, on St. Lawrence Island. Parasitology 46:395–419

Rodrigues-Silva R, Peixoto JR, de Oliveira RM, MagalhãesPinto R, Gomes DC (2002) An autochthonous case of *Echinococcus vogeli* Rausch & Bernstein, 1972 polycystic echinococcosis in the state of Rondônia, Brazil. Mem Inst Oswaldo Cruz 97(1):123–126

Romig T, Ebi D, Wassermann M (2015) Taxonomy and molecular epidemiology of Echinococcus granulosus sensu lato. Vet Parasitol 213(3–4):76–84

Rudolphi CA (1801) Beobachtungen uber die Eingeweidew urmer. Archiv für Zoologie und Zootomie 2:1–32

Scott JC, Stefaniak J, Pawlowski ZS, McManus DP (1997) Molecular genetic analysis of human cystic hydatid cases from Poland: identification of a new genotypic group (G9) of *Echinococcus granulosus*. Parasitology 114:37–43

Shul'ts RS (1962) On some epidemiological problems of echinococcosis and methods for its control. Med Parazitol 31:272–274

Sweatman GK, Williams RJ (1963) Comparative studies on the biology and morphology of *Echinococcus granulosus* from domestic livestock, moose and reindeer. Parasitology 53:339–390

Szidat L (1971) Neue Aspekte des Echinococcen-Problems. Angew Parasitol 12(3):133–143

Szldat L (1960) *Echinococcus patagonicus sp. nov.* (Cestoda) parasito del zorro Dusicyon culpaeus culpaeus(Mol.). Neotropica 6(19):13–16

Tang CT, Cui GW, Qian YC, Kang YM, Wang YH, Peng WF, Hong-Chang LU, Chen D (2007) Studies on the alveolar *Echinococcus* species in northward Daxingan mountains, Inner Mongolia, China. *Echinococcus russicensis sp. nov*. Chinese J Zoonoses 23(10):957–964

Tappe D, Stich A, Frosch M (2008) Emergence of polycystic neotropical echinococcosis. Emerg Infect Dis 14:292–297

Verstr AJ (1965) Review of *Echinococcus* species in South Africa. Onderstepoort J Vet Res 32(1):7–118

Virchow R (1856) Die multiloculäre, ulcerirende Echinokokkengeschwulst der Leber. Verhandlungen der physicalisch-medicinischen Gesellschaft zu Würzburg 6:84–95

Vogel H (1955) Über den Entwicklungszyklus und die Artzugehörigkeit des europäischen Alveolar echinococcus. Dtsch Med Wochenschr 80:931–932

Vogel H (1957a) Über den Echinococcus multilocularis Süddeutschlands I. Das Bandwurmstadium von Stämmen menschlicher und tierischer Herkunft. Zeitschrift für Tropenmedizin und Parasitologie 8:404–454

Vogel H (1957b) Über den *Echinococcus multilocularis* Süddeutschlands. I. Das Bandwurmstadium von Stammen menschlicher und tierischer Herkunft. Z Trop Parasit 8:404–454

Williams RJ, Sweatman GK (1963) On the transmission, biology and morphology of *Echinococcus granulosus equinus*, a new subspecies of hydatid tapeworm in horses in Great Britain. Parasitology 53:391–407

Xiao N, Qiu J, Nakao M, Li T, Yang W, Chen X, Schantz PM, Craig PS, Ito A (2006) *Echinococcus shiquicus*, a new species from the Qinghai-Tibet plateau region of China: discovery and epidemiological implications. Parasitol Int 55:S233–S236

# Global Epidemiology of Echinococcosis: Current Status and Future Prospects

# 2

Ying Wang, Shuai Han, and Yujuan Shen

**Abstract**

Echinococcosis is a zoonotic parasitic infection caused by tapeworms of the genus *Echinococcus*. The disease affects both humans and a wide range of animal hosts, with significant pattern in its epidemiology across different global regions. This chapter aims to explore the global epidemiology of echinococcosis, including its transmission pattern, disease burden, risk factors, control measures, and distribution, and propose the potential challenges toward control of echinococcosis.

**Keywords**

Echinococcosis · Distribution · Epidemiology · Molecular epidemiology · Control and prevention

Echinococcosis is a worldly distributed zoonotic parasitic disease caused by the larvae of *Echinococcus* spp. It is a serious threat to human and animal health and causes significant economic losses, resulting a serious public health issue (Wen

Y. Wang · S. Han · Y. Shen (✉)
National Institute of Parasitic Diseases, Chinese Center for Disease Control and Prevention, Chinese Center for Tropical Diseases Research; National Key Laboratory of Inttelligent Tracking and Forecasting for Infectious Diseases; Key Laboratory on Parastite and Vector Biology, Ministry of Health; WHO Centre for Tropical Diseases; National Center for International Research on Tropical Diseases, Ministry of Science and Technology, Shanghai, China
e-mail: shenyj@nipd.chinacdc.cn


J. Li et al. (eds.), *Echinococcus: Control and Elimination of Echinococcosis with a Focus on China and Europe*, Parasitology Research Monographs 19,
https://doi.org/10.1007/978-3-031-54479-8_2

et al. 2019). The World Health Organization (WHO) has listed it as one of the 17 neglected tropical diseases, and the Food and Agriculture Organization (FAO) of the United Nations has listed it as the second most serious foodborne parasitic disease in the world (WHO 2015; WHO and FAQ 2012). The causative agent of echinococcosis is the larval stages of *Echinococcus* spp. In taxonomy, the genus *Echinococcus* belongs to the kingdom Animalia, phylum Platyhelminthes, class Cestoda, order Cyclophyllidea, family Taeniidae. It includes *Echinococcus granulosus*, which causes cystic echinococcosis (CE), *Echinococcus multilocularis*, which causes alveolar echinococcosis (AE), *Echinococcus vogeli* and *Echinococcus oligarthra*, which cause polycystic echinococcosis, and *Echinococcus shiquicus*, which is newly discovered in the Qinghai-Tibet Plateau but not found to infect people (Xiao et al. 2005). Each species has a specific geographic distribution, life cycle, host preference, and clinical manifestations. Understanding the taxonomy and characteristics of different *Echinococcus* species is crucial for better understanding and managing echinococcosis. In this chapter, we describe the epidemiology, geographic distribution of echinococcosis, and the potential challenges toward control of echinococcosis.

## 2.1 Epidemiology

The life cycle of *Echinococcus* involves two hosts, a definitive host (usually a canid), and an intermediate host (typically a herbivore or omnivore). The adult tapeworm resides in the small intestine of the definitive host, which is usually a domestic dog or other canids. The mature tapeworm can produce thousands of microscopic eggs, which are released in the feces of the definitive host. The eggs are shed into the environment through the feces of the definitive host. This ensures the survival of the eggs outside the host's body. If the eggs are accidentally ingested by an intermediate host, such as a sheep, goat, or other grazing animals, the eggs hatch in the small intestine. Once hatched, the larvae penetrate the intestinal wall of the intermediate host to enter the bloodstream. They are then carried to various organs, most commonly the liver and lungs, where they develop into fluid-filled cysts. The hydatid cysts grow slowly over time, and within the cysts, the protoscoleces are formed, which can later develop into adult tapeworms. If a definitive host, such as a dog, consumes the infected intermediate host, the protoscoleces are released from the cysts in the digestive system and develop into adult worms in small intestine of the definitive host, and star reproducing. The cycle then continues when the definitive host sheds the eggs in its feces. Humans can also become intermediate hosts if they accidentally ingest *Echinococcus* eggs through contact with contaminated environments or through close contact with infected definitive hosts; however, humans are generally not involved in the transmission of echinococcosis (McManus et al. 2012; Wen et al. 2019; Craig et al. 2019).

### 2.1.1 Source of Infection

There are different definitive hosts such as wolves, foxes, and jackals responsible for infection but dog acts a main source of infection. It is worth noting that infected wildlife, including definitive hosts and intermediate hosts, can also cause the transmission of echinococcosis, especially alveolar echinococcosis (Craig et al. 2019).

There are many kinds of wildlife as the definitive hosts in the life cycle of *Echinococcus*, including canines and felines, and wildlife as intermediate hosts include rodents and ungulates. Among them, the hosts of canids mainly include *Vulpes vulpes*, *Vulpes ferrilata*, and *Canis lupus*, which are the main wild definitive hosts of *Echinococcus granulosus*, *Echinococcus multilocularis*, and *Echinococcus shiquicus* (Craig et al. 2019; Xiao et al. 2005). Bush dogs are the main definitive host of *Echinococcus vogeli* (Matsuo et al. 2000). Felines mainly include *Prionailurus bengalensis*, *Felis lynx*, *Puma concolor*, and *Panthera onca*, which are the main wild definitive hosts of *Echinococcus oligarthra*. Rodents as intermediate hosts include *Lagomorpha* and *Rodentia*: *Lagomorpha* include *Ochotona curzoniae* and *Lepus oiostolu*; *Rodentia* include *Microtus ilaeus*, *Arvicola terrestris*, *Ondatra zibethicus*, *Microtus brandti*, *Meriones unguiculatus*, *Citellus dauricus*, and *Clethrionomys glareolus*. These are the main wild intermediate hosts of *Echinococcus granulosus* and *Echinococcus multilocularis*. The *Ochotona curzoniae* is the wild intermediate host of *Echinococcus shiquicus*. In addition, *Dasyprocta leporina* and *Cavia porcellus* are the main wild intermediate hosts of *Echinococcus oligarthra* and *Echinococcus vogeli*, respectively. Ungulates such as *Pseudois nayaur* and *Pantholops hodgsoni* and marsupials such as *Macropus eugenii* and *Vombatus ursinus* are mainly wild intermediate hosts of *Echinococcus granulosus* (Deplazes et al. 2017; Craig et al. 2017).

Geographically, the wildlife hosts of *Echinococcus granulosus* and *Echinococcus multilocularis* are the most widely distributed in Europe, Asia, Africa, and other parts of the world while the hosts of *Echinococcus shiquicus* are mainly distributed in the Qinghai-Tibet Plateau region of China, so this species has not been found in other areas so far. The definitive hosts and intermediate hosts of *Echinococcus vogeli* and *Echinococcus oligarthra* are mainly distributed in the Americas, so these two species are currently only found in the Americas (Deplazes et al. 2017; Craig et al. 2017, 2019; Alvi and Alsayeqh 2022).

### 2.1.2 Transmission Patterns

The transmission patterns of different echinococcosis are basically similar, mainly through fecal–oral transmission. *Echinococcus* eggs are highly resistant to low temperature and dryness in the environment, and can maintain long-term infectivity in a natural state. Humans and animals are infected by eating or drinking food and water contaminated by the eggs. In dry and windy areas, the eggs are scattered with the wind, and there is also the possibility of respiratory tract infection through inhalation of dust and droplets containing eggs. People, especially children, are susceptible to infection during contact with and playing with domestic dogs, as well as when adults engage in livestock fur processing, shearing, and other processes. In addition, animal fur trading, transportation, pet dog transportation, and so on also increase the risk of indirect infection.

The definitive host canines are mainly infected by ingesting animal organs containing *Echinococcus* cysts (Craig et al. 2017; Wen et al. 2019).

### 2.1.3 Susceptibility

The species and biological characteristics of *Echinococcus*, egg status, environment, host species, and immune status can affect the susceptibility of the host. Humans are generally susceptible to echinococcosis, depending on the intensity of exposure to *Echinococcus* eggs. The susceptible population are mainly in pastoral and semipastoral areas, especially children, women who often take housework, and people engaged in animal husbandry and hunting. For example, in the Qinghai-Tibet Plateau area of China, where echinococcosis is most prevalent, women often undertake more housework than men, feeding cattle and sheep and other livestock and domestic dogs, drying cow dung as fuel, and burying domestic dog feces. Contact with infected dogs or infected livestock, and the level of exposure to *Echinococcus* eggs is higher than men, so women are more likely to be infected (Wu et al. 2018). Children often play closely with domestic dogs, and their range of activities overlaps with that of domestic dogs. In addition, in pastoral areas, there is a lack of good hygiene conditions and health awareness, resulting in a low frequency of hand washing, making them highly susceptible to infection. Due to the slow growth of *Echinococcus* cysts, most children are asymptomatic after infection and may not be diagnosed until adulthood.

For animal intermediate hosts, the host preference of different species is different. *Echinococcus granulosus* has the widest range of hosts, among livestock, sheep are the most susceptible, and are suitable hosts for *Echinococcus granulosus*, followed by cattle; however, cattle are not suitable hosts for *Echinococcus granulosus*, and the cysts formed after infection are mainly infertile cysts. Other animals such as pigs, horses, camels, and deer can also be infected. *Echinococcus multilocularis* primarily infects rodents and is rarely found in livestock. *Echinococcus shiquicus* mainly infects plateau pikas (Craig et al. 2019; Xiao et al. 2006).

Dogs, the definitive host, are susceptible to many kinds of *Echinococcus*. In the Qinghai-Tibet Plateau, *Vulpes vulpes* can be infected with *Echinococcus granulosus* and *Echinococcus multilocularis* and *Vulpes ferrilata* are mostly infected with *Echinococcus shiquicus*, while *Vulpes corsac*, which are the closest relatives to *Vulpes ferrilata*, are not susceptible to these tapeworms (Craig et al. 2019).

### 2.1.4 Disease Burden

The exact burden of echinococcosis is challenging to determine due to underreporting and variability in disease presentation. However, it is recognized as a neglected zoonotic disease with significant human health and economic consequences. CE is responsible for the majority of human cases, accounting for approximately 95% of all reported cases worldwide. It predominantly affects younger individuals, with the highest incidence observed in children and young adults. AE, although less common, has more severe clinical implications, with a higher mortality rate if left untreated. The disease burden of echinococcosis includes the physical suffering of infected individuals, loss of productivity, treatment costs, and potential fatalities (Alvi and Alsayeqh 2022; Miao et al. 2023).

### 2.1.5 Risk Factors

Several factors contribute to the risk of acquiring echinococcosis. Close contact with infected dogs or their feces increases the likelihood of human infection. In regions where dogs are used for herding or hunting, the risk is particularly high. Livestock-raising communities are also at risk due to the close association with intermediate hosts. Poor sanitation and hygiene practices, such as inadequate handwashing and consumption of contaminated food or water, play a significant role in the transmission of the parasite. Individuals engaged in agricultural and occupational activities that involve contact with animals are also more susceptible to infection (Craig et al. 2017, 2019; Wu et al. 2018; Alvi and Alsayeqh 2022).

### 2.1.6 Prevention and Control Measures

Preventing and controlling echinococcosis primarily involves education, awareness, and implementation of appropriate measures. Health education programs should focus on raising awareness about the disease, its mode of transmission, and preventive measures. Promoting proper hygiene practices, including regular handwashing with soap and clean water, is essential. Individuals should be educated about the importance of washing fruits and vegetables thoroughly before consumption and avoiding the consumption of raw or undercooked meat. In endemic regions, regular deworming of dogs and good animal husbandry practices can reduce the burden of infection in the definitive hosts. Additionally, early detection, diagnosis, and treatment of human cases are crucial for preventing the progression of the disease and reducing associated complications (Craig et al. 2017).

## 2.2 Distribution of Echinococcosis

Current data show that echinococcosis is still a serious public health problem in many countries and regions (Wen et al. 2019; Deplazes et al. 2017; Alvi and Alsayeqh 2022). In some regions, the epidemic scope and the number of threatened people are increasing.

### 2.2.1 Cystic Echinococcosis

CE is globally distributed, with cases reported on almost all continents, including polar, temperate, subtropical, and tropical regions. CE is highly endemic in parts of Eurasia, Africa, Oceania, and South America, such as western China, Uruguay, and other countries. In epidemic areas, the incidence of CE ranges from sporadic to high. Only a few countries around the world, such as New Zealand, Tasmania in Australia, and Iceland, have announced the elimination of CE, but there are still reports of animal infections on the mainland of Australia (Craig et al. 2017; Deplazes et al. 2017; Wen et al. 2019; Alvi and Alsayeqh 2022).

### 2.2.2 Alveolar Echinococcosis

AE is mainly distributed in the Northern Hemisphere, and the epidemic areas mainly include Eurasia, most parts of central Europe, north America, and parts of northern Africa. Most European countries such as France, Germany, Switzerland, Austria, Denmark, the United Kingdom, Sweden, and Russia have varying degrees of AE prevalence. Some states in the United States and central northern Canada in North America, western China, and northern Japan in Asia are considered high prevalence areas for AE. It is estimated that over 90% of the global burden of AE occurs in China (Craig et al. 2017; Deplazes et al. 2017; Wen et al. 2019; Alvi and Alsayeqh 2022).

### 2.2.3 Polycystic Echinococcosis

Polycystic echinococcosis caused by *Echinococcus vogeli* and *Echinococcus oligarthra* is only distributed in Central and South America. Reports of polycystic echinococcosis in human came from Nicaragua, Costa Rica, Panama, Colombia, Ecuador, Venezuela, Suriname, Brazil, Uruguay, Argentina, and Guyana (D'Alessandro 1997; Knapp et al. 2009). Compared with CE and AE, the report of polycystic echinococcosis is very limited, and the actual epidemic situation and harm degree of the disease have not been effectively evaluated.

### 2.2.4 *Echinococcus shiquicus*

*Echinococcus shiquicus* was first discovered in Shiqu County, Garzê Tibetan Autonomous Prefecture, Sichuan Province, China, in 2003, and later found in other areas of the Qinghai Tibet Plateau, but not found in other areas outside the Qinghai Tibet Plateau. The definitive host of *Echinococcus shiquicus* is mainly Tibetan fox, but dogs can also be infected. And the intermediate host is mainly plateau pika (Xiao et al. 2003, 2005 and 2006). At present, no cases of human infection with *Echinococcus shiquicus* have been found, and more research evidence is needed to determine whether *Echinococcus shiquicus* can infect humans.

According to the current data, it is difficult to accurately describe the epidemic and distribution map of echinococcosis, because the published data are not complete. Existing problems include inconsistent data sampling and evaluation methods, inconsistent sampling scope, misuse of prevalence and incidence rate, and few or no data in some countries. Therefore, the published data cannot accurately represent the epidemic situation in this region. In order to improve the epidemiological distribution map worldwide, there is an urgent need for unified data collection and analysis methods.

## 2.3 Molecular Epidemiology of *Echinococcus*

Before the discovery of *E. shiquicus*, four species of *Echinococcus* were recognized in the genus: *E. granulosus*, *E. multilocularis*, *E. vogeli*, and *E. oligarthra*. However, there has been controversy regarding the classification of *Echinococcus* genus, especially *E. granulosus*. According to morphological, taxonomic, epidemiological, and molecular biological data, *E. granulosus* is considered to be a multigenotype complex, *E. granulosus* sensu lato (*s.l.*), including ten genotypes (G1–G10) and lion strains. With the development of genetics and sequencing technology, the classification of *E. granulosus s.l.* has been more scientific. At present, *E. granulosus s.l.* has been split into five species, *E. granulosus* sensu stricto (*s.s.*) (G1–G3 genotypes), *E. equinus* (G4 genotype), *E. ortleppi* (G5 genotype), *E. canadensis* group (G6–G10 genotypes), and *E. felidis* (lion strain). Different species have differences in morphology, transmission mode, and epidemic characteristics (Nakao et al. 2013; Thompson and McManus 2002; Alvi and Alsayeqh 2022).

*E. granulosus s.s.* consists of G1, G2, and G3, corresponding to sheep strain, Tasmanian sheep strain, and buffalo strain originally according to epidemiological characteristics. *E. granulosus s.s.* is the main pathogen of CE, which can infect human and various animals. A variety of domestic animals such as yaks, buffalo, pigs, horses, and camels can be infected as intermediate hosts. The main transmission route is sheep to dog and cattle to dog. *E. granulosus s.s.* is distributed globally, among which G1 is the predominant genotype (Wang et al. 2023). The human CE infection rate caused by G1 exceeds 88% worldwide (Alvarez Rojas et al. 2014). G2 and G3 can also infect humans, but the reported cases are limited.

*E. equinus*, which corresponds to the G4 genotype, was first found to parasitize in horses, and is mainly transmitted between horses and dogs. *E. equinus* is the only species that has not yet been found infected humans. It is host-specific and can only be detected in horses, donkeys, mules, and zebras. *E. equinus* was once thought to only infect equines and not pathogenic to humans, but it has been detected in wild lemurs, lions, and jackals, confirming its possible zoonosis, which requires more evidence (Alvi and Alsayeqh 2022; Wang et al. 2023).

*E. ortleppi*, corresponding to the G5 genotype, was first found to parasitize cattle and was named as the cattle strain. Its morphology is different from *E. granulosus s.s.* The main intermediate host of *E. ortleppi* is cattle, and the infections in cattle are common. In addition to cattle, *E. ortleppi* can also infect buffalo, goats, camels, pigs, sheep, and other domestic animals. *E. ortleppi* can also infect human, but reported cases are rare (Alvi and Alsayeqh 2022; Wang et al. 2023).

*Echinococcus canadensis* is currently considered to include the genotype G6–G10, but it remains controversial. It has been thought that G9 is a variant of G7. And it has also been thought that *E. canadensis* should be split into three groups, *E. intermedius* composed of G6 and G7, *E. borealis* composed of G8 and G10, and *E. canadensis*. More biological, morphological, and genetic data are needed to develop a commonly accepted taxonomic theory (Alvi and Alsayeqh 2022; Wang et al. 2023).

The G6 and G7 genotypes of *E. canadensis* are distributed globally, and the infection of G6 and G7 is common in domestic animals, and the main intermediate hosts are camels and pigs. The human CE caused by G6 and G7 infection is second only to G1. The G8 and G10 genotypes of *E. canadensis* are mainly transmitted among wolves and deer, and so far only found and reported in the Northern Hemisphere. As one of the pathogens that can cause human CE, the reported cases caused by G8 and G10 genotypes are rare (Alvi and Alsayeqh 2022; Wang et al. 2023).

The geographical distribution of different species (genotypes) of *Echinococcus* genus is shown in Table 2.1.

## 2.4 Control of Echinococcosis and Current Status in China

Echinococcosis in China includes cystic echinococcosis caused by *E. granulosus* and alveolar echinococcosis caused by *E. multilocularis*. *E. shiquicus* infection has been detected in animals, but no human infection has been reported. *Echinococcus vogeli* and *Echinococcus oligarthra* have not been reported. Therefore, the control of echinococcosis in China focuses on cystic echinococcosis and alveolar echinococcosis. China is the only country that controls and monitors echinococcosis at the national level; the central government prioritized echinococcosis control, and great success has been achieved following the efforts for almost 20 years. China's experience in controlling echinococcosis is feasible and worth popularizing, especially for developing countries.

**Table 2.1** Species and genotypes within the genus *Echinococcus* and their hosts and geographical distribution (Wen et al. 2019; Thompson 2023)

| Species (genotypes) | Definitive hosts | Intermediate hosts | Disease | Human cases | Distribution |
|---|---|---|---|---|---|
| *E. granulosus* sensu stricto (G1–G3) | Dog, wolf, dingo, fox, jackal, other canids | Sheep, goat, cattle, pig, camel, buffalo, horse, wild ungulates, marsupials, etc. | CE | Yes | Worldwide |
| *E. equinus* (G4) | Dog | Horse, other equids, cervids | – | Not reported | Eurasia, Africa |
| *E. ortleppi* (G5) | Dog | Cattle, buffalo (sheep, goat) | CE | Yes | Eurasia, Africa, Central and South America |
| *E. canadensis* group (G6–G10) | Dog, wolf | Pig, sheep, camel, cervids | CE | Yes | Eurasia, Africa, North and South America |
| *E. felidis* (lion strain) | Lion | Hyena, warthog, zebra, wildebeest, bush pig, buffalo, various antelopes, giraffe | – | Not reported | Africa |
| *E. multilocularis* | All fox species, wolf, raccoon dog, dog, etc. | Rodents, domestic and wild pig, dog, monkey, horse | AE | Yes | Eurasia, North America |
| *E. vogeli* | Bush dog | Rodents | PE | Yes | Central and South America |
| *E. oligarthra* | Wild felids | Rodents | PE | Yes | Central and South America |
| *E. shiquicus* | Tibetan fox | Pika | – | Not reported | Tibetan Plateau |

### 2.4.1 Interventions of Echinococcosis in China

The complex life cycle of *Echinococcus* determines that the control and prevention of echinococcosis is a multifaceted and complex project that requires the participation of the whole society. The strategy for the control and prevention of echinococcosis in China is based on controlling the source of infection, combining the control and prevention of intermediate hosts with the screening and treatment of patients. Specific measures include monthly deworming of dogs, safe slaughter of livestock, screening of patients and free treatment, and so on to cut off the transmission from dogs to humans and from intermediate hosts to dogs. Under the leadership of the central government, many departments from the health, water resource, education, public security, and other industries have jointly participated in the prevention and control of echinococcosis to promote the control of the disease (MOH 2010; NHC 2016).

### 2.4.2 National Control Program of Echinococcosis in China

In 1989, China classified echinococcosis as a Class C infectious disease and included it in the management of the Law of the People's Republic of China on the Prevention and Treatment of Infectious Disease. Pilot prevention and control work was gradually carried out, exploring and accumulating prevention and control experience and models. Since 2005, a nationwide echinococcosis control and prevention program has been launched in China, and a large amount of dedicated funds and many targeted measures have been used to control echinococcosis. Population screening, treatment and patient management, dog deworming and testing, health education, and other work have been included in the program budget, ensuring the implementation of echinococcosis prevention and control work. In 2007, echinococcosis was listed as one of the major infectious diseases for free treatment, which further promoted the control of echinococcosis. Since 2010, the central government has formulated a control plan for echinococcosis every 5 years, specifying control objectives, control measures, and assessment indicators. At present, the national echinococcosis control program has been implemented in all echinococcosis endemic counties, with an annual special fund of more than 300 million RMB yuan.

### 2.4.3 National Epidemiological Survey on Echinococcosis in China

To identify the endemic areas and endemic situation of echinococcosis in China, a nationwide epidemiological survey on echinococcosis was carried out during 2012–2016. This is the largest, most scientific, and comprehensive epidemiological survey of echinococcosis. According to the results from survey, echinococcosis was found to be highly endemic in pasture areas of northwestern China where the prevalence of echinococcosis in humans was 0.28%, in livestock was 4.7%, and dog infection rate was 4.3%. A total of 370 counties were identified as endemic counties of echinococcosis distributed in the nine provinces and autonomous regions (Inner Mongolia, Sichuan, Tibet, Gansu, Qinghai, Ningxia, Yunnan, Shaanxi, and Xinjiang (including Xinjiang Production and Construction Corps)). China is the most vital endemic country across the globe, with more than 0.16 million patients and more than 50 million threatened people. Among them, the Qinghai-Tibet Plateau area is the most important, of which the prevalence in humans (1.28%) was significantly higher than that of non-Qinghai-Tibet Plateau area (0.13%) (Wu et al. 2018; Li et al. 2019). These survey results clarified the scope and level of echinococcosis epidemic in China, and provided a solid data basis for the echinococcosis control and prevention. Based on prevalence in humans and the infection rate of dogs, all 370 echinococcosis endemic counties were categorized into three classes (Fig. 2.1). Class I represents a prevalence of ≥1% in humans or a dog infection rate of ≥5%, Class II represents a prevalence between 0.1% and 1% or 1% in humans or a dog infection rate between 5% and 1%, while Class III represents a prevalence of <0.1% in

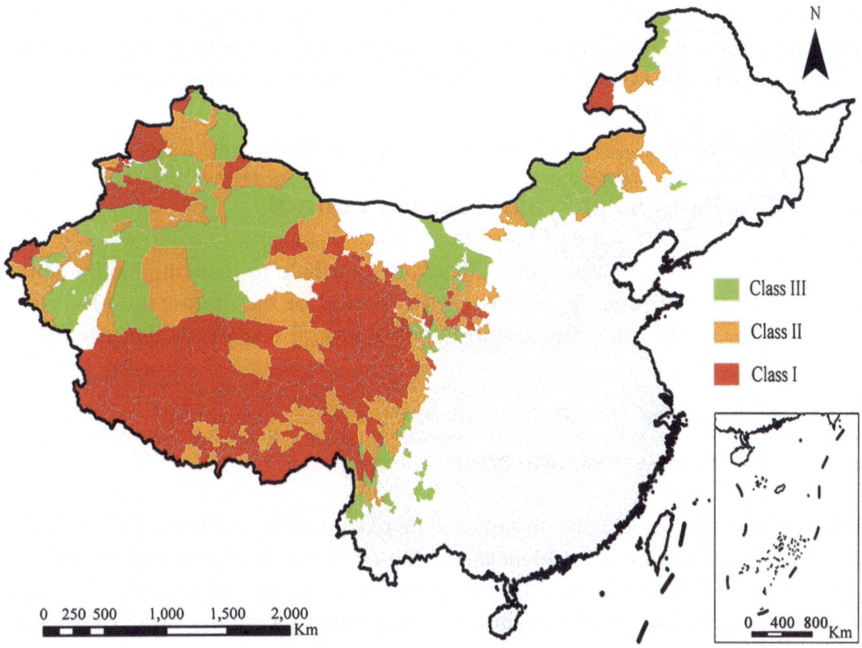


**Fig. 2.1** Epidemiological map of echinococcosis at the county level in China based on the results of the national epidemiological survey of echinococcosis from 2012 to 2016 (Tibet Autonomous Region based on the survey results of 2016), cited from the *Technical Plan for Echinococcosis Control* (2019 version) (Wang et al. 2023)

humans or a dog infection rate of less than 1%. Class I is the most severe level among them (Wang et al. 2023).

### 2.4.4 Chinese Echinococcosis Surveillance Network

Since 2016, surveillance on the prevalence of humans, livestock, and dogs has been carried out in all endemic counties. Annual surveillance can accurately evaluate the impact of control and prevention programs in addition to gaining a better understanding of the trend of prevalence of echinococcosis. Considerable progress has been made in the continuous control and prevention program of epidemic areas of echinococcosis, and the effects can be seen from the echinococcosis surveillance data. The surveillance data in recent years demonstrated that the prevalence of echinococcosis in human, livestock, and dogs have significantly dropped in most endemic areas (Wei et al. 2020; Kui et al. 2021a, b). Some areas have even attained low endemic levels. The data in 2021 showed that the prevalence of echinococcosis in humans was 0.06%, the infection rates of livestock, and dogs were 0.63% and 0.78%, respectively (Kui et al. 2023). Nevertheless, there is a gradually increasing trend that warrants attention, which is that echinococcosis patients have been identified in all nonendemic provinces. Epidemiological investigation revealed that some patients were imported cases with a history of living or traveling in endemic areas (Kui et al. 2021a, b; Han et al. 2022). However, it is challenging to determine the route of infection in locally infected cases (Wang et al. 2023).

## 2.5 Challenges for Echinococcosis Control

With advancements in various areas that can contribute to reducing the burden of echinococcosis, the future prospects for the control of the disease are promising. However, the control of echinococcosis poses several challenges that need to be addressed.

### 2.5.1 Awareness and Education

Echinococcosis is a neglected disease, and there is limited awareness about its transmission, prevention, and control measures among at-risk populations and healthcare professionals. Lack of knowledge about the disease can hinder early detection, timely treatment, and implementation of preventive measures. Increasing public awareness about echinococcosis and its transmission, prevention, and control measures is crucial. Future efforts should focus on developing targeted educational campaigns to reach at-risk populations, healthcare professionals, veterinarians, and policymakers. Utilizing various communication channels such as social media, educational materials can help disseminate accurate information and promote behavior change.

### 2.5.2 Diagnosis

Diagnosis of echinococcosis can be challenging due to its varied presentations and lack of specific symptoms. Detection and confirmation of the disease rely on a combination of clinical evaluation, imaging techniques (such as ultrasound and computed tomography), serological tests, and pathological examination (McManus et al. 2012). Access to appropriate diagnostic facilities and trained personnel may be limited in some endemic areas. Advances in diagnostic techniques can facilitate early detection and accurate diagnosis of echinococcosis. Future prospects include the development of rapid diagnostic tests that are affordable, easy to use, and suitable for resource-limited areas. These tests should be specific, sensitive, and capable of detecting different stages of the disease, enabling timely treatment initiation and monitoring.

### 2.5.3 Treatment and Vaccination

Adequate medical infrastructure, skilled doctors, and availability of drugs are crucial for the effective treatment of echinococcosis. In some resource-limited areas, access to timely and appropriate treatment options, such as surgery and antiparasitic drugs, may be limited, leading to delayed or inadequate treatment. Continuous research and development efforts should focus on improving treatment options for echinococcosis. This includes the development of new antiparasitic drugs with improved efficacy, reduced side effects, and shorter treatment durations. Additionally, innovative approaches such as targeted drug delivery systems or immunotherapeutic strategies may offer alternative treatment modalities. Meanwhile, developing an effective vaccine against echinococcosis could revolutionize control efforts (McManus et al. 2012). While progress has been made in developing vaccines for animals, vaccines for humans are still under investigation. Future research should focus on identifying promising vaccine candidates and conducting clinical trials to assess their safety, efficacy, and long-term protection.

### 2.5.4 Surveillance and Reporting

Due to underreporting, surveillance gaps, and lack of a comprehensive reporting system, the true burden of echinococcosis is often underestimated. Improved surveillance systems and enhanced reporting mechanisms are necessary to capture accurate data on the disease's incidence, prevalence, and distribution in different regions. Strengthening surveillance systems and enhancing reporting mechanisms are essential for monitoring the disease burden and evaluating control interventions. Future prospects involve adopting innovative technologies, such as digital surveillance platforms and real-time data analysis tools, to streamline data collection, analysis, and reporting. This will enable timely identification of outbreaks, hotspot mapping, and evaluation of control measures' effectiveness.

### 2.5.5 One Health Approach

The lifecycle of *Echinococcus* involves both animal and human hosts, making it a complex zoonotic disease. Dogs play a crucial role in the transmission of *Echinococcus*. Implementing effective strategies for dog population management, such as regular deworming, vaccination, and responsible pet ownership, can be challenging, especially in resource-limited areas and areas with cultural barriers. The persistence of *Echinococcus* eggs in the environment poses challenges for disease control. Controlling the disease requires a One Health approach that integrates efforts from the human health, veterinary, and environmental sectors. Future prospects involve strengthening collaboration, joint efforts should focus on integrating surveillance systems, implementing synchronized control programs, and sharing data to enhance understanding of the disease's epidemiology and improve interventions.

## 2.6 Future Prospects

Echinococcosis is a global health concern, and international collaboration is crucial for its control. Future prospects involve fostering collaboration among different countries, sharing experiences and best practices, harmonizing control strategies, and providing technical and financial support to endemic regions. Collaborative research efforts can also enhance our understanding of the disease's complexities and develop effective control measures. Continued efforts in these areas can contribute to reducing the burden of echinococcosis and improving the overall health and well-being of at-risk populations.

## References

Alvi MA, Alsayeqh AF (2022) Food-borne zoonotic echinococcosis: a review with special focus on epidemiology. Front Vet Sci 9:1072730

Alvarez Rojas CA, Roming T, Lightowlers MW (2014) Echinococcus granulosus sensu lato genotypes infecting humans-review of current knowledge. Int J Parasitol 44(1):9–18

Craig PS, Hegglin D, Lightowlers MW, Torgerson PR, Wang Q (2017) Echinococcosis: control and prevention. Adv Prasitol 96:55–158

Craig PS, Giraudoux P, Wang ZH, Wang Q (2019) Echinococcosis transmission on the Tibetan Plateau. Adv Parasitol 104:165–246

Deplazes P, Rinaldi L, Alvarez Rojas CA, Torgerson PR, Harandi MF, Romig T et al (2017) Global distribution of alveolar and cystic echinococcosis. Adv Parasitol 95:315–493

D'Alessandro A (1997) Polycystic echinococcosis in tropical America: *Echinococcus vogeli* and *E. oligarthrus*. Acta Trop 67(1–2):43–65

Han S, Kui Y, Xue C, Zhang Y, Zhang B, Li Q et al (2022) The endemic status if echinococcosis in China from 2004 to 2020. Chin J Parasitol Parasit Dis 40(4):475–480

Knapp J, Chirica M, Simonnet C et al (2009) Echinococcus vogeli infection in a hunter, French Guiana. Emerg Infect Dis 15(12):2029–2031

Kui Y, Liu B, Wang X, Xue C, Xue J, Zhang Y et al (2021a) Epidemiological characteristics of echinococcosis in non-endemic PLADs-China, 2017-2020. China CDC Wkly 3(51):1084–1088

Kui Y, Wu W, Han S, Xue C, Liu B, Xue J et al (2021b) Analysis of the results of echinococcosis surveillance in China from 2018 to 2019. J Pathog Biol 16(9):1025–1029

Kui Y, Xue CZ, Wang X, Liu BX, Wang Y et al (2023) Progress of echinococcosis control in China, 2021. Chin J Parasitol Parasit Dis. 41(2):142–148

Li B, Quzhen G, Xue CZ, Han S, Chen WQ, Yan XL et al (2019) Epidemiological survey of echinococcosis in Tibet Autonomous Region of China. Infect Dis Poverty 8(1):29

Matsuo K, Shimizu M, Nonaka N et al (2000) Development and sexual maturation of Echinococcus vogeli in an alternative definitive host, Mongolian gerbil (Meriones unguiculatus). Acta Trop 75(3):323–330

McManus DP, Gray DJ, Zhang W, Yang Y (2012) Diagnosis, treatment, and management of echinococcosis. BMJ 344:e3866

Miao L, Tian T, Zhou X (2023) Progress of international and national researches on disease burden of echinococcosis. Chin J Schisto Control 35(3):307–312

Ministry of Health of the People's Republic of China. Echinococcosis Prevention Action Plan (2010–2015). (2010). http://www.gov.cn/zwgk/2010-12/14/content_1765485.htm. Accessed 24 Dec 2010. (in Chinese)

National Health and Family Planning Commission of the People's Republic of China. National plan for echinococcosis and other key parasitic diseases prevention (2016–2020). (2016). http://www.nhc.gov.cn/jkfpwlz/zcwj1g/201902/a999e19bfae94d3ea68da8be5add58b5.shtml. Accessed 24 Oct 2016. (in Chinese)

Nakao M, Lavikainen A, Yanagida T, Ito A (2013) Phylogenetic systematic of the genus *Echinococcus* (Cestoda: Taeniidae). Int J Parasitol 43(12–13):1017–1029

Thompson RC, McManus DP (2002) Towards a taxonomic revision of the genus *Echinococcus*. Trends Parasitol 18(10):452–457

Thompson RC (2023) Zoonotic helminths—why the challenge remains. J Helminthol 97:e21,1–e2115

Wang Y, Zhang J, Wang X, Ahmed H, Shen YJ, Cao JP (2023) Molecular epidemiology and the control and prevention of cystic echinococcosis in China: what is known from current research. Zoonoses 3:24

Wei S, Wu W, Han S, Xue C, Liu B, Wang X et al (2020) Analysis of the results of echinococcosis surveillance in China from 2016 to 2017. J Pathog Biol. 15(8):924–928

Wen H, Vuitton L, Tuxun T, Li J, Vuitton DA, Zhang W et al (2019) Echinococcosis: advances in the 21st century. Clin Microbiol Rev 32:e00075–e00018

World Health Organization (2015) Investing to overcome the global impact of neglected tropical diseases: third WHO report on neglected tropical diseases. WHO, Geneva, pp 95–99

World Health Organization, Food and Agriculture Organization of the United Nations (2012) Multicriteria-based ranking for risk management of food borne parasites. FAO headquarters, Rome

Wu W, Wang H, Wang Q, Zhou X, Wang L, Zheng C et al (2018) A nationwide sampling survey on echinococcosis in China during 2012-2016. Chin J Parasitol Parasit Dis. 36(1):1–14

Xiao N, Qiu J, Nakao M, Nakaya K, Yamasaki H, Sako Y et al (2003) Short report: identification of *Echinococcus* species from a yak in the Qinghai-Tibet plateau region of China. Am J Trop Med Hyg 69(4):445–446

Xiao N, Qiu JM, Nakao M, Li TY, Yang W, Chen XW et al (2005) Echinococcus shiquicus n. sp. a taeniid cestode from Tibetan fox and plateau pika in China. Int J Paraditol 35:693–701

Xiao N, Qiu JM, Nakao M, Li TY, Yang W, Chen XW et al (2006) *Echinococcus shiquicus,* a new species from the Qinghai-Tibet plateau regions of China: discovery and epidemiological implications. Parasitol Int 55:S233–S236

# Alveolar and Cystic Echinococcosis in Central Europe: Epidemiology, Diagnosis, Therapy, and Prophylaxis

# 3

Julian Frederic Hotz, Felix Lötsch, Moritz Staudacher, Amelie Leutzendorff, Herbert Auer, Fredrik Waneck, Klaus Kaczirek, Stefan Stremitzer, and Heimo Lagler

J. F. Hotz
Department of Medicine I, Division of Infectious Diseases and Tropical Medicine, Medical University of Vienna, Vienna, Austria

Department of Internal Medicine III, Division of Infectious Diseases, Ulm University Hospital, Ulm, Germany

Department of Neurology, Evangelical Hospital Vienna, Vienna, Austria

F. Lötsch
Department of Laboratory Medicine, Division of Clinical Microbiology, Medical University of Vienna, Vienna, Austria

M. Staudacher
Department of Medicine I, Division of Infectious Diseases and Tropical Medicine, Medical University of Vienna, Vienna, Austria

Department of Clinical Pharmacology, Medical University of Vienna, Vienna, Austria

Department of Angiology, Medical University of Vienna, Vienna, Austria

A. Leutzendorff
Department of Medicine I, Division of Infectious Diseases and Tropical Medicine, Medical University of Vienna, Vienna, Austria

Department of Clinical Pharmacology, Medical University of Vienna, Vienna, Austria

H. Auer
Department of Medical Parasitology, Institute of Specific Prophylaxis and Tropical Medicine, Center of Pathophysiology, Infectiology and Immunology, Medical University of Vienna, Vienna, Austria

F. Waneck
Department of Biomedical Imaging and Image-Guided Therapy, Division of Cardiovascular and Interventional Radiology, Medical University of Vienna, Vienna, Austria



J. Li et al. (eds.), *Echinococcus: Control and Elimination of Echinococcosis with a Focus on China and Europe*, Parasitology Research Monographs 19,
https://doi.org/10.1007/978-3-031-54479-8_3

**Abstract**

Echinococcosis is a zoonotic disease caused by an infection with a tapeworm of the genus *Echinococcus (E.)*. Cystic echinococcosis results from an infection with *E. granulosus* sensu lato complex, while alveolar echinococcosis is caused by *E. multilocularis*. In Central Europe, both alveolar and cystic echinococcosis pose significant interdisciplinary challenges and costs for the healthcare system. Over the past years, the incidence of alveolar echinococcosis has notably increased in Central Europe. Due to its frequent misidentification as a malignant disease, comprehensive education on epidemiology, clinical findings, diagnostic approaches, and treatment of alveolar echinococcosis is crucial. Early detection of alveolar echinococcosis is essential for successful treatment outcomes. Furthermore, the incidence of cystic echinococcosis in Central Europe has also risen in the past decade, mainly attributed to increased immigration. This chapter presents an overview of the epidemiology, clinical manifestations, diagnostic approaches, and treatment options for both diseases in Central Europe, while underscoring the importance of implementing preventive measures.

**Keywords**

*Echinococcosis* · *Echinococcus multilocularis* · *Echinococcus granulosus* · Zoonosis · Central Europe · Epidemiology · Clinical findings · Diagnostic approaches · Treatment · Preventive strategies

## 3.1 Life Cycle of *Echinococcus multilocularis*

The definitive hosts of *Echinococcus (E.) multilocularis* in Central Europe are primarily foxes of the genus *Vulpes (Vulpes vulpes)*. Although rare, dogs and cats can also act as hosts (Fig. 3.1, No. 7) (Duscher et al. 2015). The infection usually does not cause any noticeable symptoms in the definitive hosts. Once the adult tapeworm establishes itself in the small intestine of the definitive host, it starts producing eggs (Fig. 3.1, No. 1). The eggs mature in the last proglottid (segment-like reproductive unit of a tapeworm) and contain the oncosphere equipped with three pairs of hooklets at the posterior end. Every 3–4 weeks, the last segment of the proglottids,

K. Kaczirek · S. Stremitzer
Department of General Surgery, Medical University Vienna, Vienna, Austria

H. Lagler (✉)
Department of Medicine I, Division of Infectious Diseases and Tropical Medicine, Medical University of Vienna, Vienna, Austria

Members of the Interdisciplinary Echinococcosis Board, Medical University of Vienna, Vienna, Austria
e-mail: heimo.lagler@muv.ac.at

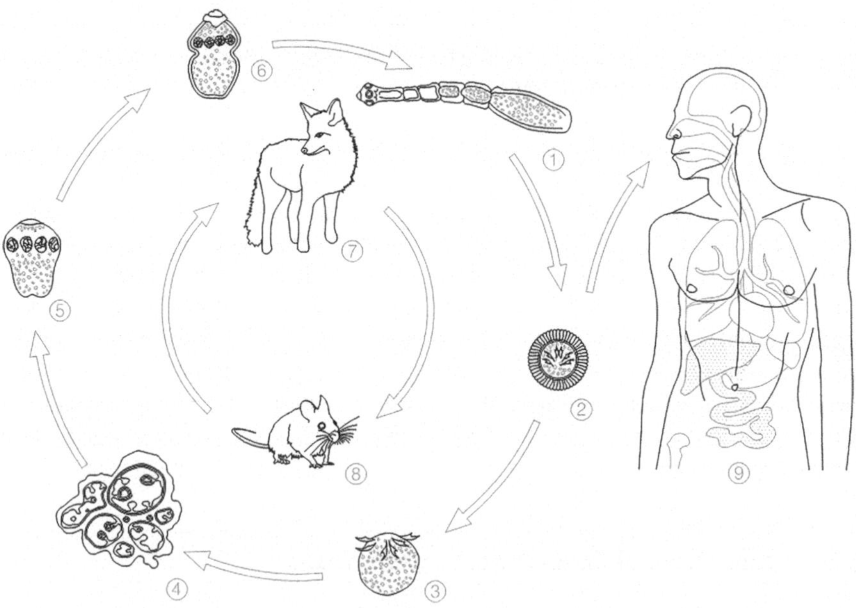


**Fig. 3.1** Life cycle of *E. multilocularis*. The figure was created by Thomas Rohrer and Dr. Dr. Julian Frederic Hotz (Division of Infectious Diseases and Tropical Medicine, Department of Medicine I, Medical University of Vienna, A-1090 Vienna, Austria)

containing mature eggs, detaches and is subsequently expelled through the feces (Fig. 3.1, No. 2). Outside the definitive host, *E. multilocularis* eggs can survive and remain infectious for a considerable period (many months) under specific conditions typical for their endemic area, such as moderately moist and cool temperatures in Central Europe. *E. multilocularis* can infect over 40 different intermediate hosts (Fig. 3.1, No. 8). An intermediate host is an essential host and vector for a parasite, serving as a temporary residence during its asexual life phase. These intermediate hosts of *E. multilocularis* primarily consist of small rodents from families such as *Arvicolidae* (voles), *Soricidae* (shrews), *Talpidae* (moles), *Sciuridae* (squirrels), *Muridae* (mice), *Dipodidae* (jerboas), and *Ochotonidae* (pikas). In Central Europe, notable intermediate hosts include the common vole (*Microtus arvalis*), water vole (*Arvicola terrestris*), and muskrat (*Ondatra zibethicus*) (Aspöck et al. 2002; Auer and Aspöck 2002; Oksanen et al. 2016).

The oncosphere possesses a keratinized embryophore that is liberated from it by the acidity of the stomach. Activation occurs upon contact with bile fluid. Utilizing its six embryonic hooks, the oncosphere attaches itself to the small intestine (Fig. 3.1, No. 3). Subsequently, the parasite is transported via the bloodstream mainly (>99%) to the liver, although less commonly it may reach the lungs, central nervous system, or other organs. Within the liver, the metacestode stage (larva) begins to develop. The metacestode comprises an outer acellular laminated layer and an inner nucleated germinal layer. At a later stage, the metacestode infiltrates the liver parenchyma (Fig. 3.1, No. 4). Through blood and lymphatic vessels, the metacestode tissue can spread to other (mainly neighboring) organs, potentially resembling metastatic dissemination. This infiltrative growth provides *E. multilocularis* with a significant advantage for survival. Due to the small size of its intermediate hosts, the formation of cysts as observed in *E. granulosus* would impose limitations on reproduction.

After several months, the growth of the metacestodes ceases and protoscoleces (invaginated head structures) are formed through asexual budding within the intermediate hosts (Fig. 3.1, No. 5). The infection weakens the intermediate host to the extent that it becomes vulnerable to potential definitive hosts. When a definitive host ingests an infected intermediate host, protoscoleces establish themselves in the small intestine of the definitive host, utilizing their scolex as an anchoring organ and the proliferative zone located behind the scolex as a production site for proglottids (Fig. 3.1, No. 6). Over the course of weeks to months, a new generation of *E. multilocularis* develops.

## 3.2 Life Cycle of *Echinococcus granulosus*

The most important definitive hosts of *E. granulosus* species and genotypes are canids (Fig. 3.2, No. 7). Due to close contact, the domestic dog is the most important definitive host for humans. Aside from the different intermediate and definitive hosts, the life cycle of *E. granulosus* follows a similar pattern to that of *E. multilocularis*. Depending on the strain, the adult *E. granulosus* begins egg production 34–53 days after oral ingestion (Fig. 3.2, No. 1). Mature eggs are released along with the terminal segment of the proglottids every 1–2 weeks and are excreted in the feces. Each detached proglottid contains approximately 1000–1500 mature eggs (Fig. 3.2, No. 2). The intermediate hosts vary greatly depending on the species and genotype (Fig. 3.2, No. 8). They include the families *Bovidae* (ruminants), *Cervidae* (deer), *Suidae* (true pigs), *Camelidae* (camels), *Giraffidae* (giraffids), *Elephantidae* (elephants), and *Hippopotamidae* (hippopotamuses). On the basis of phylogenetic analysis and parasite host specificity in their life cycles, five species belonging to the *E. granulosus* sensu lato *(s.l.)* complex can be differentiated: *E. granulosus* sensu stricto *(s.s.)* (G1–G3), *E. equinus*, *E. ortleppi*, *E. canadensis*, and *E. felidis* (Lymbery 2017). This chapter primarily concerns *E. granulosus s.s.* and secondly *E. canadensis* as those are of paramount importance in human medicine compared to other *E. granulosus* species. The most important intermediate hosts for the G1 strain in humans are sheep and goats (Auer and Aspöck 2002). We recommend referring to the list provided by Casulli et al. for a more detailed compilation of species and genotypes belonging to the *E. granulosus* family responsible for human cystic echinococcosis in Europe (Casulli et al. 2022).

The intermediate host usually ingests the infectious eggs through contaminated food or direct contact with feces (Fig. 3.2, No. 9). Similar to *E. multilocularis*, the oncosphere of *E. granulosus* frees itself from the keratinized embryophore through gastric acid. The oncospheres can then attach to the small intestine using their six

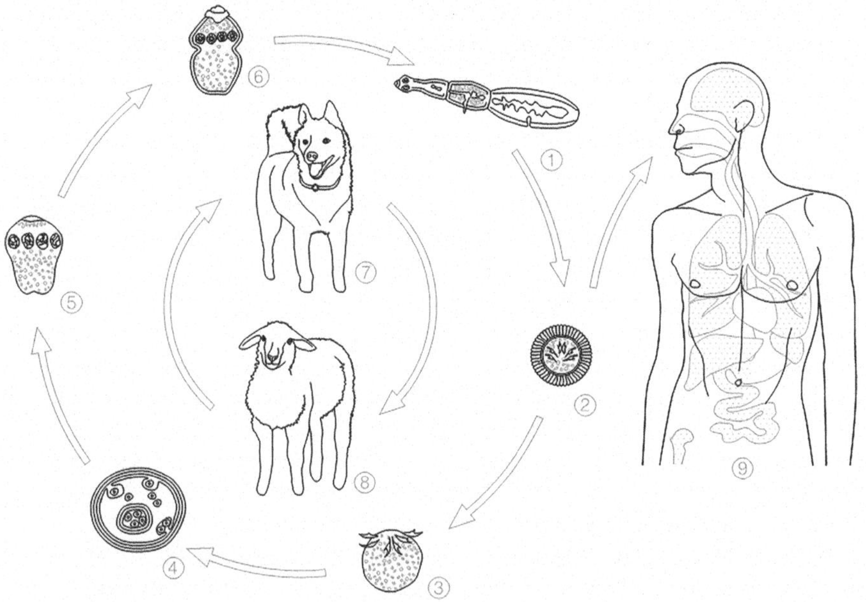


**Fig. 3.2** Life cycle of *E. granulosus* (G1). The figure was created by Thomas Rohrer and Dr. Dr. Julian Frederic Hotz (Division of Infectious Diseases and Tropical Medicine, Department of Medicine I, Medical University of Vienna, A-1090 Vienna, Austria)

embryonic hooks, penetrate it, and distribute hematogenously throughout the body (Fig. 3.2, No. 3). Through the portal circulation, the larvae primarily settle in the liver. Differently to *E. multilocularis*, *E. granulosus* often initially occurs in the lungs, central nervous system, or other organs. At the manifested location, the metacestode stage begins (Fig. 3.2, No. 4). It forms a unilocular or multilocular hydatid cyst. This fluid-filled cyst gradually increases in size and displaces the surrounding tissue. The cyst is equipped with a germinal layer in which protoscoleces develop (Fig. 3.2, No. 5, 6). The life cycle is completed when the definitive host ingests a cyst filled with protoscoleces (Mandal and Deb Mandal 2012). Infection typically occurs when the intermediate hosts are killed by definitive hosts or when they are fed to them without being cooked (Fig. 3.2, No. 7).

## 3.3 Epidemiology of Alveolar and Cystic Echinococcosis in Central Europe

Central Europe is a region in Europe that lacks a clear definition of its geographical boundaries. Traditionally, the following countries are considered part of Central Europe: Austria, Czechia, Germany, Hungary, Liechtenstein, Poland, Slovakia, Slovenia, and Switzerland.

Both alveolar echinococcosis (AE) and cystic echinococcosis (CE) are prevalent in this region. In human medicine, AE is caused by *E. multilocularis*, while CE is mainly caused by *E. granulosus s.s.* (see Sect. 3.3.10), which is part of the *E. granulosus s.l.* complex (Lymbery 2017), but also infections with *E. canadensis* are frequently reported in human medicine. Humans are accidental intermediate hosts for *E. multilocularis* and *E. granulosus s.l.* complex, whereas carnivores serve as definitive hosts. The occurrence of human infections with *Echinococcus* spp. depends on the presence of *E. granulosus s.l.* complex and/or *E. multilocularis* in definitive hosts, as well as human contact with these carnivores.

AE and CE exhibit significant differences in their epidemiology due to their preferred definitive and intermediate hosts. *E. granulosus s.l.* complex primarily involves domesticated carnivores such as dogs, and domesticated intermediate hosts such as cattle, sheep, and camels, in its life cycle (see Sect. 3.2). As a result, CE primarily affects humans who live or work with these animals, such as in rural pastoral communities. On the other hand, AE is traditionally associated with spillover from sylvatic life cycles. However, there is growing evidence that pets such as dogs and cats can also serve as vectors, contributing to human AE.

Both AE and CE are classified as "neglected tropical diseases" by the World Health Organization (WHO), but their occurrence extends beyond tropical or subtropical regions.

This chapter specifically focuses on the epidemiology of AE and CE in humans. A detailed and exhaustive discussion of the occurrence of *E. multilocularis* and *E. granulosus s.l.* complex in animals is not within the scope of this chapter.

### 3.3.1 Epidemiology of Alveolar Echinococcosis

Alveolar echinococcosis is limited to the Northern Hemisphere of the globe. Large areas of Central Europe are endemic for *E. multilocularis*. Other affected areas include Eastern Europe, Central Asia, Japan, and North America. In Central Europe, the red fox and voles are the most relevant definitive and intermediate hosts, respectively, for *E. multilocularis*, making local transmission the almost exclusive source of infection. The first recorded case of AE in a human from Central Europe dates back to 1852 when the German pathologist Ludwig Buhl described a patient with a liver tumor. Comprehensive overviews on the epidemiology of AE in Europe have been previously published (Deplazes et al. 2017; Vuitton et al. 2015). Historically, AE was considered a rare disease in Central Europe. However, in recent years, a significant increase in cases has been observed. For instance, in Switzerland, the annual number of registered AE cases ranged from 8 to 12 before 2000 (Schweiger et al. 2007). In contrast, recent years have seen an alarming rise, with 24–30 cases reported annually, representing a threefold increase. This surge is believed to be (partly) attributable to the expanding fox populations in Europe (Gottstein et al. 2015).

### 3.3.2 Austria

In Austria, the first recorded case of alveolar echinococcosis was documented in 1892 by Adolf Posselt. Since 2004, AE has been classified as a notifiable disease in Austria. The disease tends to occur more frequently in the alpine and western regions of the country (Auer and Aspöck 2002). Until 2010, Austria reported approximately 2 to 3 cases of AE each year. However, there was a dramatic increase in 2011, with 13 cases reported in a single year (Schneider et al. 2013). Since then, the incidence has remained elevated, with 67 cases reported to the Austrian Agency for Health and Food Safety (AGES) between 2017 and 2021 (AGES 2023). The reasons behind this surge are not fully understood, but it is believed that factors such as increasing fox populations and previous underreporting may contribute to the rise (Schneider et al. 2013).

### 3.3.3 Czechia

In Czechia, a total of 36 cases of AE were documented between 1998 and 2016, resulting in a mean incidence of 0.034 cases per 100,000 population (Kolářová et al. 2017). Additionally, two more cases were reported to the European Centre for Disease Prevention and Control (ECDC) in 2018 and 2020, respectively (ECDC 2020). The detection of *E. multilocularis* in wild boars and foxes provides evidence of local circulation of the parasite in Czechia (EFSA 2021; Kolářová et al. 2017).

### 3.3.4 Germany

In Germany, AE is a notifiable disease. The majority of cases are reported from the regions of Bavaria and Baden-Württemberg. According to data from EurEchinoReg between 1982 and 2000, Germany accounted for 24% of the 599 reported cases of AE (Kern et al. 2003). From 2016 to 2020, a total of 220 cases of AE were reported to the ECDC (ECDC 2020). An increase in fox populations and consequently in parasite densities has also been reported from Germany (Romig et al. 2006). Studies have also shown that dogs can carry *E. multilocularis* in Germany (Dyachenko et al. 2008). Dog-ownership, especially when dogs were fed with organic waste, has been linked to an increased risk of AE (Schmidberger et al. 2022).

### 3.3.5 Hungary

The first Hungarian case of human AE was reported in 2008 (Horváth et al. 2008). Another autochthonous case of AE was published in 2017, involving a patient from the south-western region of the country (Dezsényi et al. 2017). A systematic survey conducted between 2003 and 2018 identified a total of 16 cases of AE. The study found that nearly one in five patients died during the study period. The majority of patients originated from the western and northern parts of Hungary, where there is an overlap with infected fox populations (Dezsényi et al. 2021).

### 3.3.6 Poland

In Poland, the first confirmation of *E. multilocularis* was published in 1995 after the detection of adult worms in foxes near Gdansk (Malczewski et al. 1995). Between 1990 and 2011, a total of 121 cases of AE in humans were described, with 23 of them being fatal. North-Eastern Poland was identified as a hotspot for AE cases in humans (Nahorski et al. 2013). It was found that most patients in Poland were involved in agricultural activities or worked in animal shelters, indicating their higher risk of exposure to the parasite (Stefaniak et al. 2023). The presence of *E. multilocularis* in cats and dogs has also been documented, highlighting their potential role in transmitting AE to humans.

### 3.3.7 Slovakia

Surveillance in Slovakia has revealed a high prevalence of *E. multilocularis* infection in red foxes, reaching up to 41.6% in certain areas. Between 2000 and 2013, a total of 26 cases of AE in humans were documented (Antolova et al. 2014). Subsequent years saw additional cases reported to the European Centre for Disease Prevention and Control (ECDC), with a peak of eight cases reported in 2019 (ECDC 2020).

### 3.3.8 Slovenia

The presence of *E. multilocularis* in foxes has also been confirmed in Slovenia (Rataj et al. 2010). In a study with 1263 patients suspected of suffering from echinococcosis, 9 appeared to have specific antibodies against *E. multilocularis* (Logar et al. 2007).

### 3.3.9 Switzerland

Switzerland is recognized as an endemic region for *E. multilocularis*, with hot spots in the northern and western cantons (Deplazes et al. 2017). Between 1982 and 2000, Switzerland accounted for 21% of the 599 reported cases of AE in the European Echinococcosis Registry (EurEchinoReg) database. The incidence of AE in Switzerland is believed to be on the rise, likely due to an increase in the fox population (Schweiger et al. 2007).

### 3.3.10 Epidemiology of Cystic Echinococcosis

Cystic echinococcosis is a rare disease in Europe that disproportionately affects rural and especially pastoral communities. Although it is a notifiable disease in most of Europe (including Central Europe), it is most likely underreported. A recent survey estimated approximately 0.64 cases per 100,000 inhabitants per year for Europe cumulating to a total of 64,745 cases between 1997 and 2021. Out of these cases, 3238 were reported from Central Europe. In a survey focusing on the species responsible for CE in humans in Europe, molecular methods were employed to investigate 599 echinococcal cysts. Most of the isolates (76,8%) were attributed to *E. granulosus s.s.* The second most frequent species was the *E. canadensis* cluster (21.7%). Additionally, a few sporadic cases of CE caused by *E. ortleppi* (1.2%) and *E. vogeli* (0.3%) were identified. Geographical hotspots were identified with Central Europe being at the intersection of these hotspots (Casulli et al. 2022).

### 3.3.11 Austria

In Austria, *E. granulosus s.s.* in definitive hosts is very rare, although the available data is vastly outdated. Prior to 2001, 553 cases had been documented with approximately a third of the cases involving patients of Austrian origin and the other half being migrants from countries where *E. granulosus s.s.* is endemic, such as Turkey or the Balkans (Auer and Aspöck 2002). In recent years, the majority of CE cases in Austria have been diagnosed with migrants, with only a few sporadic cases believed to have been acquired locally. According to a recent survey, a total of 527 cases were diagnosed between 1997 and 2021, with an average of 29 annual cases of CE in Austria based on data from 2017 to 2019 (Casulli et al. 2023).

### 3.3.12 Czechia

In Czechia, 28 cases were documented between 1997 and 2021, with approximately one case per year between 2017 and 2019 (Casulli et al. 2023). Most cases are thought to be imported from elsewhere.

### 3.3.13 Germany

In Germany, 1578 cases were registered between 2001 and 2021 (Casulli et al. 2023). Most cases are imported with only few sporadic cases acquired locally. However, at least two cases have been documented in patients who had never left Germany (Richter et al. 2009).

### 3.3.14 Hungary

Hungary counted 144 cases of human cystic echinococcosis between 2000 and 2020 of which the majority is thought to have been acquired locally (Casulli et al. 2023).

### 3.3.15 Poland

In Poland, 590 cases of CE in humans were reported between 1997 and 2021, with the majority of cases thought to be acquired locally (Casulli et al. 2023). In a molecular analysis of nine patients, *E. canadensis* (G7) was detected in seven and *E. granulosus s.s.* G1 in two patients (Dybicz et al. 2015). Women and persons from rural areas were found to be more frequently affected (Gołąb and Czarkowski 2014; Waloch 2012).

### 3.3.16 Slovakia

In Slovakia, 101 cases of CE in humans were counted between 1997 and 2021. Most cases are thought to be acquired locally, with some imported cases (Casulli et al. 2023).

From Slovenia, 104 cases of human CE were reported between 1997 and 2021, with both locally acquired and imported cases (Casulli et al. 2023). In a molecular analysis of cyst material of 13 patients with CE, seven were identified as *E. granulosus s.s.* G3, five as *E. canadensis*, and one as *E. granulosus s.s.* G1 (Šoba et al. 2020). The eastern part of the country bordering Croatia and Hungary is more affected than the western part (Logar et al. 2008).

### 3.3.17 Switzerland

Switzerland counted 166 cases of human CE between 1997 and 2021, which were all imported (Casulli et al. 2023).

## 3.4 Clinical Findings

In the life cycle of *E. multilocularis* and *E. granulosus*, humans act as accidental intermediate hosts. Following infection with one of those parasites, AE or CE might develop. However, it should be noted that not every contact with infectious eggs automatically leads to a manifest disease. In the majority of exposed individuals, with intact immune systems, the disease does not develop. Genetic factors such as specific HLA characteristics or immunosuppression due to conditions such as HIV or the use of immunosuppressive medications may play a crucial role in the manifestation of the disease, especially of AE (Conraths et al. 2017; Eiermann et al. 1998).

### 3.4.1 Clinical Findings in Patients with Alveolar Echinococcosis

Alveolar echinococcosis is a zoonotic disease and one of the deadliest helminthic infections in humans. It is characterized by multiple small vesicles ranging in diameter from a few millimeters to 3 cm. Infection usually occurs through the oral ingestion of infected eggs by consuming contaminated food. After the activation of the oncosphere in the gastrointestinal tract, the parasite penetrates the intestinal mucosa. Through portal venous transport, the pathogen primarily reaches the liver. Primary extrahepatic locations are very rare (<1% of cases). Within 5–15 years (average incubation period), the metacestode tissue infiltrates the liver, leading to the manifestation of the disease. Symptomatically, the disease can be divided into initial, progressive, advanced, and stable phases. The initial phase is asymptomatic, and the infection can only be detected as an incidental finding during this phase. With an intact immune system, the larvae can be eliminated early. Imaging may reveal "died-out lesions." However, if the immune system is suppressed during this phase, a rapidly progressive or atypical course can occur (Lachenmayer et al. 2019). As the metacestode stage progresses with increasing infiltration, the progressive phase begins. The symptoms are usually nonspecific and include itching, epigastric pain, fatigue, and weight loss. Further on, jaundice and hepatomegaly may develop. The advanced phase is characterized by worsening hepatic dysfunction, leading to portal hypertension and acute liver failure. In addition to severe liver damage, *E. multilocularis* can metastasize via the blood and lymphatic vessels to adjacent tissues and also in distant organs. This represents a malignant progression of the disease. In 34% of initial diagnoses, metastasis to other organs has already occurred (Kern et al. 2003). Metastasis to other organs may cause corresponding symptoms, which can vary depending on the location of the metastases. Lung involvement is the second most probable point of infection (Brunetti et al. 2010).

The 10-year survival rate of AE without treatment is below 10%. With improved diagnostic and therapeutic tools, the 10-year survival rate with consistent treatment exceeds 90% (stable phase). The most common causes of death include septic shock, liver failure, gastrointestinal bleeding, and complications from surgical interventions.

### 3.4.2 Clinical Findings in Patients with Cystic Echinococcosis

After the oral ingestion of infectious *E. granulosus* eggs, similar to intermediate hosts, the oncosphere is activated in the gastrointestinal tract. Following penetration of the small intestine and subsequent portal venous transport, the oncospheres primarily migrate to the liver (50–75%). In contrast to *E. multilocularis*, extrahepatic primary manifestations are more common. The second most common primary site is the lung (20–25%), followed by the kidney, spleen, and central nervous system (together 15–20%) (Brunetti et al. 2010). CE is a disease that affects only one organ in approximately 85–90% of cases (Agudelo Higuita et al. 2016; Bhutani and Kajal 2018; Brunetti et al. 2010). After reaching the organ, it takes at least 10 months (up to 5 years) until the manifestation of the disease. During this phase, patients are usually asymptomatic or have only mild symptoms. After manifestation of the disease, CE is characterized by the formation of mostly unilocular cysts (>70%). Depending on the age, location, and growth rate, these cysts can vary in size and present with variable symptoms. First unspecific symptoms are usually associated with liver manifestation as right-upper quadrant discomfort, jaundice, as well as nausea and/or vomiting or with lung manifestation as cough, chest pain, dyspnea, and hemoptysis (Brunetti et al. 2010). Symptoms can abruptly occur due to complications such as cyst rupture with subsequent infection or anaphylaxis (McManus et al. 2003; Vuitton 1997). Alternatively, the disease can gradually become symptomatic as it progressively displaces tissue in the affected organ. Commonly, there is obstruction of bile ducts or blood vessels, leading to jaundice, gallbladder colic, or portal hypertension. Furthermore, different strains of the parasite demonstrate variable organ involvement and disease progression. For example, the G8 strain is often found in the lung, exhibits relatively slow growth, and rarely leads to serious complications (Moro and Schantz 2006). In contrast, the G6 strain is less frequently found in the lung but has a higher affinity for the central nervous system (Sadjjadi et al. 2013).

## 3.5 Diagnosis

### 3.5.1 Diagnosis of Alveolar Echinococcosis

Alveolar echinococcosis is a difficult-to-diagnose disease. Due to its long incubation period up to 15 years, it is usually detected by routine check-ups. Patients are typically diagnosed between the ages of 40 and 70 years (Grüner et al. 2017a; Piarroux et al. 2011). Many factors need to be considered in order to find the right diagnosis. These include epidemiological data, medical history, clinical examination, imaging, and serological testing. Obtaining useful information through anamnesis is limited due to the long incubation period. Therefore, it is only possible to make assumptions about increased exposure based on epidemiological data (e.g., working in agriculture). Clinical examination has a limited diagnostic value. Due to nonspecific symptoms such as itching, epigastric pain, fatigue, or weight loss, further imaging can sometimes detect a liver manifestation in an early stage, but due to the relatively low prevalence of AE, it is often initially misinterpreted as a malignancy. Depending on the experience and knowledge of the attending physician, it can still take months or years from the onset of symptoms to the diagnosis. Imaging plays a central role in diagnosis, and the primary suspicion is often detected during a routine abdominal ultrasound while investigating another issue. In addition to ultrasound, computed tomography (CT) and magnetic resonance imaging (MRI) have become established methods for diagnosis and follow-ups. In recent times, other methods such as contrast-enhanced ultrasound (CEUS) and combined positron emission tomography (PET) with CT (PET-CT) or MRI (PET-MRI) have also become available (Lotsch et al. 2019). Confirmation of a suspected diagnosis requires subsequent serological testing and direct pathogen detection.

### 3.5.2 PNM Classification

The low incidence of AE makes cross-country and standardized research difficult. As a therapy should be evaluated based on evidence, there is a need to classify patients according to the extent of their parasitic infection. Therefore, based on the TNM (tumor, node, metastasis) classification of hepatocellular carcinoma, the WHO Informal Working Group on Echinococcosis (WHO-IWGE) developed the PNM classification for AE (Agudelo Higuita et al. 2016; Brunetti et al. 2010; Kern et al. 2006).

The PNM classification can lead to more specific and evidence-based treatment concepts, which is particularly important due to the relatively low number of AE cases. Another classification based on the PNM classification is the stage grouping, which further divides the PNM classification into stages I–IV (Figs. 3.3 and 3.4).

### 3.5.3 Imaging

Imaging techniques play a crucial role in the initial diagnosis and follow-up of AE. The discovery of the disease is often incidental during routine ultrasound (Fig. 3.5), which can detect various morphologies. Common findings in ultrasound are juxtaposition of hyper- and hypoechogenic areas in a pseudo-tumor with irregular limits and scattered calcification and pseudo-cystic appearances due to a large area of central necrosis surrounded by an irregular hyperechogenic ring (Bresson-Hadni et al. 2006). Less typical findings include hemangioma-like lesions, which are often radiologically mistaken for hemangiomas and can therefore lead to delayed diagnosis. The metastasis-like lesions are often serologically and radiologically inconspicuous, so they may be misdiagnosed as oncological liver metastases. However, they might be differentiated from oncological liver metastases using the *E. multilocularis* Ulm Classification—Ultrasound (EMUC-US) classification (Grüner et al. 2017b; Kratzer et al. 2015).

To capture all characteristics of various lesions and possible metastasis, other imaging techniques such as CT or MRI should be used (Grüner et al. 2017a; Kodama et al. 2003). CT provides detailed anatomical and morphological information about the lesions and is very effective in visualizing the characteristic pattern of calcification (WHO/OIE 2001). When there is diagnostic uncertainty, MRI can reveal the multivesicular morphology of the lesions, which helps support the diagnosis. A PET-CT or PET-MRI scan can detect active liver or metastasized lesions even in the absence of clinical symptoms and before recurring disease becomes detectable by conventional imaging methods. Currently, PET-CT is the preferred method for evaluating the metabolic activity of parasitic lesions and is also one of the most

| PNM classification of alveolar echinococcosis | |
|---|---|
| P | Hepatic localisation of the *P*arasite |
| P X | Primary tumor cannot be assessed |
| P 0 | No detectable tumor in the liver |
| P 1 | Peripheral lesions without proximal vascular and/or biliar involvement |
| P 2 | Central lesions with proximal vascular and/or biliar involvement of one lobe[a] |
| P 3 | Central lesions with hilar vascular or biliar involvement of both lobes and/ or with involvement of two hepatic veins |
| P 4 | Any liver lesion with extension along the vessels[b] and the biliary tree |
| N | Extra hepatic involvement of neighbouring organs [diaphragm, lung, pleura, pericardium, heart, gastric and duodenal wall, adrenal glands, peritoneum, retroperitoneum, parietal wall (muscles, skin, bone), pancreas, regional lymph nodes, liver ligaments, kidney] |
| N X | Not evaluable |
| N 0 | No regional involvement |
| N 1 | Regional involvement of contiguous organs or tissues |
| M | The absence or presence of distant *M*etastasis [lung, distant lymph nodes, spleen, CNS, orbital, bone, skin, muscle, kidney, distant peritoneum and retroperitoneum] |
| M X | Not completely evaluated |
| M 0 | No metastasis[c] |
| M 1 | Metastasis |

[a] For classification, the plane projecting between the bed of the gall bladder and the inferior vena cava divides the liver in two lobes.
[b] Vessels mean inferior vena cava, portal vein and arteries.
[c] Chest X-ray and cerebral CT negative.

| PNM stage grouping of alveolar echinococcosis | | | |
|---|---|---|---|
| Stage I | P1 | N0 | M0 |
| Stage I | P2 | N0 | M0 |
| Stage IIIa | P3 | N0 | M0 |
| Stage IIIb | P1–3 | N1 | M0 |
| | P4 | N0 | M0 |
| Stage IV | P4 | N1 | M0 |
| | Any P | Any N and/or | M1 |

**Fig. 3.3** PNM classification and PNM stage grouping of alveolar echinococcosis. Kern P, Wen H, Sato N, Vuitton DA, Gruener B, Shao Y, et al. WHO classification of alveolar echinococcosis: Principles and application. Parasitology International. 2006;55:S283-S7

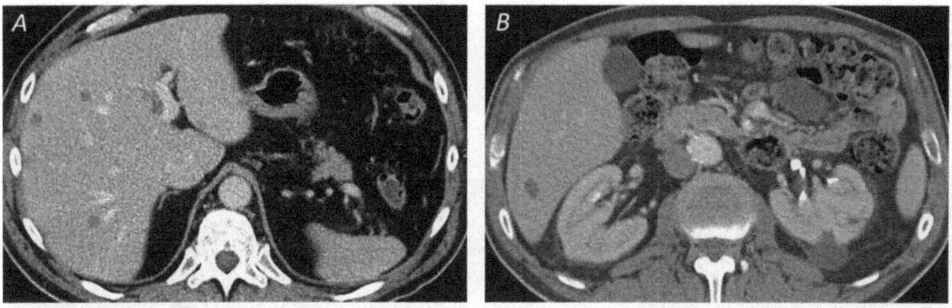


**Fig. 3.4** Especially lower stages of the PNM classification can be mistaken for benign liver diseases in imaging. (**a**) shows three small *E. multilocularis* lesions in the liver on contrast-enhanced CT, P1 N0 M0, Stage I. (**b**) depicts a small dysontogenetic liver cystic lesion on contrast-enhanced CT venous phase. Courtesy of Dr. med. univ. Fredrik Waneck (Department of Biomedical Imaging and Image-Guided Therapy, Division of Cardiovascular and Interventional Radiology, Medical University of Vienna, Währinger Gürtel 18–20, 1090 Vienna, Austria)

important methods for monitoring the progress of AE due to its availability. Given that the majority of patients undergo long-term follow-up controls from years to decades, they receive a cumulative exposure of radiation. Hence, whenever feasible, prioritizing the utilization of PET-MRI is advisable (Lotsch et al. 2019). The Fluorodeoxyglucose (FDG) uptake is often observed at the periphery of the lesion, reflecting the patient's inflammatory reaction, while the central region of the lesions usually does not show FDG uptake. So, a reduction in FDG activity over the long term does not necessarily indicate parasite death, but rather suppression of the host's inflammatory reaction. Therefore, follow-up and treatment decisions based solely on PET-CT lead to a higher recurrence rate (Reuter et al. 2004) (Fig. 3.6).

### 3.5.4 Serology

In cases of suspicious AE based on imaging, blood samples should be taken from every patient for serological testing. The serological confirmation serves to demonstrate specific antibodies and is usually performed using enzyme-linked immunosorbent assay (ELISA) as a basic test and Western blot (WB) as a confirmation test. Tests utilizing *E. multilocularis* crude antigens (ISPTM) and/or antigens such as Em2, Em18, Em alkaline phosphatase, rec II/3–10, and Em10 antigens are used (Gottstein et al. 1989, 2019; Ito et al. 1993, 1999; 2002; Sarciron et al. 1997). It has been shown that all previously used antigens (Em10, Em4, EmII/3 or EmII/3–10, and Em18) come from the same protein family as Em10. Em10 has functional similarities to human ERM (Ezrin, Radixin, and Moesin). Therefore, false positive results are more likely. Em18 is the smallest product of Em10 and shows the least functional similarity to ERM. These tests have a sensitivity ranging above 90.0%, but a low specificity. The sensitivity of the ELISAs may vary due to the different properties of the pathogen strains and the different organ localizations. To enhance specificity, a combination of Em2 and II/3–10 antigens, known as Em2+ ELISA, is widely used. It can differentiate

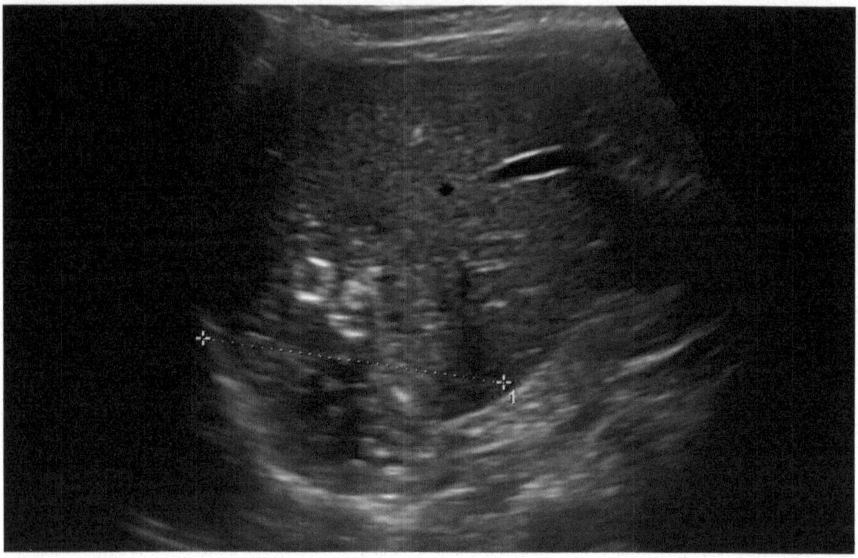

**Fig. 3.5** The patient underwent a routine ultrasound examination as part of a preventive check-up. In ultrasound, an inhomogeneous lesion with centrally hypoechoic components was observed in the liver. After further imaging: P2 N0 M0, Stage II. Courtesy of Dr. med. univ. Fredrik Waneck (Department of Biomedical Imaging and Image-Guided Therapy, Division of Cardiovascular and Interventional Radiology, Medical University of Vienna, Währinger Gürtel 18–20, 1090 Vienna, Austria)

between AE and CE with an accuracy of 90.0–95.0% and exhibits a sensitivity of >97.0% and a specificity of >98.0% (Gottstein et al. 1993). Secondary test methods include Western blots, immunoblots, ELISAs with *E. multilocularis* PSC antigens, or IgG4 subtypes (Dreweck et al. 1997; Gottstein et al. 2019; Ito et al. 1993).

However, serology should not only be seen as confirmation of imaging. As serological tests with high specificity are continuously being developed, while maintaining high sensitivity, they can increasingly compete with the importance of PET and might in the long term contribute to reducing radiation exposure (Hotz et al. 2022).

### 3.5.5 Direct Pathogen Detection

During a resection, a sample should always be taken for further examination. The sample can then be histologically processed and undergo polymerase chain reaction (PCR), allowing for direct detection of *E. multilocularis*-specific DNA (Schneider et al. 2008). If an operation is necessary, it is imperative to intraoperatively obtain a specimen for histological processing (Figs. 3.7 and 3.8).

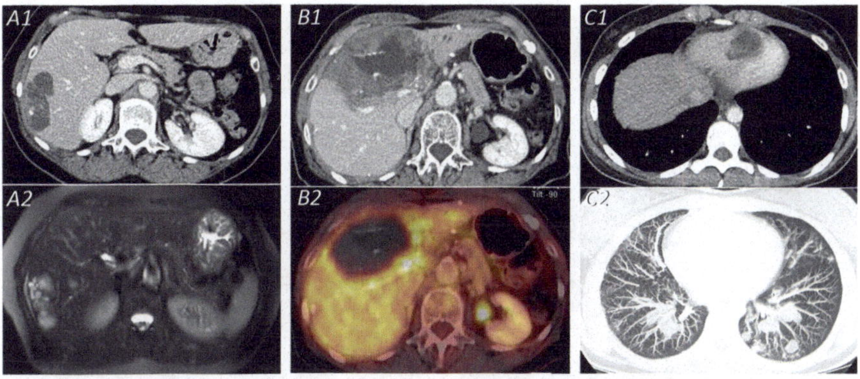


**Fig. 3.6** Left: Liver lesion without involvement of vessels or bile ducts on CT (**a1**) and T2-weighted MRI (**a2**). The lesion in segment VII is clearly visible on both CT and T2-weighted MRI, but the alveolar structure is much better differentiated on MRI. Middle: Lesion with spread along the vessels and infiltration of the peritoneum, central necrosis, and increased cellular activity at the liver hilum (**b1**, **b2**). The central necrosis and the surrounding tissue show no activity on PET-CT (**b2**). There is no vascularization or cellular activity in this area. The slight increase in activity at the medial edge of the lesion near the liver hilum supports the suspicion of hilar vascular infiltration with spread along the vessels. Right: Metastasized *E. multilocularis* in the heart (**c1**) and lung (**c2**). These lesions do not exhibit the typical central necrosis seen in most AE lesions in the liver. However, they classify as alveolar echinococcosis due to the metastasis in stage 4. Courtesy of Dr. med. univ. Fredrik Waneck (Department of Biomedical Imaging and Image-Guided Therapy, Division of Cardiovascular and Interventional Radiology, Medical University of Vienna, Währinger Gürtel 18–20, 1090 Vienna, Austria)

### 3.5.6 Case Definitions (Table 3.1)

**Table 3.1** Modified according to Brunetti E, Kern P, Vuitton DA. Expert consensus for the diagnosis and treatment of cystic and alveolar echinococcosis in humans. Acta Trop. 2010;114(1):1–16

| Case Definitions of alveolar echinococcosis | |
|---|---|
| **Confirmed** | Histopathology compatible with AE[a] *AND/OR* |
| | Detection of *E. multilocularis* nucleic acid sequence(s) in a clinical specimen. *AND* case definition probable AE |
| **Probable** | A typical organ lesion confirmed by imaging (ultrasound, computed tomography, magnetic resonance imaging) AND |
| | The detection of AE serum antibodies through highly sensitive serological tests and confirmation through serological tests with high specificity. |
| **Possible** | Appropriate clinical OR epidemiological history. |
| | Suggestive imaging OR positive AE serology. |

[a]Alveolar echinococcosis

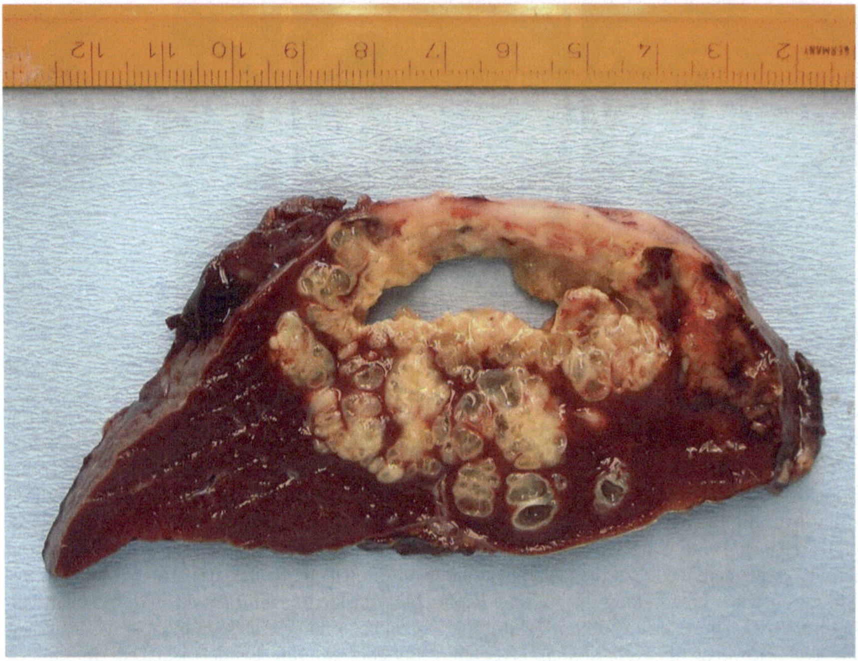

**Fig. 3.7** Macroscopic preparation of an alveolar echinococcosis of the spleen. Courtesy of Univ.-Prof. Dr. Herbert Auer (Medical Parasitology, Institute of Specific Prophylaxis and Tropical Medicine (ISPTM), Center for Pathophysiology, Infectiology and Immunology, Medical University of Vienna, Kinderspitalgasse 15, 1090 Vienna, Austria). Austrian Reference Center for Echinococcosis and other Parasitic Diseases (ÖRZP)

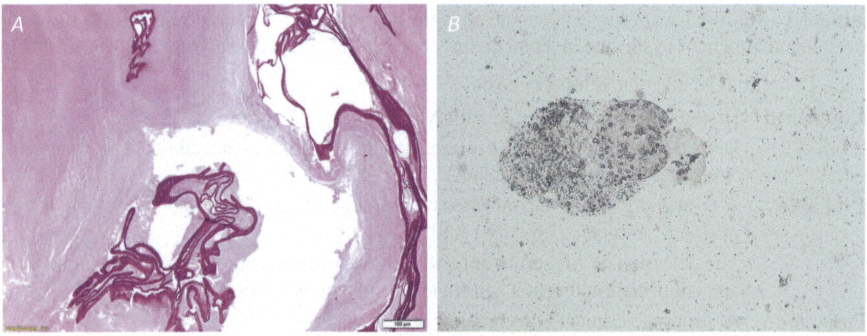


**Fig. 3.8** Histologic section of the liver of an Austrian patient, showing the PAS-stained laminated layers without protoscoleces (**a**). Degenerating protoscolex from a liver lesion (left lobe) of an Austrian AE patient—native, unstained (**b**). Courtesy of Univ.-Prof. Dr. Herbert Auer (Medical Parasitology, Institute of Specific Prophylaxis and Tropical Medicine (ISPTM), Center for Pathophysiology, Infectiology and Immunology, Medical University of Vienna, Kinderspitalgasse 15, 1090 Vienna, Austria). Austrian Reference Center for Echinococcosis and other Parasitic Diseases (ÖRZP)

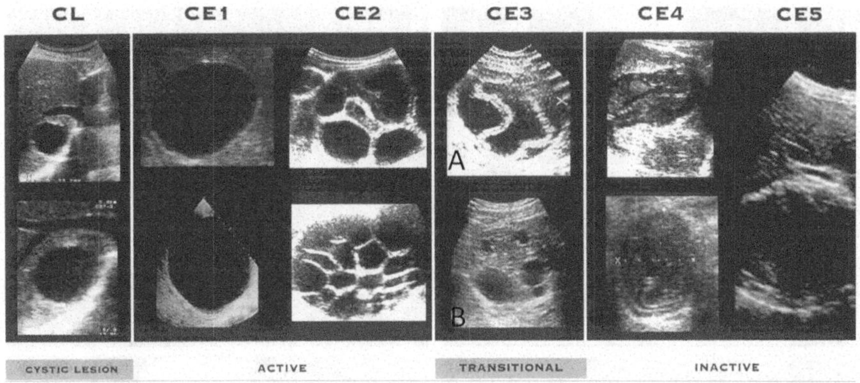


**Fig. 3.9** WHO-IWGE standardized classification. WHO Informal Working Group. International classification of ultrasound images in cystic echinococcosis for application in clinical and field epidemiological settings. Acta Tropica. 2003;85(2):253–61

## 3.6 Alveolar Echinococcosis

### 3.6.1 Diagnosis of Cystic Echinococcosis

With up to 5 years, cystic echinococcosis has a shorter incubation period than AE. For this reason, there are often more anamnestic data available. In order to find a diagnosis, epidemiological data (such as origin, travel history, etc.) should be ascertained, especially in non-endemic areas. The disease primarily targets a singular organ, with varying cyst size, location, growth rate, and activity, leading to the manifestation of multiple nonspecific symptoms following an initial asymptomatic phase. Consequently, clinical examination is in an inferior position to imaging techniques and serological testing in terms of diagnosis. However, the suspected diagnosis is often based on routine ultrasound.

### 3.6.2 Imaging

CE is easier to detect with imaging techniques compared to AE. Ultrasound is the preferred method for primary diagnosis and follow-up of CE, particularly in cases with liver manifestation, due to its frequent availability and sufficient evaluability of the cysts (Brunetti et al. 2010). In ultrasound, both the endocyst and the human pericyst are clearly visible. Gharbi et al., and subsequently the WHO-IWGE, developed a standardized classification for ultrasound (Gharbi et al. 1981; Working Group 2003) (Fig. 3.9).

This classification differentiates between active cysts (CE1, CE2), transitional cysts (CE3), and inactive cysts (CE4, CE5). The CL (cystic lesion) stage includes unilocular cysts, which are suspicious for *E. granulosus*, but other differential

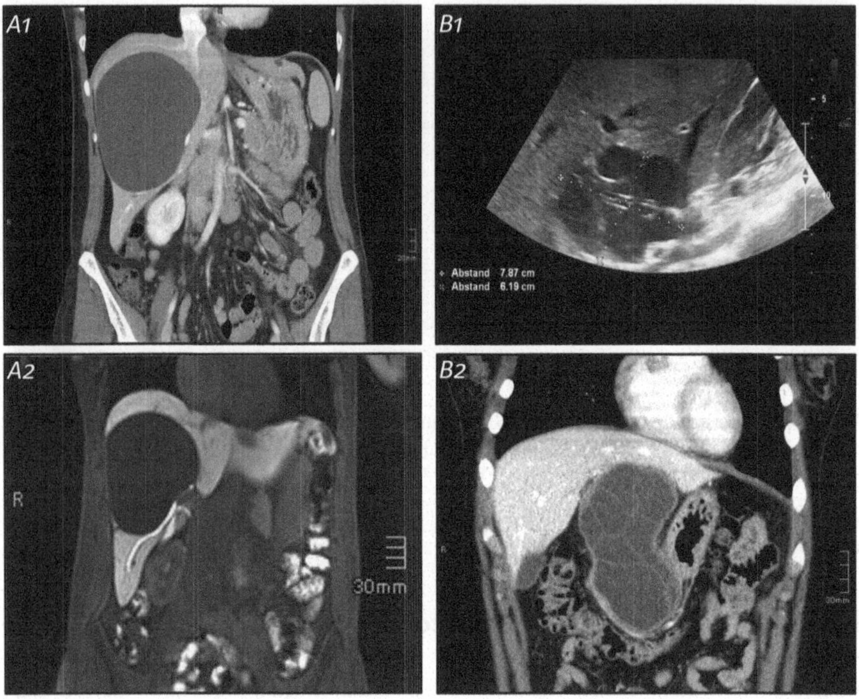


**Fig. 3.10** Left: Unilocular hypodense liver cyst CE1 in CT (**a**1) and T1-weighted MRI (**a**2). Due to the fluid content within the cyst, the lesions appear hypointense in T1-weighted imaging. Right: Ultrasound of a CE2 cyst in the liver, with septa delineating daughter cysts (**b**1). (**b**2) shows a large CE2 cyst caudal to the liver in the peritoneum, with multiple septa and daughter cysts. The strong cyst wall is clearly visible. Courtesy of Dr. med. univ. Fredrik Waneck (Department of Biomedical Imaging and Image-Guided Therapy, Division of Cardiovascular and Interventional Radiology, Medical University of Vienna, Währinger Gürtel 18–20, 1090 Vienna, Austria)

diagnoses must be excluded through further testing. The classification can also be seen in a temporal context. Stages CE1 and CE2 are usually detected in early phases, while stages CE4 and CE5 are detected in late phases of the disease. This sonography schema was primarily designed for liver involvement but can also be considered for other organs. Additionally, this schema can be used for CT and MRI. CE1 (Fig. 3.10a) is an active unilocular cyst with uniformly echo-free or low-echogenic (brood capsules, snowflake sign, hydatid sand) content, whereas stage CE2 (Fig. 3.10b) contains multiple and multiseptated cysts, possibly with daughter cysts.

CE3 can be further divided into CE3a (Fig. 3.11) and CE3b. CE3a ("water-lily sign") can be either active or inactive, whereas CE3b stage is always active. CE3b includes unilocular cysts (possibly with daughter cysts) with echo-free and echo-rich contents.

Due to the less infiltrative growth of *E. granulosus* compared to *E. multilocularis*, a PET is usually not necessary for further evaluation and follow-ups after staging using CT or MRI. Both CT and MRI can be used for the diagnosis of extrahepatic manifestations. MRI provides the best opportunity for visualizing cerebral cysts and,

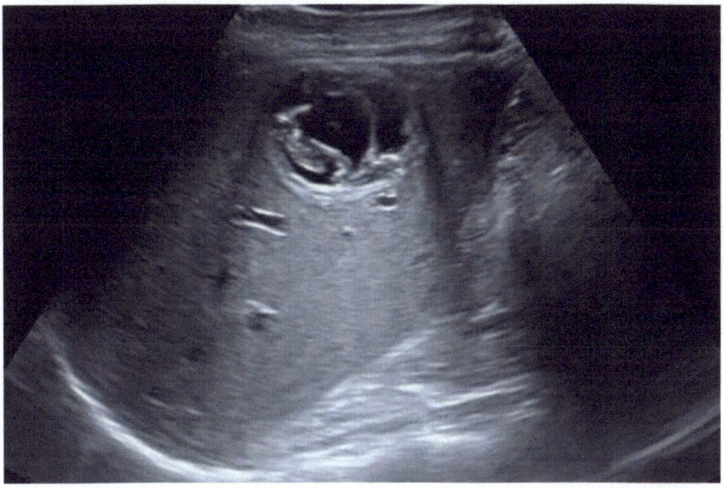

**Fig. 3.11** During a routine ultrasound examination, a patient presents a high suspicion of cystic echinococcosis. Sonographically, the lesion appears as a well-defined, highly suggestive CE3a stage, exhibiting an echogenic "water-lily"-shaped structure within an otherwise hypoechoic content. Courtesy of Dr. med. univ. Fredrik Waneck (Department of Biomedical Imaging and Image-Guided Therapy, Division of Cardiovascular and Interventional Radiology, Medical University of Vienna, Währinger Gürtel 18–20, 1090 Vienna, Austria)

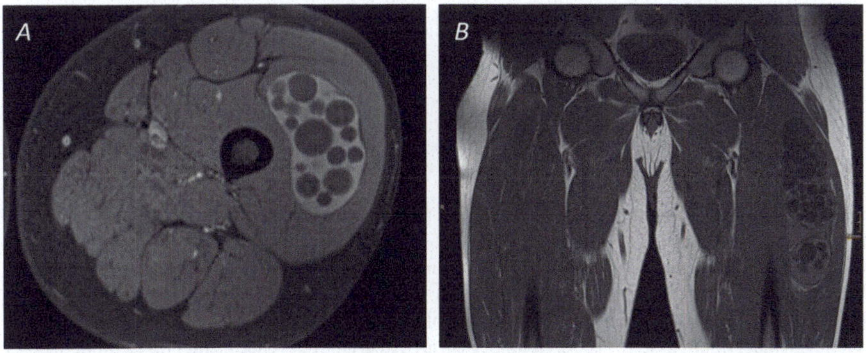


**Fig. 3.12** Atypical primary manifestation of cystic echinococcosis (CE3b) in the thigh in a patient from Syria. In axial T2-weighted (**a**) and sagittal T1-weighted (**b**) sections. Courtesy of Dr. med. univ. Fredrik Waneck (Department of Biomedical Imaging and Image-Guided Therapy, Division of Cardiovascular and Interventional Radiology, Medical University of Vienna, Währinger Gürtel 18–20, 1090 Vienna, Austria)

due to its high soft tissue contrast, offers optimal visualization of cardiac or pelvic cysts. A strongly T2-weighted MRCP (magnetic resonance cholangiopancreatography) should be performed for cysts larger than 5 cm prior to planned percutaneous-interventional therapy procedures to exclude a relevant bile fistula (Fig. 3.12).

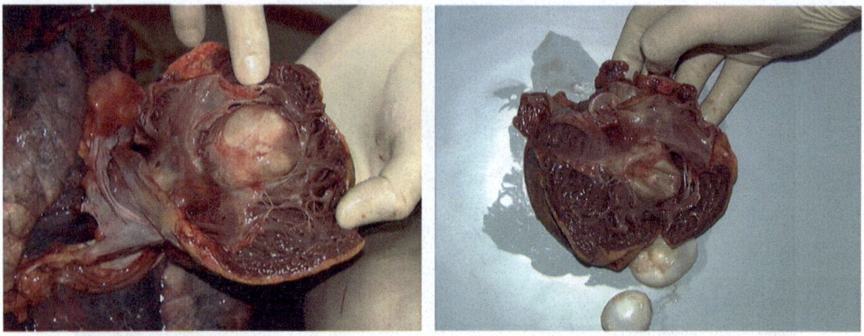

**Fig. 3.13** The image depicts the intraoperative macroscopic presentation of a cystic echinococcosis in an atypical location within the cardiac region. Courtesy of Univ.-Prof. Dr. Herbert Auer (Medical Parasitology, Institute of Specific Prophylaxis and Tropical Medicine (ISPTM), Center for Pathophysiology, Infectiology and Immunology, Medical University of Vienna, Kinderspitalgasse 15, 1090 Vienna, Austria). Austrian Reference Center for Echinococcosis and other Parasitic Diseases (ÖRZP)

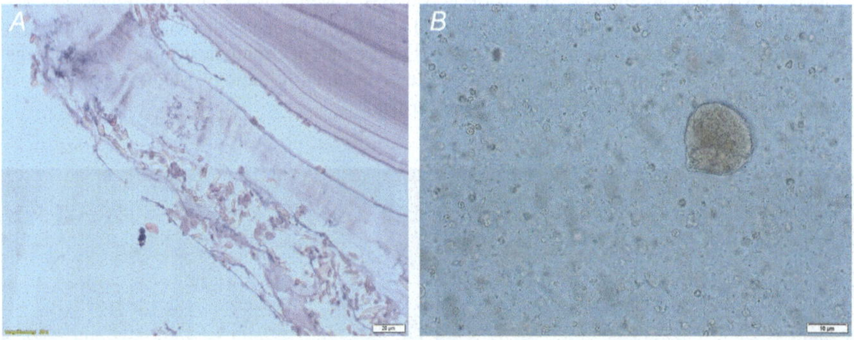


**Fig. 3.14** Histologic section of the liver of a CE patient showing the typical laminated layer of an *Echinococcus granulosus* cyst—HE staining (**a**). Isolated and degenerated protoscolex of a cerebral *Echinococcus granulosus* cyst of a Philippinian CE patient (**b**). Courtesy of Univ.-Prof. Dr. Herbert Auer (Medical Parasitology, Institute of Specific Prophylaxis and Tropical Medicine (ISPTM), Center for Pathophysiology, Infectiology and Immunology, Medical University of Vienna, Kinderspitalgasse 15, 1090 Vienna, Austria). Austrian Reference Center for Echinococcosis and other Parasitic Diseases (ÖRZP)

### 3.6.3 Serology

The sensitivity for ELISA (Eg-ELISA, based on *E. granulosus* hydatid fluid antigen EgHF), indirect hemagglutination inhibition test (IHA), or latex agglutination test is between 85.0 and 98.0% for liver cysts, 50.0–60.0% for pulmonary cysts, and 90.0–100% for multiple organ involvement (Brunetti et al. 2010; Ito and Craig 2003). However, specificity is limited by cross-reactions with other parasites such as *E. multilocularis* or *T. solium*, malignant tumors, liver cirrhosis, or other liver diseases. Therefore, an Eg-Western blot should be used for confirmation (55). If this is not sufficient to differentiate between *E. granulosus* and *E. multilocularis*, the diagnosis can only be confirmed through PCR of biopsy or surgical specimens. However, because of unambiguously epidemiological data and/or imaging, there is often no need to differ between AE and CE.

### 3.6.4 Direct Pathogen Detection

PCR testing of the affected tissue has proven to be a valuable tool for the detection and characterization of *E. granulosus* and its various species and genotypes associated with cystic echinococcosis (Schneider et al. 2008). Distinguishing between different *Echinococcus granulosus* species and genotypes is of scientific importance. However, at the present time, species that cannot be attributed to *E. granulosus s.s.* are treated in the same manner as *E. granulosus s.s.* Additionally, it should be ensured that a histological analysis is consistently performed during surgical procedures (Figs. 3.13 and 3.14).

### 3.6.5 Case Definitions (Table 3.2)

**Table 3.2** Modified according to Brunetti E, Kern P, Vuitton DA. Expert consensus for the diagnosis and treatment of cystic and alveolar echinococcosis in humans. Acta Trop. 2010;114(1):1–16

| Case Definitions of Cystic Echinococcosis | |
|---|---|
| **Confirmed** | Probable CE[a] plus EITHER (1) demonstration of protoscoleces or their components, using direct microscopy or molecular biology, in the cyst contents aspirated by percutaneous puncture or at surgery, OR |
| | (2) changes in US[b] appearance, e.g., detachment of the endocyst in a CE1 cyst, thus moving to a CE3a stage, or solidification of a CE2 or CE3b, thus changing to a CE4 stage, after administration of ABZ[c] (at least 3 months) or spontaneous. |
| **Probable** | Any patient with the combination of clinical history, epidemiological history, imaging findings |
| | AND serology positive for CE on two tests |
| **Possible** | Any patient with a clinical OR epidemiological history |
| | AND imaging findings OR serology positive for CE |

[a]Cystic Echinococcosis
[b]Ultrasound
[c]Albendazole

## 3.7 Treatment

### 3.7.1 Treatment of Alveolar Echinococcosis

Alveolar echinococcosis is a very rare disease, which is why affected individuals should be treated at experienced centers with interdisciplinary boards. The treatment involves pharmacological interventions, typically administered with benzimidazoles (especially albendazole), as well as surgical approaches. However, in stable patients, a watch-and-wait strategy can also be considered after initial treatment. For all newly diagnosed patients, initiating temporary or lifelong pharmacological therapy is recommended, and an R0 resection should be pursued if operability is achievable. To enable comparability among individual patients, Brunetti et al. developed a treatment scheme based on the WHO PNM classification (see Sect. 3.5.2), providing recommendations for acute therapy and subsequent follow-up assessments (Brunetti et al. 2010).

## 3.8 Benzimidazole

With the introduction of pharmacological therapy in the 1980s, the treatment of AE patients underwent a significant change, leading to a shift in treatment focus from surgery to internal medicine. According to the current pharmacological treatment scheme (Table 3.3), pharmacological therapy with benzimidazoles (BMZ) is indicated in all stages of the WHO PNM classification (Brunetti et al. 2010; Kern et al. 2006). For the stages P1N0M0 and P2N0M0, a BMZ therapy for 2 years is recommended following an R0 resection. Subsequently, in the presence of normal imaging and serology, an attempt to discontinue pharmacological therapy with regular monitoring should be initiated. From stage P3N0M0 onward, lifelong palliative BMZ therapy is envisaged (Brunetti et al. 2010). However, even in these cases,

**Table 3.3** Stage-specific approach to alveolar echinococcosis. Modified according to Brunetti E, Kern P, Vuitton DA. Expert consensus for the diagnosis and treatment of cystic and alveolar echinococcosis in humans. Acta Trop. 2010;114(1):1–16

| WHO classification | Surgery | Interventional treatment | Drug therapy | Suggested |
|---|---|---|---|---|
| P1N0M0 | Yes | No | Yes | Optimal: Radical resection (R0); BMZ[a] for 2 years; PET-CT controls<br>Minimal: Radical resection (R0); BMZ for 3 months |
| P2N0M0 | Yes | No | Yes | Optimal: Radical resection (R0); BMZ for 2 years; minimal: Radical resection (R0);BMZ for 3 months |
| P3N0M0 | No | No | Yes | Optimal: BMZ continuously; PET-CT or -MRI initially and in 2 years intervals; minimal: BMZ continuously |
| P3N1M0 | No | Yes | Yes | Optimal: BMZ continuously; PET-CT or -MRI initially and in 2 years intervals;minimal: Surgery, in indicated |
| P4N0M0 | No | Yes | Yes | Optimal: BMZ continuously; PET-CT or -MRI initially and in 2 years intervals;minimal: Surgery, in indicated |
| P4N1M1 | No | Yes | Yes | Optimal: BMZ continuously; PET-CT or -MRI initially and in 2 years intervals;minimal: Surgery, in indicated |

[a]benzimidazole

under strict risk–benefit assessment and in the absence of abnormal PET-CT and/or serological findings, a structural treatment interruption (STI) of BMZ therapy may be considered (Ammann et al. 2015; Grüner et al. 2017a; Hotz et al. 2022).

The drug of choice is albendazole (ABZ), an oral benzimidazole. Treatment with ABZ is usually performed with 10–15 mg/kg per day divided in two doses resulting in a typical dose of 400 mg twice daily. Alternatively, mebendazole (MBZ) can be used divided in three daily doses with a total dose of 40–50 mg/kg per day. The most common side effects include gastrointestinal complaints such as diarrhea, nausea, and vomiting (Grüner et al. 2017a). Furthermore, hair loss, neutropenia, or elevated liver enzymes can occur less frequently. Among those experiencing side effects, in a well-characterized cohort at the university hospital of Ulm, 6.9% had to discontinue ABZ therapy due to pronounced hepatotoxicity (Grüner et al. 2017a). Rarely, if BMZ cannot be taken due to intolerance or exceptionally rarely due to ineffectiveness, amphotericin B or nitazoxanide might be treatment options with usually a lack of effect and high side effects. ABZ therapeutic drug monitoring (TDM) should be determined to evaluate treatment success or to confirm a true treatment failure. BMZ should not be taken during pregnancy or while breastfeeding due to their teratogenic effects. Women of childbearing age should be advised about strict contraception methods.

Unlike the pharmacological therapy scheme with BMZ, surgical treatment approaches offer a potential curative option. Therefore, interdisciplinary exchange is crucial to select the optimal therapeutic approach for the affected patient. Surgical interventions should always involve an experienced AE team to avoid further dissemination during the operation. Surgical treatment of AE of the liver is similar to surgical treatment of malignancies. The aim is to remove all disease with a clear margin (1 mm if possible). An R0 resection provides the only curative approach for AE and is the best possible treatment option, particularly for low PNM classifications (Brunetti et al. 2010; Kern et al. 2006). However, in cases of good operability and low postoperative risk, an R0 resection can also be considered for higher stages. An R1 resection margin (microscopic disease at the surgical margin) is acceptable at vital structures, e.g., hepatic vein, portal vein, bile duct, if they are not invaded and all macroscopic abnormalities can be removed. To assess resectability, MRI with hepatocyte-specific contrast agents, e.g., gadolinium ethoxybenzyl diethylenetriamine pentaacetic acid (Gd-EOB DTPA) to identify all lesions and 18F-FDG-PET-CT with delayed images to evaluate resection margins, should be routinely performed prior to surgery. General principles of liver surgery also apply to surgical treatment of AE, such as conserving a sufficient size of the future liver remnant, arterial and portal venous inflow, hepatic venous outflow, and biliary drainage. Additional methods to preoperatively augment the future liver remnant such as portal vein embolization should be considered. However, especially in elderly patients or those with severe comorbidities, an interdisciplinary board should take place to determine whether surgery or conservative management should be preferred. In rare cases where a patient suffers from irreversible and severe liver damage, liver transplantation may be considered.

### 3.8.1 Treatment of Cystic Echinococcosis

The recommended therapy for cystic echinococcosis depends on the localization, size of the cyst(s), and activity stage of the disease. The decision regarding the best approach should be made individually for each patient. Similar to AE, the WHO has developed treatment recommendations for uncomplicated CE of the liver. It should be noted here that the minimal and optimal treatment recommendations have also been adjusted according to the medical resources available in each respective country. Furthermore, it is important to know that treatment schemes are developed by expert recommendations without randomized controlled trials (Table 3.4).

The goal of all surgical and percutaneous interventional procedures is to remove the active germinal wall and the metacestodes. It is always necessary to consider whether resection is possible without the leakage of active cyst fluid, which could lead to further dissemination of *E. granulosus* and to anaphylaxis. The more radical the intervention, the higher the surgical risk. Surgery such as pericystectomy, closed total pericystectomy, open total pericystectomy, and laparoscopic surgery (combined non-PAIR) are indicated for complicated cysts of stages CE2–CE3b with multiple daughter cysts or when percutaneous procedures are not available. Closed pericystectomy (removal of the entire cyst including the germinal wall) is preferred because it has a lower risk of seeding compared to open pericystectomy and laparoscopic surgery is just indicated for selected cases due to missing evidence (Brunetti et al. 2010; Filice et al. 2000). At Medical University of Vienna, liver resection is also performed for CE1 and CE3a if the cyst size is >10 cm or if interventional methods, e.g., PAIR, are contraindicated due to biliary fistulas. The aim is to

**Table 3.4** Suggested stage-specific approach to uncomplicated cystic echinococcosis of the liver. Modified according to Brunetti E, Kern P, Vuitton DA. Expert consensus for the diagnosis and treatment of cystic and alveolar echinococcosis in humans. Acta Trop. 2010;114(1):1–16

| WHO classification | Surgery | Percutaneous treatment | Drug therapy | Suggested |
|---|---|---|---|---|
| CE1 | No | Yes | Yes | Optimal: <5 cm ABZ[a]; >5 cm PAIR[b] + ABZ<br>Minimal: <5 cm PAIR; >5 cm PAIR |
| CE2 | Yes | Yes | Yes | Optimal: Other PT[c] + BZ<br>minimal: Other PT |
| CE3a | No | Yes | Yes | Optimal: <5 cm ABZ; >5 cm PAIR + ABZ<br>Minimal: <5 cm PAIR; >5 cm PAIR |
| CE3b | Yes | Yes | Yes | Optimal: Non-PAIR PT + ABZ<br>minimal: Non-PAIR PT |
| CE4 | No | No | No | Optimal: Watch and wait |
| CE5 | No | No | No | Optimal: Watch and wait |

[a]Albendazole
[b]Puncture, Aspiration, Injection, Reaspiration
[c]Percutaneous treatment

perform a pericystectomy without opening the cyst to avoid spillage of cyst content. In cases where major vital structures of the liver are at risk, cysts can be opened, evacuated, and irrigated with protoscoicidal agents. Orifices of biliary fistulas should be closed with sutures to avoid postoperative bile leaks. 20% hypertonic saline solution should be avoided in case of a biliary fistula as it could reach and damage the bile ducts leading to secondary biliary sclerosis. In case of involvement of the biliary tree, disease clearance should be performed during the surgical procedure or alternatively by endoscopic means (ERCP). During all surgical procedures cyst content spillage should be avoided at all costs by covering other organs with surgical cloths soaked in 20% hypertonic saline.

The noninvasive PAIR (Puncture, Aspiration, Injection, Reaspiration) therapeutic procedure, developed in 1985, involves the aspiration of cyst fluid under local anesthesia using ultrasound (or CT) guidance. In Austria, PAIR is usually under general anesthesia. It is crucial to exercise caution in the detection and identification of cystobiliary fistulas, as the likelihood of their presence is approximately 80% when the cyst diameter exceeds 7.5 cm (Filice et al. 2000). Therefore, preoperative magnetic resonance cholangiopancreatography (MRCP) can be performed. Bilirubin rapid tests should be performed for further evaluation. If cystobiliary fistulas are present, the use of protoscolicidal substances should be avoided due to the risk of sclerosing cholangitis. If no relevant cystobiliary fistulas are present, injection of a protoscolicidal solution (20% NaCl solution or 95% ethanol) is initiated for 10–15 min, followed by reaspiration at the end of the intervention. For uncomplicated large (>5 cm) CE1 and CE3a cysts, the combination therapy of PAIR and ABZ is the first-line treatment (Khuroo et al. 1993). It should not be used for CE2, CE3b, inactive CE4 and CE5 cysts, as well as lung cysts (Brunetti et al. 2010). Larger cysts (>10 cm) should be catheterized after injection until the drainage decreases to less than 10 mL/day (Fig. 3.15).

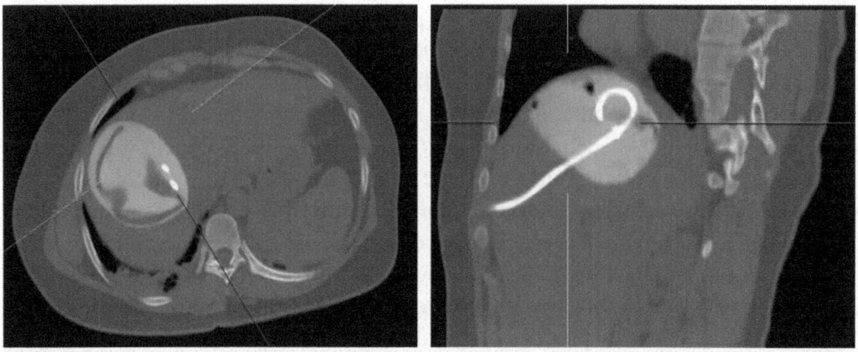

**Fig. 3.15** Catheterization under CT-guidance of a CE liver cyst with instillation of 20% NaCl. Courtesy of Dr. med. univ. Fredrik Waneck (Department of Biomedical Imaging and Image-Guided Therapy, Division of Cardiovascular and Interventional Radiology, Medical University of Vienna, Währinger Gürtel 18–20, 1090 Vienna, Austria)

Concurrent anthelmintic therapy with ABZ or MBZ should be initiated before scheduled surgery to avoid secondary infestation of hydatid cysts and disease recurrence. Furthermore, perioperative treatment leads to cyst softening, which facilitates surgical removal (Bildik et al. 2007; Mahmoodi et al. 2023). The duration of preoperative treatment remains controversial and can vary depending on several factors, including the size and location of the cyst, the expertise of the interdisciplinary team, and the individual patient's condition. Guidelines suggest treatment initiation 4–30 days prior to surgery (Manterola et al. 2005; WHO/OIE 1996). However, opinions on this matter diverge (Brunetti et al. 2010). For example, in Austria, ABZ treatment usually begins 7 days before the planned operation. In Germany, it starts at least 1 day and up to 7 days prior. However, small CE1 cysts, for which PAIR could be considered, are exempted from ABZ therapy for more than 1 week, as prolonged treatment could lead to membrane collapse. Furthermore, large pulmonary CE1 cysts should only be treated with ABZ approximately 24 h before the operation to minimize the intraoperative rupture risk. Nevertheless, the rationale suggests that ABZ therapy should commence upon diagnosis, and the pre- and perioperative ABZ management should occur in interdisciplinary collaboration with the surgery department.

According to a recent RCT, where the efficacy of ABZ as an adjuvant therapy along with the main surgical treatment for hepatic hydatid cysts was investigated, optimal duration of medical treatment before surgery seems to be 4 weeks or more (Vahidirad et al. 2022).

BMZ should be used for at least 2 years after surgery or PAIR to prevent recurrence. Due to the superior medical care available in Central Europe, non-PAIR procedures are generally preferred, as the interventional complication risk is considered low compared to the risk of a lower outcome in PAIR procedures. However, in regions with limited medical resources, where access to advanced interventions may be challenging, the PAIR procedure is often utilized even for higher CE stages (Fig. 3.16).

In inoperable cases, therapy with ABZ is permanent. Similar to AE, treatment with ABZ is performed with 10–15 mg/kg per day divided in two doses, resulting in a typical dose of 400 mg twice daily, and the treatment with the alternative MBZ can be used divided in three daily doses with a total dose of 40–50 mg/kg per day. Praziquantel might be another alternative, although there is a lack of sufficient evidence and effectiveness seems to be low (Cobo et al. 1998). Despite the elevated therapeutic efficacy of benzimidazoles, the occurrence of treatment intolerance due to side effects or primary contraindications against these agents are prevalent. Consequently, these patients face limited alternatives in terms of therapeutic interventions, as benzimidazoles remain the sole authorized pharmaceutical agent for this specific indication. Thus, to date no second-line treatment option can be clearly recommended and further pharmacological studies are urgently needed to pave the way for new therapy opportunities against this zoonotic disease. As for AE, ABZ therapeutic drug monitoring (TDM) should be determined to evaluate treatment success or to confirm a treatment failure.

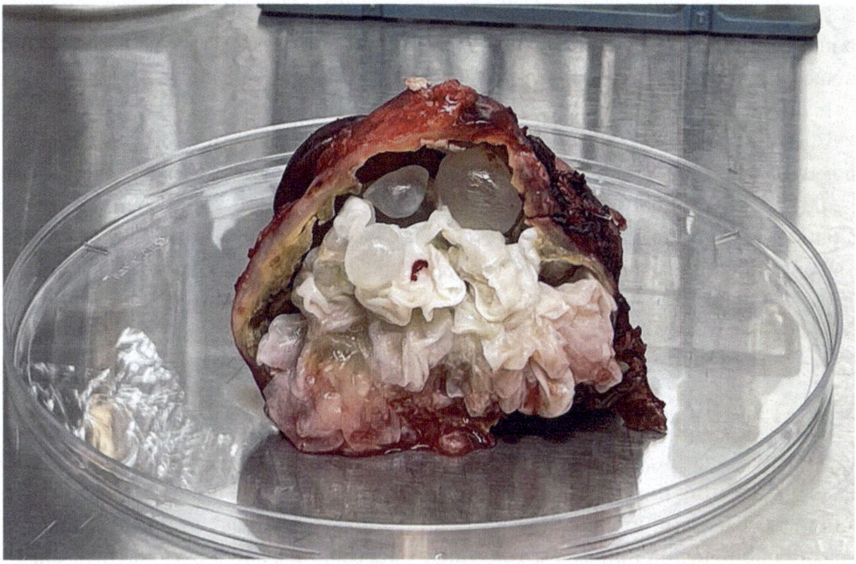

**Fig. 3.16** Typical liver specimen of a CE3b cyst from a South Italian patient with cystic echinococcosis. Courtesy of Assoc.-Prof. Dr. Heimo Lagler MPH DTM (Department of Medicine I, Division of Infectious Diseases and Tropical Medicine, Medical University of Vienna, Währinger Gürtel 18–20, 1090 Vienna, Austria)

### 3.8.2 Pharmacokinetics of Albendazole

Albendazole, as a broad-spectrum anthelmintic agent, is commonly used for the treatment of various parasitic infections such as filariosis, giardiosis, cysticercosis, toxocariosis, and echinococcosis.

The knowledge of its pharmacokinetic profile is crucial for the optimization of treatment efficacy, which is strongly determined by the drugs absorption, distribution, metabolization, and elimination within the body.

After oral administration, ABZ undergoes extensive first-pass metabolism in the liver and rapidly converts into its active metabolite ABZ-sulfoxide (ABZ-SO). The agent is characterized by poor aqueous solubility and high lipid solubility, leading to its low oral bioavailability (max 5%) if taken in a fasted state. According to studies investigating the dietary impact on bioavailability of ABZ, the intake of ABZ concomitant with a fatty meal can lead to a two- to five-fold enhancement of the mean plasma area under the concentration–time curve (AUC) of ABZ-SO (Lange et al. 1988; Ochoa et al. 2021). Therefore, administration with fatty meals is highly recommended.

ABZ-SO is approximately 70% bound to plasma proteins and is widely distributed throughout the body tissues. The agent can cross the blood–brain barrier, allowing it to effectively target parasites within the central nervous system. The pharmacokinetic transfer rate from plasma into cystic tissue after oral administration is found to be between 13 and 22% (Al Jalali and Zeitlinger 2020).

First-pass metabolization in the liver is primarily conducted through enzymatic oxidation mediated by cytochrome P450 (CYP) enzymes. ABZ-SO, the major metabolite formed, is responsible for exhibition of anthelmintic activity. Further metabolism of ABZ-SO leads to the formation of ABZ sulfone and other metabolites.

Elimination of ABZ and its metabolites is primarily performed through the bile into the feces. Only a small portion of the drug and its metabolites are excreted in the urine. The elimination half-life of ABZ-SO is 8.5 h. In individuals with impaired hepatic function, the elimination of ABZ may be delayed, leading to higher plasma concentrations. Dose adjustments are necessary in these patients.

## 3.9 Strategies for Prevention

Given that echinococcosis is a zoonotic disease, it primarily affects animals such as dogs, wolves, and foxes. However, humans can inadvertently serve as intermediate hosts by ingesting the eggs released in the feces of infected animals. Consequently, echinococcosis is a preventable disease for humans. Various approaches, encompassing public health initiatives, veterinary interventions, and individual awareness, can effectively mitigate the risk of infection.

Furthermore, it should be noted that surgical interventions, imaging, and ABZ therapy for AE and CE, significantly burden the healthcare system, partly due to unexpectedly high drug prices during long-term treatment. Hence, preventive measures, accompanied by cost-effectiveness analyses, should be given much greater emphasis in the management of AE and CE (Lotsch et al. 2019).

### 3.9.1 General Interventions

Enhancing public consciousness among high-risk populations and children, who are in frequent contact with animals, about the disease and its modes of transmission is crucial in the prevention of echinococcosis (Craig et al. 2017). The practice of meticulous hygiene and food safety measures, such as thorough and mindful hand washing after animal contact, as well as refraining from consuming unwashed fruits and vegetables, plays a significant role in the control of the disease.

### 3.9.2 Preventive Strategies for Alveolar Echinococcosis

Due to the engagement of wild animal species in the life cycle of AE (see Sect. 3.1), the management of disease transmission and control becomes complicated. Research endeavors, exploring the consequences of anthelmintic interventions targeted at prospective wild hosts in Europe and Japan, have noted noteworthy reductions in disease prevalence with these measures (Umhang et al. 2019). Furthermore, preventive measures play a crucial role in combating AE. Individuals engaged in agricultural activities should be educated about the life cycle of *E. multilocularis*, especially its intermediate and definitive hosts. Additionally, they should be informed about preventive measures such as washing hands after any contact with soil or grass, cleaning fallen fruits, berries, or garden vegetables, and avoiding direct contact with foxes or fox feces. It should also be considered that serological screening examinations (e.g., every 2–3 years) in highly exposed individuals could be beneficial (Auer and Aspöck 2014).

### 3.9.3 Preventive Strategies for Cystic Echinococcosis

Given that the transmission of this entity occurs through domestic animals, particularly domestic dogs, the principal focus of prevention lies in implementing deworming interventions for these animals (see Sect. 3.2). Praziquantel represents a widely employed agent for this specific purpose. Further, proper disposal of dog's feces can significantly reduce transmission of infectious eggs.

The utilization of a recombinant antigen (EG95) derived from *E. granulosus* for immunizing sheep presents promising possibilities in terms of prevention and control measures. Argentinian trials have demonstrated significant benefits of vaccinating sheep, while China has been extensively implementing the vaccine as part of their disease management strategy (Larrieu et al. 2015). This vaccine is presently undergoing commercial production and has obtained official registration in both China and Argentina. Furthermore, an educational campaign on the life cycle of *E. granulosus* and its associated intermediate and definitive hosts should be conducted in endemic areas. Special attention should be given to individuals in the agricultural sector or those with close contact with dogs. In many regions, animals are slaughtered under unhygienic conditions, and the raw meat residues are fed to domestic dogs. After infection of these animals with *E. granulosus*, anthelminthic therapy is usually not initiated. Therefore, in 2001, the WHO recommended the implementation of comprehensive awareness programs in endemic areas (WHO/OIE 2001).

**Competing Interests** The authors have no conflicts of interest to declare that are relevant to the content of this chapter.

**Institutional Review Board Statement:**The study was conducted in accordance with the 1964 Declaration of Helsinki.

## References

Austrian Agency for Health and Food Safety (AGES) (2023) Report 2023. Kleiner Fuchsbandwurm. Available: https://www.ages.at/mensch/krankheit/krankheitserreger-von-a-bis-z/kleiner-fuchsbandwurm. Accessed 18 Jul 2023

Agudelo Higuita NI, Brunetti E, M cCloskey C (2016) Cystic echinococcosis. J Clin Microbiol 54:518–523

Al Jalali V, Zeitlinger M (2020) Systemic and target-site pharmacokinetics of Antiparasitic agents. Clin Pharmacokinet 59:827–847

Ammann RW, Stumpe KD, Grimm F, Deplazes P, Huber S, Bertogg K, Fischer DR, Müllhaupt B (2015) Outcome after discontinuing long-term Benzimidazole treatment in 11 patients with non-resectable alveolar echinococcosis with negative FDG-PET/CT and anti-EmII/3-10 serology. PLoS Negl Trop Dis 9:e0003964

Antolova D, Miterpakova M, Radoňak J, Hudačkova D, Szilagyiova M, Začek M (2014) Alveolar echinococcosis in a highly endemic area of northern Slovakia between 2000 and 2013. Euro Surveill 19

Aspöck H, Auer H, Walochnik J (2002) Parasitenzyklen: Die (manchmal) verschlungenen Wege von Wirt zu Wirt

Auer H, Aspöck H (2002) Die zystische und die alveoläre Echinokokkose–Die gefährlichsten Helminthosen Mitteleuropas. Denisia 6:333–353

Auer H, Aspöck H (2014) Helminths and helminthoses in Central Europe: diseases caused by nematodes (roundworms). Wien Med Wochenschr (1946) 164:414–423

Bhutani N, Kajal P (2018) Hepatic echinococcosis: a review. Ann Med Surg (Lond) 36:99–105

Bildik N, Cevik A, Altintaş M, Ekinci H, Canberk M, Gülmen M (2007) Efficacy of preoperative albendazole use according to months in hydatid cyst of the liver. J Clin Gastroenterol 41:312–316

Bresson-Hadni S, Delabrousse E, Blagosklonov O, Bartholomot B, Koch S, Miguet JP, Mantion GA, Vuitton DA (2006) Imaging aspects and non-surgical interventional treatment in human alveolar echinococcosis. Parasitol Int 55(Suppl):S267–S272

Brunetti E, Kern P, Vuitton DA (2010) Expert consensus for the diagnosis and treatment of cystic and alveolar echinococcosis in humans. Acta Trop 114:1–16

Casulli A, Abela-Ridder B, Petrone D, Fabiani M, Bobić B, Carmena D, Šoba B, Zerem E, Gargaté MJ, Kuzmanovska G, Calomfirescu C, Rainova I, Sotiraki S, Lungu V, Dezsényi B, Herrador Z, Karamon J, Maksimov P, Oksanen A, Millon L, Sviben M, Shkjezi R, Gjoni V, Akshija I, Saarma U, Torgerson P, Šnábel V, Antolová D, Muhovic D, Besim H, Chereau F, Belhassen García M, Chappuis F, Gloor S, Stoeckle M, Müllhaupt B, Manno V, Santoro A, Santolamazza F (2023) Unveiling the incidences and trends of the neglected zoonosis cystic echinococcosis in Europe: a systematic review from the MEmE project. Lancet Infect Dis 23:e95–e107

Casulli A, Massolo A, Saarma U, Umhang G, Santolamazza F, Santoro A (2022) Species and genotypes belonging to Echinococcus granulosus sensu lato complex causing human cystic echinococcosis in Europe (2000–2021): a systematic review. Parasit Vectors 15:109

Cobo F, Yarnoz C, Sesma B, Fraile P, Aizcorbe M, Trujillo R, Diaz-de-Liano A, Ciga MA (1998) Albendazole plus praziquantel versus albendazole alone as a pre-operative treatment in intra-abdominal hydatisosis caused by Echinococcus granulosus. Trop Med Int Health 3:462–466

Conraths FJ, Probst C, Possenti A, Boufana B, Saulle R, La Torre G, Busani L, Casulli A (2017) Potential risk factors associated with human alveolar echinococcosis: systematic review and meta-analysis. PLoS Negl Trop Dis 11:e0005801

Craig PS, Hegglin D, Lightowlers MW, Torgerson PR, Wang Q (2017) Echinococcosis: control and prevention. Adv Parasitol 96:55–158

Deplazes P, Rinaldi L, Alvarez Rojas CA, Torgerson PR, Harandi MF, Romig T, Antolova D, Schurer JM, Lahmar S, Cringoli G, Magambo J, Thompson RC, Jenkins EJ (2017) Global distribution of alveolar and cystic echinococcosis. Adv Parasitol 95:315–493

Dezsényi B, Dubóczki Z, Strausz T, Csulak E, Czoma V, Káposztás Z, Fehérvári M, Somorácz Á, Csilek A, Oláh A, Almási K, Patonai A, Görög D, Széll Z, Tolnai Z, Sréter T, Danka J, Auer H, Grüner B, Barth TFE, Casulli A (2021) Emerging human alveolar echinococcosis in Hungary (2003–2018): a retrospective case series analysis from a multi-Centre study. BMC Infect Dis 21:168

Dezsényi B, Strausz T, Makrai Z, Csomor J, Danka J, Kern P, Rezza G, Barth TF, Casulli A (2017) Autochthonous human alveolar echinococcosis in a Hungarian patient. Infection 45:107–110

Dreweck CM, Luder CG, Soboslay PT, Kern P (1997) Subclass-specific serological reactivity and IgG4-specific antigen recognition in human echinococcosis. Trop Med Int Health 2:779–787

Duscher GG, Leschnik M, Fuehrer HP, Joachim A (2015) Wildlife reservoirs for vector-borne canine, feline and zoonotic infections in Austria. Int J Parasitol Parasites Wildl 4:88–96
Dyachenko V, Pantchev N, Gawlowska S, Vrhovec MG, Bauer C (2008) Echinococcus multilocularis infections in domestic dogs and cats from Germany and other European countries. Vet Parasitol 157:244–253
Dybicz M, Borkowski PK, Dąbrowska J, Chomicz L (2015) Cases of Echinococcus granulosus Sensu Stricto isolated from polish patients: imported or indigenous? Biomed Res Int 2015:728321
ECDC (2020) Echinococcosis. Annual epidemiological report for 2020. Available: https://www.ecdc.europa.eu/sites/default/files/documents/echinococcosis-annual-epidemiolpgical-report-2020.pdf. Accessed 18 Jul 2023
EFSA (2021) The European Union one health 2019 Zoonoses report. EFSA J 19:e06406
Eiermann TH, Bettens F, Tiberghien P, Schmitz K, Beurton I, Bresson-Hadni S, Ammann RW, Goldmann SF, Vuitton DA, Gottstein B, Kern P (1998) HLA and alveolar echinococcosis. Tissue Antigens 52:124–129
Filice C, Brunetti E, Bruno R, Crippa FG (2000) Percutaneous drainage of echinococcal cysts (PAIR–puncture, aspiration, injection, reaspiration): results of a worldwide survey for assessment of its safety and efficacy. WHO-Informal Working Group on Echinococcosis-Pair Network. Gut 47:156–157
Gharbi HA, Hassine W, Brauner MW, Dupuch K (1981) Ultrasound examination of the hydatic liver. Radiology 139:459–463
Gołąb E, Czarkowski MP (2014) Echinococcosis and cysticercosis in Poland in 2012. Przegl Epidemiol 68(279–82):379–381
Gottstein B, Jacquier P, Bresson-Hadni S, Eckert J (1993) Improved primary immunodiagnosis of alveolar echinococcosis in humans by an enzyme-linked immunosorbent assay using the Em2plus antigen. J Clin Microbiol 31:373–376
Gottstein B, Lachenmayer A, Beldi G, Wang J, Merkle B, Vu XL, Kurath U, Müller N (2019) Diagnostic and follow-up performance of serological tests for different forms/courses of alveolar echinococcosis. Food Waterborne Parasitol 16:e00055
Gottstein B, Stojkovic M, Vuitton DA, Millon L, Marcinkute A, Deplazes P (2015) Threat of alveolar echinococcosis to public health–a challenge for Europe. Trends Parasitol 31:407–412
Gottstein B, Tschudi K, Eckert J, Ammann R (1989) Em2-ELISA for the follow-up of alveolar echinococcosis after complete surgical resection of liver lesions. Trans R Soc Trop Med Hyg 83:389–393
Grüner B, Kern P, Mayer B, Gräter T, Hillenbrand A, Barth TEF, Muche R, Henne-Bruns D, Kratzer W, Kern P (2017a) Comprehensive diagnosis and treatment of alveolar echinococcosis: a single-center, long-term observational study of 312 patients in Germany. GMS Infect Dis 5:Doc01
Grüner B, Schmidberger J, Drews O, Kratzer W, Gräter T (2017b) Imaging in alveolar echinococcosis (AE): comparison of Echinococcus multilocularis classification for computed-tomography (EMUC-CT) and ultrasonography (EMUC-US). Radiol Infect Dis 4:70–77
Horváth A, Patonay A, Bánhegyi D, Szlávik J, Balázs G, Görög D, Werling K (2008) The first case of human alveolar echinococcosis in Hungary. Orv Hetil 149:795–799
Hotz JF, Peters L, Kapp-Schwörer S, Theis F, Eberhardt N, Essig A, Grüner B, Hagemann JB (2022) Evaluation of serological markers in alveolar echinococcosis emphasizing the correlation of PET-CTI tracer uptake with RecEm18 and Echinococcus-specific IgG. Pathogens 11:239
Ito A, Craig PS (2003) Immunodiagnostic and molecular approaches for the detection of taeniid cestode infections. Trends Parasitol 19:377–381
Ito A, Ma L, Schantz PM, Gottstein B, Liu YH, Chai JJ, Abdel-Hafez SK, Altintas N, Joshi DD, Lightowlers MW, Pawlowski ZS (1999) Differential serodiagnosis for cystic and alveolar echinococcosis using fractions of Echinococcus granulosus cyst fluid (antigen B) and E. Multilocularis protoscolex (EM18). Am J Trop Med Hyg 60:188–192
Ito A, Nakao M, Kutsumi H, Lightowlers MW, Itoh M, Sato S (1993) Serodiagnosis of alveolar hydatid disease by western blotting. Trans R Soc Trop Med Hyg 87:170–172

Ito A, Xiao N, Liance M, Sato MO, Sako Y, Mamuti W, Ishikawa Y, Nakao M, Yamasaki H, Nakaya K, Bardonnet K, Bresson-Hadni S, Vuitton DA (2002) Evaluation of an enzyme-linked immunosorbent assay (ELISA) with affinity-purified Em18 and an ELISA with recombinant Em18 for differential diagnosis of alveolar echinococcosis: results of a blind test. J Clin Microbiol 40:4161–4165

Kern P, Bardonnet K, Renner E, Auer H, Pawlowski Z, Ammann RW, Vuitton DA, Kern P (2003) European echinococcosis registry: human alveolar echinococcosis, Europe, 1982-2000. Emerg Infect Dis 9:343–349

Kern P, Wen H, Sato N, Vuitton DA, Gruener B, Shao Y, Delabrousse E, Kratzer W, Bresson-Hadni S (2006) WHO classification of alveolar echinococcosis: principles and application. Parasitol Int 55:S283–S2S7

Khuroo MS, Dar MY, Yattoo GN, Zargar SA, Javaid G, Khan BA, Boda MI (1993) Percutaneous drainage versus albendazole therapy in hepatic hydatidosis: a prospective, randomized study. Gastroenterology 104:1452–1459

Kodama Y, Fujita N, Shimizu T, Endo H, Nambu T, Sato N, Todo S, Miyasaka K (2003) Alveolar echinococcosis: MR findings in the liver. Radiology 228:172–177

Kolářová L, Matějů J, Hozáková L, Stejskal F, Hrdý J, Kolářová H, Leissová M, Skála V, Dundr P (2017) Human alveolar echinococcosis and an overview of the occurrence of Echinococcus multilocularis in animals in The Czech Republic. Epidemiol Mikrobiol Imunol 66:163–172

Kratzer W, Gruener B, Kaltenbach TE, Ansari-Bitzenberger S, Kern P, Fuchs M, Mason RA, Barth TF, Haenle MM, Hillenbrand A, Oeztuerk S, Graeter T (2015) Proposal of an ultrasonographic classification for hepatic alveolar echinococcosis: echinococcosis multilocularis Ulm classification-ultrasound. World J Gastroenterol 21:12392–12402

Lachenmayer A, Gebbers D, Gottstein B, Candinas D, Beldi G (2019) Elevated incidence of alveolar echinococcosis in immunocompromised patients. Food Waterborne Parasitol 16:e00060

Lange H, Eggers R, Bircher J (1988) Increased systemic availability of albendazole when taken with a fatty meal. Eur J Clin Pharmacol 34:315–317

Larrieu E, Mujica G, Gauci CG, Vizcaychipi K, Seleiman M, Herrero E, Labanchi JL, Araya D, Sepúlveda L, Grizmado C, Calabro A, Talmon G, Poggio TV, Crowley P, Cespedes G, Santillán G, García Cachau M, Lamberti R, Gino L, Donadeu M, Lightowlers MW (2015) Pilot field trial of the EG95 vaccine against ovine cystic echinococcosis in Rio Negro, Argentina: second study of impact. PLoS Negl Trop Dis 9:e0004134

Logar J, Soba B, Kotar T (2008) Serological evidence for human cystic echinococcosis in Slovenia. BMC Infect Dis 8:63

Logar J, Šoba B, Lejko-Zupanc T, Kotar T (2007) Human alveolar echinococcosis in Slovenia. Clin Microbiol Infect 13:544–546

Lotsch F, Waneck F, Groger M, Auer H, Kaczirek K, Rausch I, Wadsak W, Hacker M, Lagler H, Ramharter M, Karanikas G (2019) FDG-PET/MRI imaging for the management of alveolar echinococcosis: initial clinical experience at a reference Centre in Austria. Trop Med Int Health 24:663–670

Lymbery AJ (2017) Phylogenetic pattern, evolutionary processes and species delimitation in the genus Echinococcus. Adv Parasitol 95:111–145

Mahmoodi S, Ebrahimian M, Mirhashemi SH, Soori M, Rashnoo F, Oshidari B, Shadidi Asil R, Zamani A, Hajinasrollah E (2023) A 20 years retrospective descriptive study of human cystic echinococcosis and the role of Albendazole concurrent with surgical treatment: 2001-2021. Iran J Parasitol 18:100–106

Malczewski A, Rocki B, Ramisz A, Eckert J (1995) Echinococcus multilocularis (Cestoda), the causative agent of alveolar echinococcosis in humans: first record in Poland. J Parasitol 81:318–321

Mandal S, Deb Mandal M (2012) Human cystic echinococcosis: epidemiologic, zoonotic, clinical, diagnostic and therapeutic aspects. Asian Pac J Trop Med 5:253–260

Manterola C, Mansilla JA, Fonseca F (2005) Preoperative albendazole and scolices viability in patients with hepatic echinococcosis. World J Surg 29:750–753

McManus DP, Zhang W, Li J, Bartley PB (2003) Echinococcosis Lancet 362:1295–1304

Moro P, Schantz PM (2006) Cystic echinococcosis in the Americas. Parasitol Int 55(Suppl):S181–S186

Nahorski WL, Knap JP, Pawłowski ZS, Krawczyk M, Polański J, Stefaniak J, Patkowski W, Szostakowska B, Pietkiewicz H, Grzeszczuk A, Felczak-Korzybska I, Gołąb E, Wnukowska N, Paul M, Kacprzak E, Sokolewicz-Bobrowska E, Niścigorska-Olsen J, Czyrznikowska A, Chomicz L, Cielecka D, Myjak P (2013) Human alveolar echinococcosis in Poland: 1990-2011. PLoS Negl Trop Dis 7:e1986

Ochoa D, Saiz-Rodríguez M, González-Rojano E, Román M, Sánchez-Rojas S, Wojnicz A, Ruiz-Nuño A, García-Arieta A, Abad-Santos F (2021) High-fat breakfast increases bioavailability of Albendazole compared to low-fat breakfast: single-dose study in healthy subjects. Front Pharmacol 12:664465

Oksanen A, Siles-Lucas M, Karamon J, Possenti A, Conraths FJ, Romig T, Wysocki P, Mannocci A, Mipatrini D, La Torre G, Boufana B, Casulli A (2016) The geographical distribution and prevalence of Echinococcus multilocularis in animals in the European Union and adjacent countries: a systematic review and meta-analysis. Parasit Vectors 9:519

Piarroux M, Piarroux R, Giorgi R, Knapp J, Bardonnet K, Sudre B, Watelet J, Dumortier J, Gérard A, Beytout J, Abergel A, Mantion G, Vuitton DA, Bresson-Hadni S (2011) Clinical features and evolution of alveolar echinococcosis in France from 1982 to 2007: results of a survey in 387 patients. J Hepatol 55:1025–1033

Rataj AV, Bidovec A, Žele D, Vengušt G (2010) Echinococcus multilocularis in the red fox (Vulpes vulpes) in Slovenia. Eur J Wildl Res 56:819–822

Reuter S, Buck A, Manfras B, Kratzer W, Seitz HM, Darge K, Reske SN, Kern P (2004) Structured treatment interruption in patients with alveolar echinococcosis. Hepatology 39:509–517

Richter J, Orhun A, Grüner B, Müller-Stöver I, Reuter S, Romig T, Häussinger D, Kern P (2009) Autochthonous cystic echinococcosis in patients who grew up in Germany. Eur Secur 14:19229

Romig T, Dinkel A, Mackenstedt U (2006) The present situation of echinococcosis in Europe. Parasitol Int 55(Suppl):S187–S191

Sadjjadi SM, Mikaeili F, Karamian M, Maraghi S, Sadjjadi FS, Shariat-Torbaghan S, Kia EB (2013) Evidence that the Echinococcus granulosus G6 genotype has an affinity for the brain in humans. Int J Parasitol 43:875–877

Sarciron EM, Bresson-Hadni S, Mercier M, Lawton P, Duranton C, Lenys D, Petavy AF, Vuitton DA (1997) Antibodies against Echinococcus multilocularis alkaline phosphatase as markers for the specific diagnosis and the serological monitoring of alveolar echinococcosis. Parasite Immunol 19:61–68

Schmidberger J, Uhlenbruck J, Schlingeloff P, Maksimov P, Conraths FJ, Mayer B, Kratzer W (2022) Dog ownership and risk for alveolar echinococcosis, Germany. Emerg Infect Dis 28:1597–1605

Schneider R, Aspöck H, Auer H (2013) Unexpected increase of alveolar echincoccosis, Austria, 2011. Emerg Infect Dis 19:475–477

Schneider R, Gollackner B, Edel B, Schmid K, Wrba F, Tucek G, Walochnik J, Auer H (2008) Development of a new PCR protocol for the detection of species and genotypes (strains) of Echinococcus in formalin-fixed, paraffin-embedded tissues. Int J Parasitol 38:1065–1071

Schweiger A, Ammann RW, Candinas D, Clavien PA, Eckert J, Gottstein B, Halkic N, Muellhaupt B, Prinz BM, Reichen J, Tarr PE, Torgerson PR, Deplazes P (2007) Human alveolar echinococcosis after fox population increase, Switzerland. Emerg Infect Dis 13:878–882

Šoba B, Gašperšič Š, Keše D, Kotar T (2020) Molecular characterization of Echinococcus granulosus sensu lato from humans in Slovenia. Pathogens 9

Stefaniak M, Derda M, Zmora P, Nowak SP (2023) Risk factors and the character of clinical course of the Echinococcus multilocularis infection in patients in Poland. Pathogens 12

Umhang G, Possenti A, Colamesta V, d'Aguanno S, La Torre G, Boue F, Casulli A (2019) A systematic review and meta-analysis on anthelmintic control programs for Echinococcus multilocularis in wild and domestic carnivores. Food Waterborne Parasitol 15:e00042

Vahidirad A, Mansouri M, Shamshirian A, Berenji F, Motie MR (2022) Efficacy of preoperative Albendazole on Protoscoleces viability in hydatid cyst of the liver. Iran J Parasitol 17:483–487

Vuitton DA (1997) The WHO informal working group on echinococcosis. Coordinating Board of the WHO-IWGE Parassitologia 39:349–353

Vuitton DA, Demonmerot F, Knapp J, Richou C, Grenouillet F, Chauchet A, Vuitton L, Bresson-Hadni S, Millon L (2015) Clinical epidemiology of human AE in Europe. Vet Parasitol 213:110–120

Waloch M (2012) Cystic echinococcosis in Poland in 2010. Przegl Epidemiol 66:311–313

WHO/OIE (1996) Guidelines for treatment of cystic and alveolar echinococcosis in humans. WHO informal working group on echinococcosis. Bull World Health Organ 74:231–242

WHO/OIE (2001) Manual on echinococcosis. Echinococcosis in humans and animals: a public health problem of global concern. World Organisation for Animal Health (Office International des Epizooties) and World Health Organization

Working Group, W. H. O. I (2003) International classification of ultrasound images in cystic echinococcosis for application in clinical and field epidemiological settings. Acta Trop 85:253–261

# Echinococcosis in China: Epidemiology, Burden of Disease, and Economic Impacts

4

Wei Wang, Lilei Wang, and Xiumin Han

**Abstract**

Echinococcosis, also known as hydatidosis or hydatid disease, is a neglected, tropical, zoonotic parasitic disease caused by infection with larvae of the genus *Echinococcus*. China is one of the countries with the highest disease burden of echinococcosis in the world, and the disease is particularly prevalent in western and northwestern pastoral areas of China where the highest prevalence of both alveolar echinococcosis (AE) and cystic echinococcosis (CE) is detected in the world. Results from the nationwide epidemiological survey on echinococcosis in China from 2012 to 2016 showed that the overall detection of echinococcosis was 0.51% among the study participants, and the prevalence of human echinococcosis was estimated to be 0.28% in endemic foci, with more than 50 million people at risk of infections. The disease burden due to CE in China accounts for 40% of global burden of CE, and the economic loss to human patients with CE consists of 30% of global economic losses; in addition, 91% of global new cases diagnosed with AE occur in China, which is responsible for 95% of total DALYs worldwide. Since the national echinococcosis control program was launched in 2006, an integrated strategy based on control of the source of echinococcus

W. Wang (✉)
National Health Commission Key Laboratory on Parasitic Disease Prevention and Control, Jiangsu Provincial Key Laboratory on Parasites and Vector Control Technology, Jiangsu Institute of Parasitic Diseases, Wuxi, Jiangsu Province, China
e-mail: wangwei@jipd.com

L. Wang
Shandong Institute of Parasitic Diseases, Shandong First Medical University & Shandong Academy of Medical Sciences, Jining, Shandong Province, China

X. Han
Qinghai Provincial People's Hospital, Xining, Qinghai Province, China





J. Li et al. (eds.), *Echinococcus: Control and Elimination of Echinococcosis with a Focus on China and Europe*, Parasitology Research Monographs 19,
https://doi.org/10.1007/978-3-031-54479-8_4

infections has been implemented, including health education, chemotherapy of infected patients, periodical deworming of dogs, and improved hygiene in the slaughtering of livestock, and great strides have been achieved for echinococcosis control in China (Xiao 2021). In this chapter, we describe the evolution of the national echinococcosis control program, and summarize the advances in the epidemiology, disease burden, and economic impacts of CE and AE in China. In addition, we recommend the implementation of One Health approach to eliminate echinococcosis in China, and collaboration with other disease-affected countries to walk together for joint control of echinococcosis, thereby contributing to achievement of the target set in the WHO new roadmap.

**Keywords**

Echinococcosis · Alveolar echinococcosis · Cystic echinococcosis · Epidemiology · Disease burden · Economic burden · One Health · China

## 4.1 Introduction

Echinococcosis, also known as hydatidosis or hydatid disease, is a neglected, tropical, zoonotic parasitic disease caused by infection with larvae of the genus *Echinococcus* (McManus et al. 2003). As an old disease, echinococcosis had been recognized for centuries, and the description of hydatid cysts had been documented in Miraculous Pivot (Lingshu Jing), one of the earliest medical classics in China (460 to 379 BC), and had been described in aphorisms by ancient Greek physician Hippocrates (460 to 377 BC) (Moro and Schantz 2006). Nevertheless, it was not until in the seventeenth century that the Italian physician and poet Francesco Redi, for the first time, described the biological nature of the hydatid cysts, and Peter Simon Pallas, a German zoologist and botanist, hypothesized hydatid cysts as larval stages of tapeworms in 1760 (Eckert and Thompson 2017). Subsequently, Karl Asmund Rudolphi, a Swedish-born German botanist and naturalist who is widely accepted as "father of helminthology," first introduced the word *Echinococcus* to the world of science (Eckert 2007).

Echinococcosis, which is widespread across the world, seriously threatens human health, causes huge economic losses to livestock industry and hinders socioeconomic developments, remains a major global health problem, and is a leading cause of poverty among affected individuals and families (Wen et al. 2019). Currently, there are two main forms of human echinococcoses, including cystic echinococcosis (CE) caused by *E. granulosus* and alveolar echinococcosis (AE) caused by *E. multilocularis*, with CE found in each continent except Antarctica and distribution of AE restrained in the northern hemisphere (Cadavid Restrepo et al. 2016) (Fig. 4.1). Worldwide, more than 2 million people are estimated to suffer from echinococcosis and approximately 200,000 new cases are diagnosed with echinococcosis each year, which cause economic losses of more than 4 billion USD due to treatment of infected patients and livestock (Alvi and Alsayeqh 2022; Atkinson et al. 2013). In endemic regions, the annual incidence of CE is $1/10^4$ to

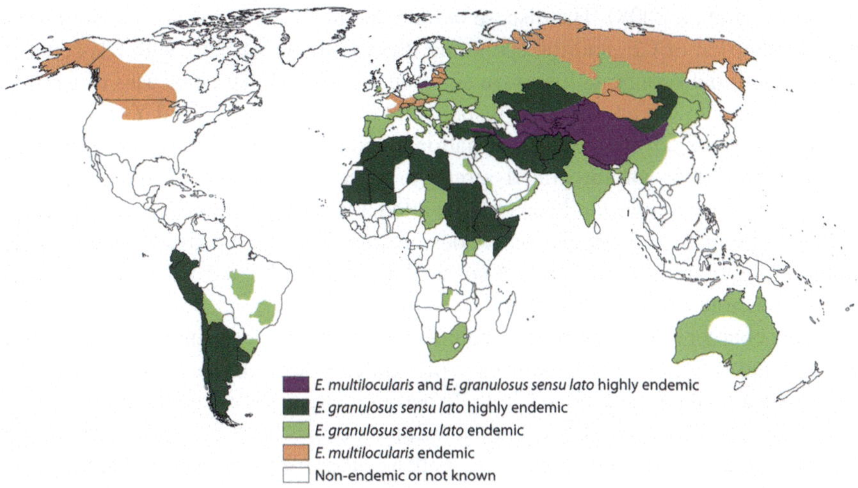


**Fig. 4.1** Worldwide distribution of cystic echinococcosis and alveolar echinococcosis

$200/10^4$, with global burden of more than 1 million disability-adjusted life years (DALYs) due to CE, and the annual incidence of AE is $0.03/10^4$ to $1.2/10^4$, with global burden of 0.67 million DALYs due to AE (Deplazes et al. 2017; Budke et al. 2006; Torgerson et al. 2010). However, echinococcosis has been always a neglected tropical disease (NTD), with underestimated incidence, prevalence, and burden of disease. Even today there is a lack of policy and financial supports in countries where the disease is prevalent (Molyneux et al. 2017).

To accelerate the progress toward global elimination of NTDs, the 73th World Health Assembly endorsed the *Ending the neglect to attain the Sustainable Development Goals: a road map for neglected tropical diseases 2021–2030* on November 13, 2020, and this new roadmap was formally initiated since January 28, 2021 (World Health Organization 2020). Global targets for eradication, elimination, and control of 20 NTDs are set in the new roadmap, and echinococcosis is targeted for control in the world by 2030 (World Health Organization 2020). Since then, One Health approach and interventions have been recommended by the World Health Organization (WHO) for mobilization of resources from health, agriculture, environment, livestock, and wildlife sectors for joint management of NTDs (Laing et al. 2021; Laing 2023). Successful experiences from countries where echinococcosis has been under control or eliminated have proven that an integrated strategy including provision of primary healthcare services, improving human education and health levels, promoting healthy production and living styles, intensified monitoring and management of animal hosts, access to safe water, improved food products quarantine, and improving the quality control of control interventions is effective to control and eliminate echinococcosis (Fu et al. 2020; Craig et al. 2017; Anisuzzaman et al. 2023).

Echinococcosis is highly prevalent in China, notably in western and northwestern pastoral areas where the highest prevalence of both AE and CE is detected in the world (Wang et al. 2008). The disease burden due to CE in China accounts for 40% of global burden of CE, and the economic loss to human patients with CE consists of 30% of global economic losses; in addition, 91% of global new cases diagnosed with AE occur in China, which is responsible for 95% of total DALYs worldwide (Qian et al. 2017, 2019a). Since the national echinococcosis control program was launched in 2006, an integrated strategy based on control of the source of *Echinococcus* infections has been implemented, including health education, chemotherapy of infected patients, periodical deworming of dogs, improved hygiene in the slaughtering of livestock, and great strides have been achieved for echinococcosis control in China (Yu et al. 2020).

In this chapter, we describe the evolution of the national echinococcosis control program and summarize the advances in the epidemiology, disease burden, and economic impacts of CE and AE in China. In addition, we recommend the implementation of One Health approach to eliminate echinococcosis in China, and collaboration with other disease-affected countries to walk together for joint control of echinococcosis, thereby contributing to achievement of the target set in the WHO new roadmap.

## 4.2 Evolution of National Echinococcosis Control Program in China

Since 1980s, the high endemicity and harmfulness of echinococcosis in western regions of China was paid much attention to by the government and a Chinese expert panel, sponsored by WHO, visited Australia and New Zealand for echinococcosis control in 1985 (Chai 2009). Then, the Training Base for Echinococcosis Control was built in Xinjiang, where echinococcosis was highly prevalent, by Ministry of Health, the People's Republic of China in 1988, and the First National Conference for Echinococcosis Control in China was held in Urumqi, Xinjiang in 1989 (Abudusalamu et al. 2019). As a consequence, the first and second training courses for echinococcosis control were held with aims to improve the capability building in China, and pilot projects for echinococcosis control were performed in western regions of China (Chai 2009). The 16th World Congress of Echinococcosis was held in Beijing, China, in 1993, which urged the implementation of interventions against zoonotic echinococcosis. Prior to 2006 when the national echinococcosis control program was initiated (Yu et al. 2020), epidemiological surveys, basic scientific researches, and clinical studies were conducted, which were supported from grants from central, provincial, and municipal governments, and the outcomes provided baseline data and useful tools for echinococcosis control in China.

Since early 2000s, the rapid economic and social developments and reform of the national public health system urged containment of infectious diseases in China (Wang et al. 2019). In 2005, the Ministry of Health proposed intensified echinococcosis control in endemic foci of China. Then, the *National Control Program for Major Parasitic Diseases in China from 2006 to 2015* was issued by Ministry of Health in 2016, and pilot projects for echinococcosis control, supported by the central government, were implemented in ten counties endemic for echinococcosis in Sichuan Province, a highly endemic areas in China (Yan et al. 2015). Since 2007, the coverage of the national echinococcosis control program gradually expanded, and currently, all countries endemic for echinococcosis have been included in Chinese national echinococcosis control program. The *2010–2015 Action for Echinococcosis Control* was jointly released by 14 ministries of China in 2010, in which a working mechanism of government-led, multisectoral collaboration and joint participation of the whole society was proposed (Wang 2017). In this action, an integrated strategy based on controlling the source of *Echinococcus* infections was employed for echinococcosis control, and the integrated interventions included health education campaigns, management of the source of *Echinococcus* infections, intermediate hosts prevention and control, and chemotherapy of infected patients, and regional collaboration, joint prevention and control, scientific and technical innovation, and international collaboration were encouraged for echinococcosis control (Wang 2017). In 2012, the State Council of the People's Republic of China issued the *National Mid- and Long-term Control Programs for Animal Diseases*, and the central government included echinococcosis in the list of top priority for animal diseases control and set the target for control of echinococcosis in animals by 2020 (Chen and Pu 2015). Subsequently, an ambitious goal of transmission control of echinococcosis in China by 2030 was set in Healthy China 2030 (Tan et al. 2019) and the *National Control Program for Major Parasitic Diseases in China from 2016 to 2020* (Qian et al. 2017). In 2021, echinococcosis was made required for compulsory immunization in domestic livestock, as proposed in the 2021 National Compulsory Immunization for Animal Diseases in China (Ministry of Agriculture and Rural Affairs 2021). Recently, the 2023 version of the National Echinococcosis Control Program was released, with aims to facilitate the progress toward echinococcosis control in China through case screening and identification, chemotherapy and surgical therapy of cases, containment of the source of *Echinococcus* infections, surveillance response, and capability building (Fig. 4.2).

## 4.3 Epidemiology of Echinococcosis in China

Understanding the epidemiology of echinococcosis is a prerequisite to formulation of echinococcosis control programs and implementation of effective interventions (Ministry of Agriculture and Rural Affairs 2021). To investigate the epidemiological characteristics of echinococcosis and associated influencing factors, a large number of epidemiological surveys have been conducted across endemic foci in China (Zhang et al. 2015) (Figs. 4.3, 4.4, and 4.5). In addition, two nationwide surveys have been conducted to investigate the epidemiology of echinococcosis in China, including the nationwide survey on the status of major human parasitic diseases in China from 2001 to 2004 (Fu et al. 2021) and the nationwide epidemiological survey on echinococcosis in China from 2012 to 2016 (Coordinating Office of the National Survey on the Important Human Parasitic Diseases 2005; Ma et al. 2023).

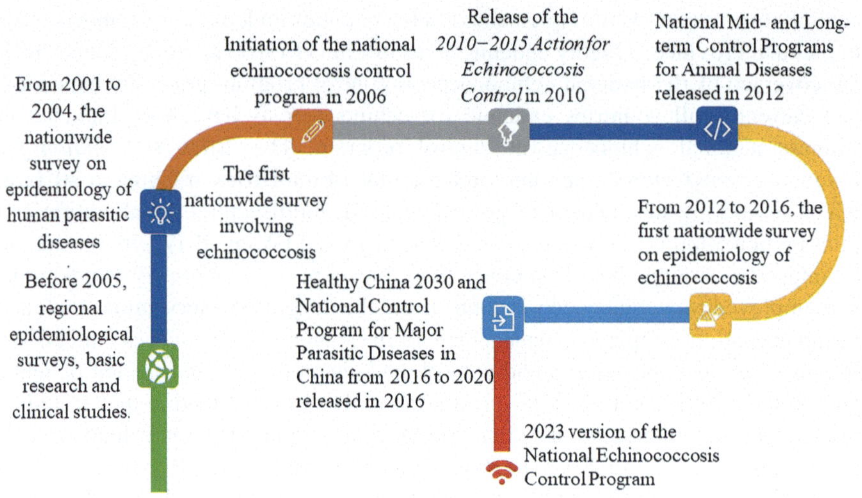


**Fig. 4.2** National echinococcosis control program in China

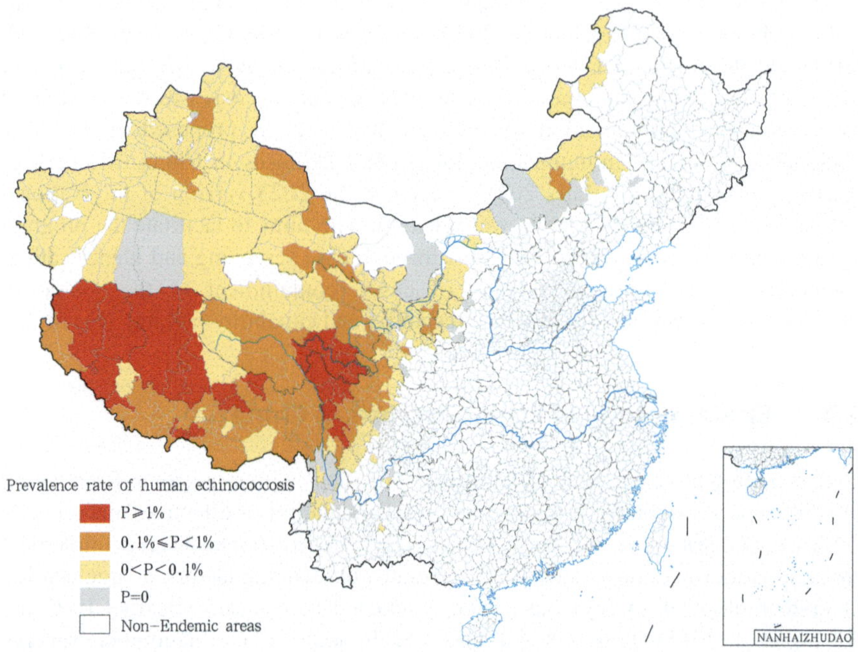


**Fig. 4.3** Geographical distribution of human echinococcosis prevalence in China in 2018

During the national survey on major human parasitic diseases conducted across 31 provinces (municipalities, autonomous regions) in China, the seroprevalence and prevalence of echinococcosis was investigated in 12 provinces (autonomous regions)

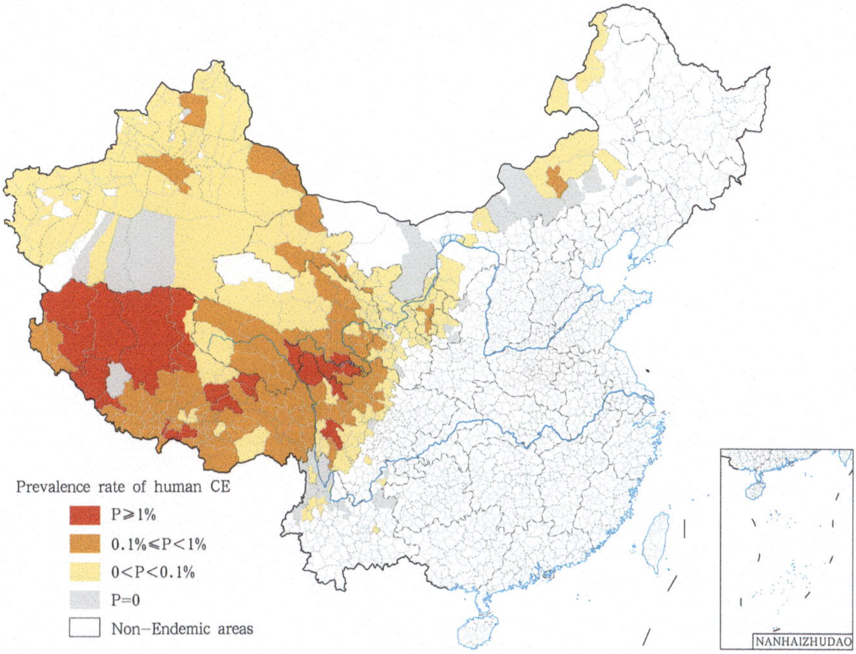


**Fig. 4.4** Geographical distribution of human cystic echinococcosis prevalence in China in 2018

of Inner Mongolia, Jilin, Yunnan, Sichuan, Guizhou, Yunnan, Shaanxi, Gansu, Qinghai, Ningxia, Xinjiang, and Tibet, and this was the first national survey on the epidemiology of echinococcosis (Fu et al. 2021). In this survey, patients with echinococcosis were detected in eight provinces (autonomous regions) of Xinjiang, Qinghai, Tibet, Sichuan, Gansu, Ningxia, Inner Mongolia, and Shaanxi, with echinococcosis seroprevalence of 12.04% and prevalence of 1.08, and 380,000 residents were estimated to have the disease, with more than 10 million at risk of infections; in addition, echinococcosis was found to be predominantly occur in Sichuan Province, Qinghai Province, Gansu Province, Ningxia Hui Autonomous Region, and Tibetan Autonomous Region, where both AE and CE were prevalent (Fu et al. 2021). Data captured from the national notifiable disease-reporting system showed that a total of 10,790 cases with echinococcosis were reported in China from 2004 to 2008, with a male to female ratio of approximately 1:1 and a mean age of 38 years (range: 8 months to 95 years), and 98.2% of the cases were found in 377 counties of six provinces (autonomous regions) of Xinjiang, Sichuan, Qinghai, Gansu, Ningxia, and Inner Mongolia (Wang et al. 2010b).

The nationwide epidemiological survey on echinococcosis was conducted using stratified randomly sampling in nine provinces (autonomous regions) of Inner Mongolia, Sichuan, Tibet, Gansu, Qinghai, Ningxia, Yunnan, Shaanxi, and Xinjiang in China from 2012 to 2016, and echinococcosis was detected in 368 out of the 413 counties (Fig. 4.6), including 115 counties endemic for both CE and AE and without

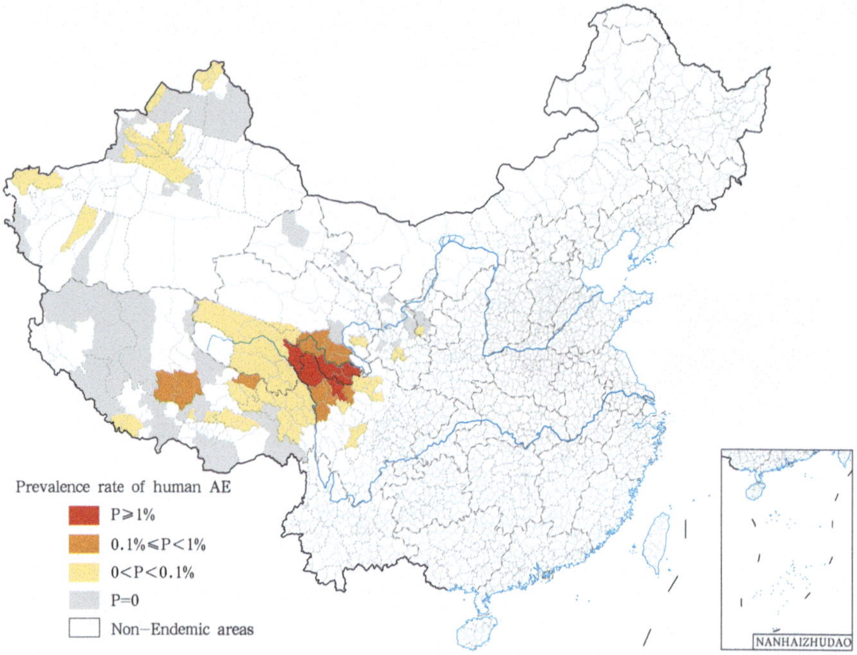


**Fig. 4.5** Geographical distribution of human alveolar echinococcosis prevalence in China in 2018

counties endemic for AE alone. The overall detection of echinococcosis was 0.51% among the study participants, and the prevalence of human echinococcosis was estimated to be 0.28% in endemic foci, with more than 50 million people at risk of infections (Coordinating Office of the National Survey on the Important Human Parasitic Diseases 2005). There were significant differences detected in the detection of human echinococcosis in terms of gender ($\chi^2 = 123.27$, $P < 0.05$), age ($\chi^2 = 247.99$, $P < 0.05$), production type ($\chi^2 = 5278.60$, $P < 0.05$), region ($\chi^2 = 2435.70$, $P < 0.05$), occupation ($\chi^2 = 6138.90$, $P < 0.05$), and educational level ($\chi^2 = 3381.40$, $P < 0.05$). More importantly, B-mode ultrasonography detected 0.10% prevalence of echinococcosis among 385,878 primary school students included in the survey, and ELISA showed 3.03% seroprevalence of echinococcosis. In addition, the positive rate of *Echinococcus* coproantigens was 4.25% in 111,832 dog samples, and the overall detection of *Echinococcus* infection was 4.68% in 217,774 livestock (sheep, bovine, and pigs) captured from 342 counties endemic for echinococcosis (Coordinating Office of the National Survey on the Important Human Parasitic Diseases 2005). Since then, the echinococcosis-endemic counties were classified into four categories according to the prevalence of echinococcosis in humans and dogs: Category A: human prevalence of 1% and higher, or dog prevalence of 5% and higher; Category B: human prevalence of 0.1–1%, or dog prevalence of 1–5%; Category C: human prevalence of 0–0.1%, or dog prevalence of 0–1%; and Category D: no local cases during the recent 3 years and no infected

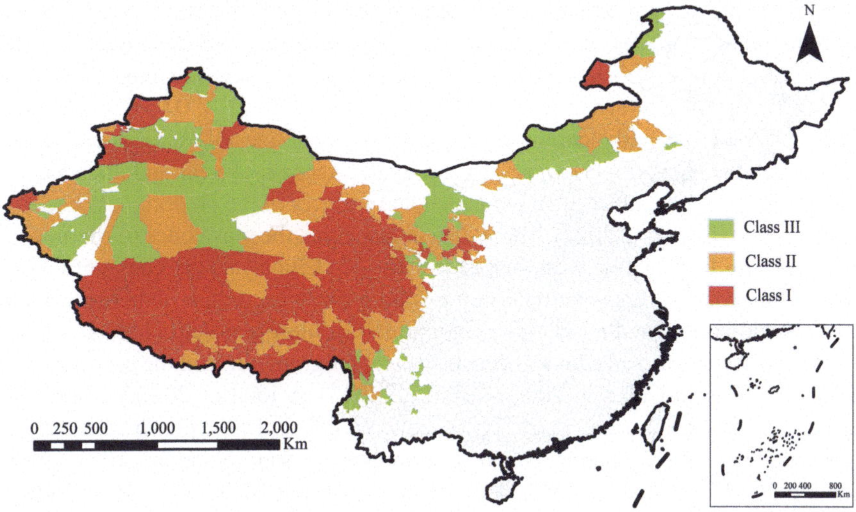


**Fig. 4.6** Nationwide distribution of human echinococcosis in China as revealed by the nationwide epidemiological survey on echinococcosis in China from 2012 to 2016

dogs. This survey provides baseline data for formulation of prevalence-specific echinococcosis control interventions and provides insights into development of national echinococcosis control programs for policymakers (Ma et al. 2023).

Epidemiological data have shown that approximately 95% of the cases with echinococcosis are detected in western pastoral areas of China (Fu et al. 2021; Coordinating Office of the National Survey on the Important Human Parasitic Diseases 2005; Ma et al. 2023); however, a tendency toward a rise has been observed in more than 20 echinococcosis non-endemic provinces of China (Qian et al. 2019a). Data captured from the national notifiable disease-reporting system showed that a total of 462 cases with echinococcosis were reported in 22 non-endemic provinces of China, including 234 men and 228 women, with a mean age of (41.42 ± 16.03) years (range: 4–86 years), and the highest proportion of cases with echinococcosis was seen at ages of 20–50 years (63.20%) among farmers and herdsmen (36.15%); in addition, 50.43% were imported from echinococcosis-endemic provinces and 97.40% were reported by hospitals (Yu et al. 2021). Because of low awareness of echinococcosis control knowledge, intensified health education pertaining to echinococcosis prevention, diagnosis and treatment, and improved screening of echinococcosis among farmers and herdsmen at ages of 20–50 years returning from echinococcosis-endemic areas are strongly recommended for timely identification and treatment of patients. As a Category C notifiable disease, echinococcosis has been included in the National Notifiable Disease Report Information System since 2004, and data from the National Notifiable Disease Report Information System showed that a total of 66,040 cases with echinococcosis were reported across 30 provinces (municipalities, autonomous regions) in China from 2004 to 2020, with 98.9% reported in nine provinces (autonomous regions) of Xinjiang, Sichuan,

Qinghai, Gansu, Ningxia, Tibet, Inner Mongolia, Shaanxi, and Yunnan and 1.1% in 22 non-endemic provinces, and including 31,494 men and 34,546 women, and the cases with echinococcosis appeared a tendency toward a rise in China ($Z = 2.27$, $P < 0.05$) and in non-endemic areas of China ($Z = 4.168$, $P < 0.05$); in addition, 91.3% of the cases with echinococcosis were found at ages of 20 years and older, and more importantly, there were 11 cases under ages of <1 year and 200 cases at ages of 85 years and older, with herdsmen and farmers as the predominant occupation (72.8%) (Han et al. 2022). These data indicate that the prevalence of echinococcosis remains high in nine western pastoral areas of China, and the number of cases with echinococcosis appear a tendency toward a rise with year, with herdsmen and farmers and young- and middle-aged individuals as high-prevalent populations.

Recently, the national echinococcosis control program released the prevalence of echinococcosis in China in 2021 (Kui et al. 2023). A total of 26,773 cases with echinococcosis were detected in 370 endemic counties of nine provinces (autonomous regions) of Xinjiang, Sichuan, Qinghai, Gansu, Ningxia, Tibet, Inner Mongolia, Shaanxi, and Yunnan, with mean prevalence of 0.06%, and 1346 new cases were diagnosed with echinococcosis (Kui et al. 2023). In addition, the positive rate of *Echinococcus* coproantigens was 0.47% in domestic dogs and 0.78% in wild canines, and the prevalence of echinococcosis was 0.63% in livestock and 1.07% in wild rodents in 2021, suggesting the transmission of echinococcosis has been almost under control; however, there are still multiple challenges for the national echinococcosis control program, and the management of the sources of *Echinococcus* infections remains to be intensified (Kui et al. 2023). Optimization of the echinococcosis control strategy, improving the capability building in grassroots disease control and prevention institutions, and establishment of a sensitive surveillance-response system are required to facilitate the progress toward echinococcosis control in China (Kui et al. 2023).

## 4.4 Disease Burden Due to Echinococcosis in China

As a global public health concern, echinococcosis causes huge public health concerns, and estimating the disease burden due to echinococcosis may facilitate the progress toward elimination. WHO estimated that there were 19,300 deaths and more than 1.67 million DALYs due to echinococcosis annually in the world (WHO 2007). China is among the top-burden echinococcosis countries in the world (Wen et al. 2019). During the period from 2008 to 2016, a total of 367 cases died from echinococcosis in China, with annual mean crude mortality of 129.29/105, and the dead cases had a mean age of 56 ± 18 years and a male to female ratio of 1:0.94; among the dead cases, Han ethnicity was the predominant ethnicity (52.0%), farmers as the predominant occupation (46.6%), junior high school and below as the predominant educational level (57.2%), and home as the predominant site of death (59.9%) (Tian et al. 2018).

In addition, data from the Global Burden of Disease Study were employed to assess the disease burden due to echinococcosis in China. Cui and colleagues evaluated the trends in disease burden of CE in China from 1990 to 2016 based on the GBD 2016 datasets, and the standardized DALY rate of CE reduced from $2.0/10^5$ to $0.3/10^5$, with a reduction of 87.2%, while the mortality increased from $143.6/10^5$ to $214.2/10^5$, with a rise of 49.1% (Cui et al. 2019). Based on the disability weight (DW) estimated using the five-level EuroQol five-dimensional questionnaire (EQ-5D-5L), the disease burden of human echinococcosis showed a remarkable decline following the implementation of the interventions in Shiqu County, Sichuan Province (Tian 2018). A multicenter hospital-based study during the period from December 2016 to March 2017 showed that the disease burden of human echinococcosis caused 322,400 DALYs, 28,800 years lived with disability (YLDs), and 293,600 years of life lost (YLLs) (Zhang et al. 2018). Wang and colleagues measured annual echinococcosis DALYs of 126,159 [95% uncertainty interval (*UI*): 122,415–137,675], annual AE DALYs of 105,829 (95% *UI*: 101,969–117,090), and CE DALYs of 20,330 (95% *UI*: 19,690–21,581) person-years in Tibetan communities (Wang et al. 2020). In addition, the DALYs were 4258.61 person-years due to echinococcosis in China in 2019 (Fig. 4.7), and there was no significant decrease in the DALYs of echinococcosis from 1990 to 2019 [average annual percent change (AAPC) = −0.5%, $P = 0.264$] (Zeng et al. 2023) (Table 4.1, Fig. 4.7). Currently, a few studies have been performed to evaluate the disease burden due to echinococcosis in China, and most studies were conducted based on field survey. Establishment of the DW system for echinococcosis is of urgent need in China (Fig. 4.8).

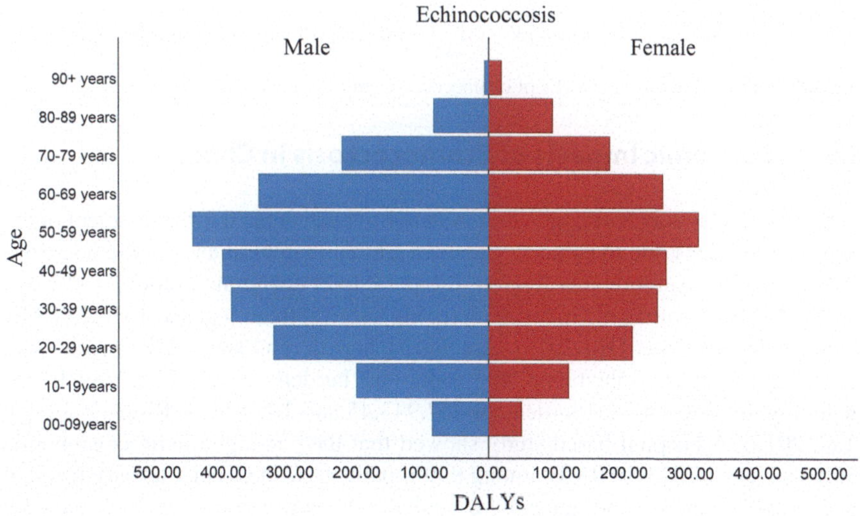


**Fig. 4.7** Gender- and age-specific DALYs of human echinococcosis in China in 2019

**Table 4.1** Disease burden of echinococcosis in China from 1990 to 2019 by Joinpoint regression analysis

| Time period | APC (95% *CI*) | *t* value | AAPC (95% *CI*) | *z* value | *P* value |
|---|---|---|---|---|---|
| 1990–2003 | −3.2% (−3.7% to −2.6) | −11.6 | | | <0.001 |
| 2003–2016 | −1.4% (−2.3% to −0.5%) | −3.3 | | | <0.003 |
| 2016–2019 | 17.1% (9.6% to 25.2%) | 4.9 | | | <0.001 |
| 1990–2019 | | | −0.5% (−1.2% to 0.3%) | −1.1 | 0.264 |

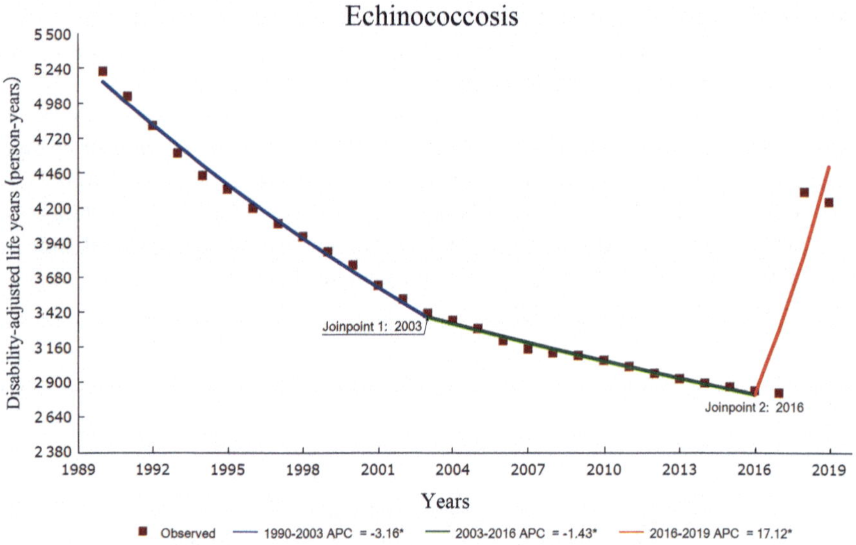


**Fig. 4.8** Temporal trends in DALYs of echinococcosis in China from 1990 to 2019

## 4.5 Economic Impacts of Echinococcosis in China

Currently, a relatively entire indicator system has been created for assessment of the economic burdens due to echinococcosis in China. In central and southern regions of Ningxia Hui Autonomous Region, the per capita economic burden of CE was 10,842.18 RMB among inpatients undergoing surgical treatment and 1318.13 RMB among non-inpatients (Hu 2010). Among 51 hepatic end-stage AE patients with autologous liver transplantation, the economic burdens were 447,128.89 RMB, including direct economic burdens of 257,943.45 and 189,834.06 RMB (Zulifeiye et al. 2017). A hospital-based study showed that the per capita indirect economic burdens were 12,044.58 RMB among 999 inpatients with CE in a tertiary hospital in Urumqi, Xinjiang, and appeared a tendency toward a rise with gross national product from 2004 to 2008 (Wang et al. 2010a). In a tertiary hospital in the

Qinghai-Tibetan Plateau, the mean expense of hospital stay was 47,100 RMB among hepatic echinococcosis patients with surgical treatment, and the mean expense of hospital stay was higher among AE patients than among CE patients (60,800 vs. 42,200 RMB/person; $t = 3.697$, $P < 0.05$) (Zhang et al. 2018). In addition, the mean expense of hospital stay was 10,726.83 RMB among 64 echinococcosis patients without surgical treatment and 55,026.71 RMB among 88 echinococcosis patients with surgical treatment (Tian 2018).

Because of data availability, most studies on economic burdens of echinococcosis have a small sample size and focus on inpatients; in addition, most studies only evaluate direct medical expenses. Since there is a lack of economic assessment of echinococcosis among untreated patients and patients receiving chemotherapy, the economic impact of echinococcosis cannot be accurately evaluated (Miao et al. 2023).

## 4.6 Concluding Remarks

Echinococcosis control involves humans, livestock, wild animals, and ecological environments. As a Category C notifiable disease, echinococcosis control has been given a high priority in China. Following concerted efforts for years, great strides have bene achieved for the national echinococcosis control program; however, the number of echinococcosis cases has recently appeared a tendency toward a rise in non-endemic areas of China (Yu et al. 2021). Therefore, One Health approach has been proposed, which mobilizes multisectoral sources to walk together to combat echinococcosis.

Furthermore, China is active to participate in international collaboration and global health governance and supports echinococcosis control in Mongolia, including assisting formulation of national echinococcosis control program, training disease control professionals, and implementing joint control activities (Qian et al. 2019b). On November 2017, international experts from 13 countries along the Belt and Road Initiative were invited in Chengdu, China to discuss how to jointly respond to the high disease burden due to echinococcosis and signed the Chengdu Declaration, with aims to establish a network for global cooperation and capacity-building to accelerate the control and elimination of echinococcosis, in line with WHO's roadmap on NTDs and the global momentum harnessed to support the targets of the United Nations Sustainable Development Goals in achieving universal health coverage (Qian and Zhou 2018). In addition, a remote medical service platform was created based on the WHO Collaborating Centre for Prevention and Care Management of Echinococcosis in the First Affiliated Hospital of Xinjiang Medical University, which provides remote medical services guidance and training on echinococcosis for middle Asian countries (Abudusalamu et al. 2019).

China is one of the countries with the highest disease burden of echinococcosis in the world. China's progress in echinococcosis control started with limited scientific research projects, followed by nationwide epidemiological surveys, and then launched a national echinococcosis control program based on pilot projects. During this process, science-based control and technological innovations have been integrated into echinococcosis control in China. The concept and mechanism innovations-based echinococcosis control models and successful experiences in highly endemic foci promote the sustainable echinococcosis control achievements in China (Xiao 2021). In addition, the joint echinococcosis prevention and control and sharing of successful experiences with other echinococcosis-endemic countries will provide China's contributions to global echinococcosis control program.

**Acknowledgments** Figures presented in this chapter were kindly provided by Miss Ting Zeng from China CDC. The study was supported by the grants from Qinghai Provincial Department of Science and Technology (grant no. 2019-HZ-803), Wuxi Municipal Health Commission (grant no. Z202116), and Wuxi Municipal Department of Science and Technology (grant no. Y20212012).

**Conflicts of Interest** The authors declare no conflicts of interest.

## References

Abudusalamu AN, Shao YM, Tuerganaili A, Zhang WB, Lin RY, Hou YY, Zhang ZZ, Wen H (2019) Establishment and innovative practice of integrative system of prevention, diagnosis and management for echinococcosis in China. Chin J Parasitol Parasit Dis. 37(4):388–394

Alvi MA, Alsayeqh AF (2022) Food-borne zoonotic echinococcosis: a review with special focus on epidemiology. Front Vet Sci 9:1072730

Anisuzzaman HMS, Hatta T, Labony SS, Kwofie KD, Kawada H, Tsuji N, Alim MA (2023) Food- and vector-borne parasitic zoonoses: global burden and impacts. Adv Parasitol 120:87–136

Atkinson JA, Gray DJ, Clements AC, Barnes TS, McManus DP, Yang YR (2013) Environmental changes impacting *Echinococcus* transmission: research to support predictive surveillance and control. Glob Chang Biol 19(3):677–688

Budke CM, Deplazes P, Torgerson PR (2006) Global socioeconomic impact of cystic echinococcosis. Emerg Infect Dis 12(2):296–303

Cadavid Restrepo AM, Yang YR, McManus DP, Gray DJ, Giraudoux P, Barnes TS, Williams GM, Soares Magalhães RJ, Hamm NA, Clements AC (2016) The landscape epidemiology of echinococcoses. Infect Dis Poverty 5:13

Chai JJ (2009) Echinococcosis control in China: challenges and research needs. Chin J Parasitol Parasit Dis. 27(5):379–383

Chen FZ, Pu YL (2015) Challenges and suggestions for implementing the National mid- and long-term control programs for animal diseases. Chi J Vet Med 51(1):110–112

Coordinating Office of the National Survey on the Important Human Parasitic Diseases (2005) A national survey on current status of the important parasitic diseases in human population. Chin J Parasitol Parasit Dis 23(5 Suppl):332–340

Craig PS, Hegglin D, Lightowlers MW, Torgerson PR, Wang Q (2017) Echinococcosis: control and prevention. Adv Parasitol 96:55–158

Cui FF, Zhao J, Yu CH, Song XQ, He XY, Wang LL, Gao JH, Zhai YK (2019) Trend analysis on the disease burden of cystic echinococcosis in China from 1990 to 2016. Modern Prev Med 46(2):193–197,220

Deplazes P, Rinaldi L, Alvarez Rojas CA, Torgerson PR, Harandi MF, Romig T, Antolova D, Schurer JM, Lahmar S, Cringoli G, Magambo J, Thompson RC, Jenkins EJ (2017) Global distribution of alveolar and cystic echinococcosis. Adv Parasitol 95:315–493

Eckert J (2007) Historical aspects of echinococcosis—an ancient but still relevant zoonosis. Schweiz Arch Tierheilkd 149(1):5–14
Eckert J, Thompson RC (2017) Historical aspects of echinococcosis. Adv Parasitol 95:1–64
Fu M, Han S, Xue C, Wang X, Liu B, Wang Y, Wang L, Wei S, Cui X, Zhang T, Zhang HB, Zheng B, Tian T, Yang S, Gao CH, Dang Z, Xu B, Yu Q, Wu W (2020) Contribution to the echinococcosis control programme in China by NIPD-CTDR. Adv Parasitol 110:107–144
Fu MH, Wang X, Han S, Guan YY, Bergquist R, Wu WP (2021) Advances in research on echinococcoses epidemiology in China. Acta Trop 219:105921
Han S, Kui Y, Xue CZ, Zhang YL, Zhang BG, Li QY, Wang T, Dai Y, Yuan Y, Tao W, Zhu AY, Li ZX, Wang ZJ, Tang L, Wu WT, Xie HG, Yao LN, Yang YC, Wu WP (2022) A national survey on current status of the important parasitic diseases in human population. Chin J Parasitol Parasit Dis 40(4):475–480
Hu XM (2010) Study on disease burden of cystic echinococcosis in central and southern Ningxia. China CDC, Beijing
Kui Y, Xue CZ, Wang X, Liu BX, Wang Y, Wang LY, Yang SJ, Han S, Wu WP, Xiao N (2023) Progress of echinococcosis control in China, 2021. Chin J Parasitol Parasit Dis. 41(2):142–148
Laing G (2023) A year of one health and neglected tropical diseases. Int Health 15(5):477–479
Laing G, Vigilato MAN, Cleaveland S, Thumbi SM, Blumberg L, Salahuddin N, Abdela-Ridder B, Harrison W (2021) One health for neglected tropical diseases. Trans R Soc Trop Med Hyg 115(2):182–184
Ma T, Wang Q, Hao M, Xue C, Wang X, Han S, Wang Q, Zhao J, Ma X, Wu X, Jiang X, Cao L, Yang Y, Feng Y, Gongsang Q, Scheffran J, Fang L, Maude RJ, Zheng C, Ding F, Wu W, Jiang D (2023) Epidemiological characteristics and risk factors for cystic and alveolar echinococcosis in China: an analysis of a national population-based field survey. Parasit Vectors 16(1):181
McManus DP, Zhang W, Li J, Bartley PB (2003) Echinococcosis Lancet 362(9392):1295–1304
Miao LY, Tian T, Zhou XN (2023) Progress of international and national researches on disease burden of echinococcosis. Chin J Schisto Control. 35(3):307–312
Ministry of Agriculture and Rural Affairs (2021) National compulsory immunization for animal diseases in China. http://www.moa.gov.cn/govpublic/xmsyj/202101/t20210111_6359742.htm
Molyneux DH, Savioli L, Engels D (2017) Neglected tropical diseases: progress towards addressing the chronic pandemic. Lancet 389(10066):312–325
Moro PL, Schantz PM (2006) Echinococcosis: historical landmarks and progress in research and control. Ann Trop Med Parasitol 100(8):703–714
Qian MB, Abela-Ridder B, Wu WP, Zhou XN (2017) Combating echinococcosis in China: strengthening the research and development. Infect Dis Poverty 6(1):161
Qian MB, Chen J, Bergquist R, Li ZJ, Li SZ, Xiao N, Utzinger J, Zhou XN (2019a) Neglected tropical diseases in the People's Republic of China: progress towards elimination. Infect Dis Poverty 8(1):86
Qian YJ, Ding W, Wu WP, Bandikhuu A, Damdindorj T, Nyamdorj T, Bold B, Dorjsuren T, Sumiya G, Guan YY, Zhou XN, Li SZ, Don Eliseo LP 3rd. (2019b) A path to cooperation between China and Mongolia towards the control of echinococcosis under the belt and road initiative. Acta Trop 195:62–67
Qian MB, Zhou XN (2018) Walk together to combat echinococcosis. Lancet Infect Dis 18(9):946
Tan X, Zhang Y, Shao H (2019) Healthy China 2030, a breakthrough for improving health. Glob Health Promot 26(4):96–99
Tian T (2018) Study on disease burden of echinococcosis in high-risk regions of the Qinghai-Tibetan plateau. Beijing, China CDC
Tian T, Cao CL, Wu WP, Jia TW, Qian MB, Hao YW, Zhu ZL, Wang Q, Li SZ, Zhou XN (2018) Epidemiological characteristics of echinococcosis death cases in China from 2008 to 2016. Chin J Schisto Control. 30(3):282–288
Torgerson PR, Keller K, Magnotta M, Ragland N (2010) The global burden of alveolar echinococcosis. PLoS Negl Trop Dis 4(6):e722
Wang LY (2017) The final evaluation of the 12th five-year plan for echinococcosis control and the 13th five-year plan. Chin Anim Health 19(7):13–19

Wang L, Feng XH, Duan XY, Wen H (2010a) Direct and indirect economic burden of 999 cystic echinococcosis patients in a tertiary hospital. Chin J Epidemiol 31(7):835–836
Wang Z, Wang X, Liu X (2008) Echinococcosis in China, a review of the epidemiology of *Echinococcus* spp. EcoHealth 5(2):115–126
Wang L, Wang Z, Ma Q, Fang G, Yang J (2019) The development and reform of public health in China from 1949 to 2019. Glob Health 15(1):45
Wang LY, Wu WP, Zhu XH (2010b) The endemic status of hydatidosis in China from 2004 to 2008. Chin J Zoon 26(7):699–702
Wang Q, Yang L, Wang Y, Zhang GJ, Zhong B, Wu WP, Zheng CJ, Liao S, Yu WJ, He W, Wang Q, Chen F, Li RR, Huang Y, Yao R, Zhou XN (2020) Disease burden of echinococcosis in Tibetan communities-a significant public health issue in an underdeveloped region of western China. Acta Trop 203:105283
Wen H, Vuitton L, Tuxun T, Li J, Vuitton DA, Zhang W, McManus DP (2019) Echinococcosis: advances in the 21st century. Clin Microbiol Rev 32(2):e00075–e00018
WHO. WHO estimates of the global burden of foodborne diseases: foodborne diseases burden epidemiology reference group 2007—2015. https://www.who.int/publications/i/item/9789241565165
World Health Organization (2020) Ending the neglect to attain the sustainable development goals: a road map for neglected tropical diseases 2021–2030. World Health Organization, Geneva. https://iris.who.int/handle/10665/338565
Xiao N (2021) Concept and mechanism innovations provide a sustainable driver for echinococcosis control in China. Chin J Schisto Control. 33(4):329–333
Yan J, Hu T, Lei ZL (2015) The endemic situation and challenges of major parasitic diseases in China. Chin J Parasitol Parasit Dis. 33(6):412–417
Yu Q, Han S, Xue JB, Xia S (2021) Epidemiological profiles of echinococcosis cases reported in the National Notifiable Disease Report System in non-endemic areas of China from 2004 to 2016. Chin J Schisto Control. 33(1):48–53
Yu Q, Xiao N, Han S, Tian T, Zhou XN (2020) Progress on the national echinococcosis control programme in China: analysis of humans and dogs population intervention during 2004-2014. Infect Dis Poverty 9(1):137
Zeng T, Lü S, Tian L, Li S, Sun L, Jia T (2023) Temporal trends in disease burden of major human parasitic diseases in China from 1990 to 2019. Chin J Schisto Control 35(1):7–14
Zhang MY, Wu WP, Guan YY, Wang LY, Hou YY, Feng Y, Ma JY, Wang Q, Wang Y, Han S, Xue CZ, Wu WT (2018) Analysis of disease burden of hydatid disease in China. Chin J Parasitol Parasit Dis. 36(1):15–20
Zhang W, Zhang Z, Wu W, Shi B, Li J, Zhou X, Wen H, McManus DP (2015) Epidemiology and control of echinococcosis in central Asia, with particular reference to the People's Republic of China. Acta Trop 141(Pt B):235–243
Zulifeiye A, Guo M, Wang L, Liu C, Tuerganaili A, Shao YM, Wen H (2017) Economic burden of hepatic alveolar echinococcosis patients with autologous liver transplantation. Chin J Public Health 33(7):1105–1109

# Biology and Hosts of *Echinococcus* Species in China

5

Baoping Guo, Kalibixiati Aimulajiang, and Rongsheng Mi

**Abstract**

Echinococcosis, is a serious zoonotic disease with severe prevalence in the vast western and northwestern pastoral areas of China, one of the highest prevalence rates in the world. Cystic echinococcosis (CE) and alveolar echinococcosis (AE), are endemic in some areas simultaneously, posing a serious threat to the health and socio-economic development of local populations. *Echinococcus* species (spp.) can infect both domestic and wild animals through the domestic cycle, wild cycle and mixed domestic–wild cycle, completing the cycle chain. In the pastoral areas of western China, the extensive distribution of pasture and forest and the abundance of domestic and wildlife resources provide a good environment for the growth and egg success of *Echinococcus* spp. This, together with the unique human factors in endemic areas, poses a great challenge to the prevention and control of *Echinococcus* spp. nowadays, resulting in a large number of population infections in the West in recent years, especially in the Qinghai-Tibet Plateau and Yili, Xinjiang. Based on previous epidemiological studies, this chapter reviews the important factors contributing to the spread of *Echinococcus* in China, focusing on three aspects: biological (parasite genotype and host species),

B. Guo (✉) · K. Aimulajiang
State Key Laboratory of Pathogenesis, Prevention and Treatment of High Incidence Diseases in Central Asia, Clinical Medicine Institute, The First Affiliated Hospital of Xinjiang Medical University, Urumqi, Xinjiang, PR China

Xinjiang Key Laboratory of Echinococcosis, Clinical Medicine Institute, The First Affiliated Hospital of Xinjiang Medical University, and WHO Collaborating Centre on Prevention and Case Management of Echinococcosis, Urumqi, Xinjiang, PR China

R. Mi
Key Laboratory of Animal Parasitology of Ministry of Agriculture, Laboratory of Quality and Safety Risk Assessment for Animal Products on Biohazards (Shanghai) of Ministry of Agriculture, Shanghai Veterinary Research Institute, Chinese Academy of Agricultural Sciences, Shanghai, PR China



J. Li et al. (eds.), *Echinococcus: Control and Elimination of Echinococcosis with a Focus on China and Europe*, Parasitology Research Monographs 19,
https://doi.org/10.1007/978-3-031-54479-8_5

environmental (landscape and climate) and social (age, sex, ethnicity, education, occupation, lifestyle, cultural practices, living conditions and hygiene habits of populations in endemic areas), which interact with each other to determine the transmission of *Echinococcus* tapeworm. In view of the current epidemiological situation of echinococcosis in China, and in view of the above-mentioned challenges in the prevention and control of echinococcosis, in order to implement precise prevention and control, and to achieve a shift from the previous 'crude' prevention and control to 'precise' prevention and control, it is more important to highlight science and precision, and it is urgent to use scientific theories and methods to grasp the many epidemiological factors of echinococcosis, to accurately identify and assess the future risk of echinococcosis transmission and its changing trends, and then to identify potential high-risk areas and develop efficient control strategies.

**Keywords**

Echinococcosis · Serious zoonotic disease · Western China · Living conditions · Hygiene habits · Potential high-risk areas · Prevention and control

## 5.1 Introduction

Echinococcosis is a zoonotic disease that lead to a serious risk to human health and livestock production and can be a global public health problem (Wen et al. 2019). Echinococcosis caused by the larvae of the tapeworm stage of the *Echinococcus*, which includes *Echinococcus granulosus sensu lato* (*s.l.*), *E. multilocularis*, *E. oligarthtus*, *E. vogeli* and *E. shiquicus* (Eckert and Thompson 2017; Vuitton et al. 2020). *Echinococcus* cestodes essentially require two mammalian hosts to complete their life history, with intermediate hosts being hoofed herbivores and rodents and definitive hosts being carnivores (canids, felines, etc.), with humans being incidental hosts and not involved in the cycle (Fig. 5.1). Cystic echinococcosis (CE) and alveolar echinococcosis (AE) infections cause very severe and even life-threatening symptoms (Stey et al. 1993; Wang et al. 2021a).

CE and AE were once neglected zoonotic parasitic diseases (Cai et al. 2019; Giraudoux et al. 2006), with CE being widely distributed globally (Torgerson et al. 2010; Wen et al. 2019) (Fig. 5.2), with endemic areas on all five continents (Budke et al. 2006; Torgerson 2003; Torgerson and Dowling 2001; Torgerson et al. 2010), and contributing to a substantial disease burden, although AE is predominantly endemic in the Northern Hemisphere, with high pathogenicity and mortality rates (Torgerson et al. 2002, 2010). Approximately 90% of global echinococcosis is distributed in China (Noguera et al. 2022; Zhang et al. 2009), with the most severe cases in the western pastoral areas of Xinjiang, Qinghai, Tibet and Inner Mongolia (Shumuye et al. 2023; Wang et al. 2021b). Since the implementation of the Action Plan for the Prevention and Control of Echinococcosis (2010–2015) and the central transfer payment programme (Al-Jawabreh et al. 2017; Loi et al. 2019), the China government has arranged special funds and achieved initial results in the prevention

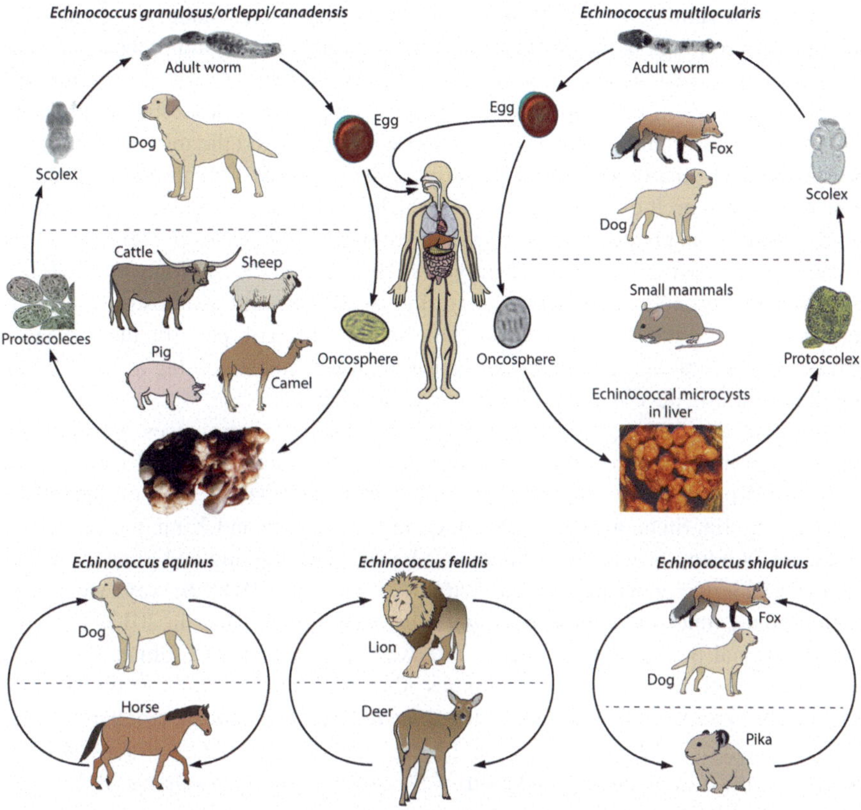


**Fig. 5.1** The life cycle of *E. granulosus* and *E. multilocularis*

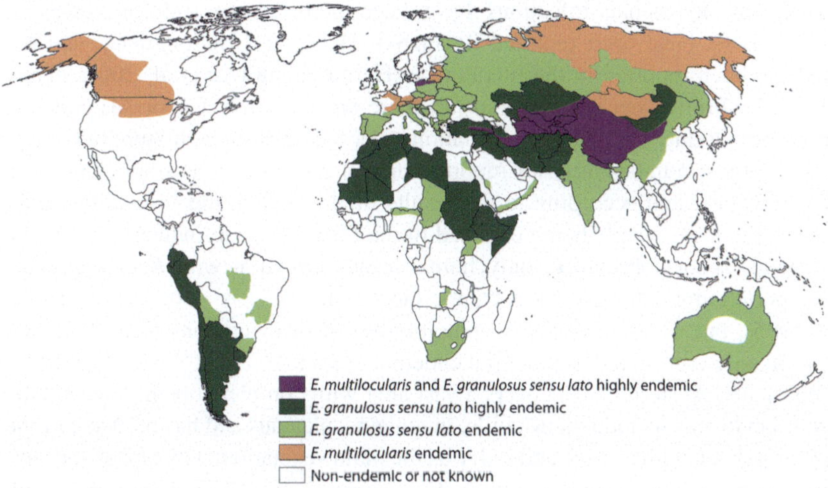


**Fig. 5.2** Global distribution of CE and AE

and control of echinococcosis. While achievements have been made, we should also see the shortcomings in the work, such as the national control and prevention measures for large livestock and domestic dogs (*Canis familiaris*) in the prevention and control objectives of CE, but the intermediate hosts of CE – domestic animals and definitive hosts – canids (unowned dogs and wild foxes) and the intermediate hosts of AE disease – small rodents and definitive hosts – canids. This has led to a lack of host-related research on echinococcosis, which has led to a shortage of prevention tools, resulting in a large number of people still being infected with echinococcosis in the western pastoral areas in recent years (Huang et al. 2018; Wang et al. 2022). In summary, echinococcosis causes suffering to patients and also restricts agricultural and pastoral production, increasing the social burden of 'poverty due to disease and poverty due to disease' in western pastoral areas, and becoming a factor limiting the economic development of western pastoral areas.

From 2004 to 2008, echinococcosis were reported in 27 provinces, autonomous regions and municipalities directly under the Central Government of China, with 98.2% of reported cases in seven provinces and autonomous regions, including Inner Mongolia, Sichuan, Tibet, Gansu, Qinghai, Ningxia and Xinjiang, as well as in the surrounding provinces of Shaanxi, Shanxi, Heilongjiang and Yunnan, where the main endemic areas are located (Wu 2017; Wu et al. 2018). Surveys were conducted on a county-by-county basis, with the exception of Tibet in 2012. The projected national population prevalence rate was 0.24%, with an estimated 110,000 people affected. The number of counties identified and the number of endemic counties has increased by 2 in Inner Mongolia and by 24 in Yunnan, compared to the 350 estimated in 2012, giving a national prevalence of at least 368 counties and 119 counties with a threatened population of 50 million and a prevalence of 170,000 (Wu et al. 2018). *E. shiquicus* was the first identified in Shiqu County, Ganzi Prefecture, Sichuan Province (Xiao et al. 2005) and is currently only found in the eastern part of the Tibetan Plateau, including western Sichuan and southeastern Qinghai, but no human infections have been reported. In addition, cases of *Echinococcosis* have been reported in Shanxi, Hebei and Heilongjiang, and AE cases have been reported in the literature in Heilongjiang (Xu et al. 2004; Zhang et al. 2001). From what is known so far, there are no purely CE or AE endemic counties on a county-by-county basis nationwide, but basically a mixture of CE and AE. The largest number of cases reported in non-endemic areas since 2004 is in Henan province, and according to the results of the survey, the cases are mainly related to the local people being engaged in the fur processing industry. In Liyang and Yixing, Jiangsu Province, only human cases infection with *Echinococcosis* were reported and definitive hosts canine faecal antigen tests were positive (Shen et al. 1995), but no more in-depth epidemiological investigations were done and they were only considered as potential endemic areas (Wu 2017; Wu et al. 2018).

Central and western China covers a vast area with complex topography, and the types of landforms include plains, hills, mountains, plateaus and basins. Most of the area is grassland, with grasses and sedge as the main vegetation, with alpine and subalpine meadows being the main types (Giraudoux et al. 2013b; Giraudoux et al. 2013a; Shang et al. 2014). In total, over 190 species of hoofed (Smith and Xie 2009), rodent and carnivorous wild mammals are found, providing a rich wildlife host for the transmission of *Echinococcosis* (Jiang et al. 2012a). The natural range of wildlife

(Wang et al. 2002, 2018) intersects with human grazing activity (Li et al. 2005; Wang et al. 2002, 2020), providing a pathway for the spread of echinococcosis to the human environment and causing a significant increase in the prevalence of echinococcosis in the population of western China (Li et al. 2010; Li et al. 2005). However, most epidemiological studies on echinococcosis have focused on domestic animals and there is a relative lack of studies on wild animals (Craig et al. 2019; Wu et al. 2018). In addition, in recent years, with the development of the western region and the 'Belt and Road Initiative', the country has invested in a large number of infrastructures that have changed the geography of the West, indirectly affecting the habitat of different organisms (Qian et al. 2017, 2019), and by conducting satellite images of the West it was found that the habitat has also changed, such as changes in grassland, cultivated land, natural forest belts and the increase in roads. The alteration of these factors has led to a serious lack of current research on wildlife, and secondly, with the increasing impact of human activities on the natural environment, understanding the mechanisms of the role of wild and domestic animals in the transmission of echinococcosis and understanding the prevalence of echinococcosis in fauna (especially wildlife) in western China is an important theoretical basis for establishing an efficient integrated control strategy. In this chapter, we have reviewed the mechanisms of echinococcosis transmission in western China. To this end, this chapter reviews the progress of host and ecological research on echinococcosis in western China and analyses the animal ecological factors influencing the spread of echinococcosis to provide data to support the prevention and control of echinococcosis, and to provide recommendations for disease control and ecological protection of the pathogen.

## 5.2 *Echinococcus granulosus*

### 5.2.1 The risk and distribution of *E. granulosus*

The main pathogens of CE include *E. granulosus sensu strict* (*s.s.*), *E. equinus*, *E. felidis*, *E. ortleppi* and *E. canadensis*. CE is widely prevalent worldwide (Table 5.1) (Wen et al. 2019) and is associated with human behavioural risk the risk and pathogenicity of zoonotic diseases caused by CE varies due to the genetic diversity of animal hosts (Otero-Abad and Torgerson 2013). The economic losses to the livestock industry are estimated to be more than 3 billion yuan, causing great harm to the health of the inhabitants and the production and development of the livestock industry in endemic areas (Agudelo Higuita et al. 2016). The clinical presentation depends on the infected organ, the size of the cyst and the location of the infection (Mcmanus et al. 2003), with rupture of the cyst causing systemic or local allergic reactions (Engelke et al. 2014) and even death (Khachatryan 2017).

### 5.2.2 *E. granulosus* genotypes in China

The existing known genotypes of *E. granulosus* s.l. have been found in China, except for *E. equinus* (G4), and the genetic diversity of *E. granulosus s.l.* is very rich, especially in the Qinghai-Tibetan Plateau. According to the literature, the

**Table 5.1** Current recognized species within the genus *Echinococcus* and their preferential hosts and geographic distribution in China

| Species | Definitive host(s) | Intermediate host(s) | Human cases | Distribution |
|---|---|---|---|---|
| *Echinococcus granulosus sensu strict* (*s.s.*) | Dog, *Canis lupus* (wolf), *Vulpes vulpes* (Red fox), Cat, etc. | Sheep, goat, cattle, domestic pig, camel, buffalo, horse, wild ungulates, marsupials, Bactrian camel, *Bos grunniens*, *Pseudois nayaur*, *Procapra picticaudata*, *Rodentia*, *Ochotona curzoniae*, *Lasiopodomys fuscus*, etc. | Reported | Qinghai, Tibet, Sichuan, Gansu, Xinjiang, Inner Mongolia, Ningxia, Heilongjiang and Yunnan |
| *Echinococcus canadensis* | Not reported | Sheep, *Bos grunniens* | Reported | Gansu, Heilongjiang |
| *Echinococcus ortleppi* | Not reported | Not reported | Reported | Guangxi |
| *Echinococcus felidis* | Not reported | Not reported | Not reported | Not reported |
| *Echinococcus equinus* | Not reported | Not reported | Not reported | Not reported |
| *Echinococcus multilocularis* | *Vulpes vulpes* (Red fox), *Canis lupus* (wolf), Tibetan dog, Dog, etc. | *Ochotona curzoniae*, *Lepus oiostlus*, *Cricetulus migratorius* Pallas, *Microtus fucus*, *Mus musculus*, Tibetau sheep, *Bos grunniens*, sheep, domestic pig, *Pitymys irene*, *Ochatona daurica* Pallas, *Myospalax fontanierii* Milne *Edwards*, *Apodemus sylvaticus*, *Rhombomys sopimus* Pallas, *Ellobius trancrei*, *Meriones tamariscinus* Pallas, *Arvicola amphibious*, etc. | Reported | Qinghai, Tibet, Sichuan, Gansu, Xinjiang, Inner Mongolia, Ningxia and Yunnan |
| *Echinococcus oligarthra* | Not reported | Not reported | Not reported | Not reported |
| *Echinococcus vogeli* | Not reported | Not reported | Not reported | Not reported |
| *Echinococcus shiquicus* | Tibetan fox (*Vulpes ferrilata*) | *Ochotona curzoniae*, *Lasiopodomys fuscus*, *Microtus limnophilus*, *Cricetulus longicaudatus*, *Phaiomys leucurus*, etc. | Not reported | Qinghai and Sichuan |

genetic composition of *E. granulosus s.l.* in China: *E. granulosus s.s.* (G1 and G3) accounts for 98.3% (Hua et al. 2022); *E. ortleppi* (G5) for 0.1%; *E. canadensis* (G6, G7, G8 and G10) for 1.6%. Of these, 97.7% of host infections and geographical distribution were G1. In Xinjiang, Sichuan and Qinghai endemic provinces, we found *E. granulosus s.l.* infections in sheep and yaks (García-Méndez et al. 2022;

Guo et al. 2019), with the most stable genetic composition, and a higher proportion of fertile cysts infected in sheep 70.69% (287/406) than in yaks 15.22% (28/184) (Hua et al. 2022). In China, a great deal of surveillance has been done on our intermediate hosts such as yaks, cattle, goats and camels (Ohiolei et al. 2020; Ohiolei et al. 2019; Yang et al. 2005a), but surveillance of pigs is lacking, mostly concentrated in slaughterhouses, so we also advocate for more CE surveillance of pigs (Anvari et al. 2021; Cucher et al. 2016), as they play an important role in the transmission and conservation of *E. granulosus s.l.*, which are often overlooked.

### 5.2.3 Transmission of *E. granulosus*

Two types of transmission routes exist for both *E. granulosus* and *E. multilocularis*, namely the wild environment route, which relies on the wildlife food chain, and the human environment route, which relies on domestic animals (for example, domestic dogs) in conjunction with wildlife hosts living in close proximity to the human environment. A wide range of intermediate host species are known for *E. granulosus*, including some medium and large ungulates of the families Bovidae, Deer, Pig, Equidae, Camelidae, Giraffidae, Elephantidae and Hippopotamidae (Eckert et al. 2001; Hua et al. 2022; Tigre et al. 2016; Zait et al. 2016), as well as some species of Leporidae, Primates and Marsupials; the definitive hosts are mainly a variety of canids (Alvi et al. 2023; Schmidberger et al. 2022). The high prevalence of both types of echinococcosis in intermediate hosts, cattle, sheep and yaks (He and Wang 2001; Ji et al. 1995), as well as instances of simultaneous infection with both types of echinococcosis in the same host, not only severely affects the economic development of the livestock industry in the area but more importantly, the involvement of domestic animals in the life cycle of *E. multilocularis* increases the risk of human infection, making it difficult and complex to develop and implement control measures. Both plateau pika (*Ochotona curzoniae*) and *Lepus oiostlus* are dominant wildlife on the Tibetan Plateau and are very widely distributed (Wang et al. 2018; Xiao et al. 2005; Chiu et al. 2000). They aren't only hunted by wild carnivores but are also preferred predators by domestic dogs, which has the potential to become a bridge animal for the presence of echinococcosis circulating between domestic and wild animals, forming a mixed cycle (Karamon et al. 2017; Wang et al. 2022). The control of echinococcosis should therefore be a two-pronged approach, with deworming management of domestic dogs accompanied by an emphasis on intervention in the wild cycle (Deplazes et al. 2017).

### 5.2.4 The intermediate hosts of *E. granulosus*

CE is mainly found in domestic animals, such as cattle, sheep and yaks, and it has been reported in various provinces in western China. For example, Chen et al. investigated yaks and sheep in Guinan County, Qinghai Province, and the infection rates were 58.00% (58/100) and 43.73% (48/567) in yaks and sheep, respectively (Chen et al. 2005); He et al. investigated yaks and sheep in southern Qinghai using ocular autopsy and lesion sectioning, with infection rates as high as 84.60% (489/578) in yaks and 79.90% (306/383) in sheep (He and Wang 2001); Schantz et al. found a high rate of 71.60% (521/728) in yaks in Qinghai (Schantz et al. 2003); Yu et al. investigated yaks and sheep in Qinghai and the infection rates were 78.52% (106/135) and 82.61% (95/115) (Yu et al. 2008), respectively. The high infection rate of domestic animals in Qinghai may be related to the backward farming practices and health education in the region; however, there is also a high infection rate in wild animals in Qinghai. He et al. conducted an epidemiological survey on wild rock sheep and Tibetan antelope (*Pantholops hodgsonii*) in Qingnan area, and the infection rates were 76.42% (210/327) and 66.66% (130/198), respectively, all of which indicate the presence of natural epidemic sites of CE in Qinghai. Plateau pika is an important intermediate host for AE, but as an intermediate host for CE is rarely reported. He et al. found up to 71.43% (5/7) infection in plateau pika in Qingnan, and Jiang et al. also reported up to 71.43% (5/7) in Qinghai. Sichuan province, which borders Qinghai, is also hardest hit by CE, especially in yaks, as reported by Guo et al. and He et al. (Guo et al. 2011; He et al. 2000), while Weng et al. not only found a 10.00% (7/70) infection rate in plateau pika in Shiqu County by mitochondrial sequencing but also found that *Lasiopodomys fuscus* and Rodentia were also infected with CE, with infection rates of 4.03% (11/273) and 3.69% (11/298), respectively, and these wild animals are rarely reported to be infected with CE in China (Weng et al. 2019). Livestock is dominated by yaks, cattle, sheep and fine wool sheep reports, mainly in the Gannan region Gansu province. For example, Niu et al. conducted an epidemiological survey of yaks, native breeds and fine wool sheep in Sunan County by dissecting and observing the touch method, and the infection rates were 20.83% (20/96), 23.68% (9/38) and 23.37% (86/368) (Niu and An 2012), respectively; Zhao et al. conducted a field survey of yaks and yaks in Gannan Autonomous Prefecture, and the infection rates were 10.47% (451/4309) and 12.37% (451/3645), respectively (Zhao et al. 2008). CE in Xinjiang is mainly concentrated in the pastoral areas of northern Xinjiang, for example, Liu et al. reported that the summary and analysis of CE epidemiological data of domestic animals (sheep, goats, cattle, etc.) in Xinjiang from 1957 to 1991 showed that the infection rate of sheep CE in Tekes County from 1963 to 1990 was 14.7–87.3% (Liu et al. 1992); Nuslaiti et al. reported that in Tekes County cattle/sheep fixed-point slaughterhouse and the results showed 76.42% infection rate of sheep and 32.76% infection rate of cattle (Nuslaiti et al. 2016); unlike the above provinces, Meng et al. investigated and found CE infection in camels and horses in northern Xinjiang by dissecting and observing touch, with 7.32% (6/82) and 4.17% (8/192) infection, respectively (Meng et al. 2014). In Inner Mongolia, Romig et al. found CE infection in Bactrian camels with a prevalence of 24.63% (67/272) (Romig et al. 2017).

There are relatively few reports of infection in pigs compared to domestic animals, both domestically and internationally, and there are some reports in China (Hua et al. 2022), such as in Tibetan areas where most domestic pigs are free-ranging and susceptible to ingesting food or drinking water contaminated with infected *E. granulosus* eggs by dog faeces, resulting in pigs being infected (Craig et al. 2019; Sierra-Ramos and Valderrama-Pome 2017). Li et al. showed through epidemiological data that the infection rate in pigs in slaughterhouses on the Tibetan Plateau was 9.92% (37/373) and a positive serological test result of 7.27% (33/454) (Li et al. 2017a, b), and overall, the number of infections in pigs in Tibetan areas is low compared to other major intermediate hosts, sheep or yaks (Craig et al. 2019). There are also some reports on pigs infected with *Echinococcus*, for example, Tian reported one case of pigs infected with *Echinococcosis* in a feedlot in Tieling city, Liaoning Province (Tian 2001); Sun et al. found 9751 pigs affected with *Echinococcosis* in 25,069 pigs kept by 6250 pig farmers in six townships in Jilin Province at the end of 1989, with an incidence rate of 38.8% (Sun 1992); An et al. found that according to the *Animal Disease Journal of Bozhou*, in Xinjiang, the average infection rate of pigs in Boertala Mongolian Autonomous Prefecture in 1988 was 23.71%, and the intensity of liver infection was 1–24 cysts (An and Shan 2013); Hong et al. found one pig kidney infected with echinococcosis in a pig farmer in Quanzhou City (Hong et al. 2000); Li et al. found one pig infected with echinococcosis in a pig designated slaughterhouse in Minqin County, Gansu Province (Li 2017); Wan et al. found one case of pig lung infected with echinococcosis in a slaughterhouse in Qinghai in 2001 (Wan et al. 2001). Guo et al. for the first identified E. granulosus G1 infection in wild boar (*Sus scrofa*) in Xinjiang, China, the infection rate was 3.70% (1/27), suggesting that this pathogen is maintained through circulation between intermediate and definitive hosts. The finding is epidemiologically significant and suggests a transmission route through wild animals (unpublished). These reports above indicated the infections for echinococcosis based on the morphology of the infected envelope and epidemiological investigations, it was concluded that the cause of infection was CE. These epidemiological data above reflect that pigs can be disseminated nationwide as an intermediate host for echinococcosis, thus suggesting that the role of pigs as an intermediate host for CE can't be ignored (Cardona and Carmena 2013; Dán et al. 2018; Zhang et al. 2014).

### 5.2.5 The definitive hosts of *E. granulosus*

On account of the difficulty of collecting samples and investigating evidence from definitive hosts, there are fewer reports of definitive hosts compared to intermediate hosts. In Qinghai, Chen et al. investigated dogs in Guinan County by faecal antigen, and the infection rate was 28.33% (17/60) (Chen et al. 2005); in 2002, Wang et al. investigated dogs and foxes in the Qingnan area, and the infection rates were 62.50% (60/96) and 16.67% (1/6), respectively (Wang et al. 2002). The prevalence of infection was 52.34% (56/107), 100% (3/3) and 20.00% (3/15), respectively. Surprisingly, Wang et al. found that a cat in the Qilian Mountains and Hehuang Valley was also infected with *E. granulosus*, which is the first report in China (Wang et al. 2002). Luo Zhaohui et al. conducted an epidemiological survey of domestic and wild dogs in Gêrzê County, Tibet, with infection rates of 17.65% (6/34) and 20.00% (9/45), respectively (Luo et al. 2022). Budke et al. and He et al. investigated dogs in Sichuan Province, with infection rates of 8.36% (31/371) and 43.75% (7/16), respectively

(Budke et al. 2005b; He et al. 2000). Niu et al. and Li et al. investigated dogs in Sunan and Jingyuan counties using the faecal antigen method, with infection rates of 5.15% (32/621) and 6.61% (20/303) (Li et al. 2006; Niu and An 2012), respectively, and Zhao et al. also investigated dogs in Gannan Autonomous Prefecture using the inducation diarrhea of arecoline hydrobromide (AH) and dissection method, with an infection rate of 22.97% (17/74) (Zhao et al. 2008). In Tekes County, Xinjiang, Liu et al. found infection rates of up to 70% in dogs in 1986 using the inducation diarrhea of (AH) and examination method (Liu et al. 1992), which decreased by 1988 but was still as high as 32.08%. The main distribution of *E. granulosus* is in Xiji County, Ningxia Autonomous Region. Li et al. in 2003 detected a high infection rate of 54.54% (6/11) in dogs in Xiji County using the faecal antigen method. The infection rate in dogs was 6.10% (40/650) and 7.00% (54/771), respectively (Li et al. 2008).

## 5.3 *Echinococcus multilocularis*

### 5.3.1 The risk and distribution of AE

E. *multilocularis* is mainly distributed in the Northern Hemisphere (Deplazes et al. 2017; Nahorski et al. 2013; Wu et al. 2017). E. *multilocularis* is an important zoonotic parasitic disease with a high prevalence in western China, and AE patients account for more than 91% of the world's cases. Seven provinces (autonomous regions) in western China are highly endemic (Issued jointly by the National Health Planning Commission and twelve departments 2010): Sichuan, Xinjiang, Ningxia, Qinghai, Gansu, Tibet and Inner Mongolia are particularly affected (Torgerson et al. 2010; Zhang et al. 2015), accounting for more than 91% of the world's cases (Torgerson et al. 2010), and AE infection can cause 'tumour-like' lesions, known as 'worm cancer' (Chong et al. 2022; Gürler et al. 2019). Most infections are predominantly in the liver and can spread to the lungs, brain and other organs through cellular infiltration, circulation and lymphatic metastases (Jenkins et al. 2005), and according to the World Health Organization, the 10-year mortality rate of untreated AE patients is as high as 95% (Brunetti et al. 2010).

There are three main endemic areas in China: (1) the northwestern part of the Xinjiang Uygur Autonomous Region, mainly in the nomadic or semi-nomadic areas of the Kazakhs in the Yili and Al regions (Guo et al. 2021b) (Fig. 5.3); (2) southern Ningxia and south-central Gansu, with southern Ningxia mainly endemic in Guyuan, Xiji and Jinyuan counties, and south-central Gansu mainly in the Dingxi region and Gannan Tibetan Autonomous Prefecture (Yang et al. 2008; Yang et al. 2005b); (3) the eastern part of the Qinghai-Tibet Plateau, including northwestern Sichuan (mainly Aba and Ganzi Tibetan Autonomous Prefecture) and southwestern Qinghai (mainly Golog, Yushu and Huangnan Tibetan Autonomous Prefecture). The number of people at risk for AE in Xinjiang reached 5.8 million, with an infection rate of approximately 0.2% in the exposed population (Guo et al. 2017) and approximately 811 new cases per year (Torgerson et al. 2010). AE cases were disseminated in northern Xinjiang, mainly in the Yili, Aletai and Tacheng regions, and an epidemiological survey of echinococcosis disease in four counties of Ganzi Tibetan Autonomous Prefecture in Sichuan found that the average AE prevalence at six survey sites in four counties, Baiyu, Sedar, Batang and Litang. The mean prevalence of AE was 1.23% (Yu et al. 2005); in Xiji County, Ningxia, the mean prevalence of AE was 5.9% (Wang et al. 1991). The main endemic areas for AE tend to be in economically deprived and remote areas, and although local government surveys are intensive, there are still many unreported AE cases and the actual number of cases may be underestimated (Otero-Abad and Torgerson 2013). In China it is now common to investigate the prevalence of AE infection by ultrasound means, according to ultrasound findings, a large number of AE cases exist in China (Table 5.2) (Torgerson et al. 2010)), accounting for at least 90% of the global incidence, with the prevalence of AE similar to that of tuberculosis in some densely populated areas (Yang et al. 2009). Since the large population base and high-risk factors in western China, the prevalence shows some variation between 1% and 12%, with a very high prevalence in Aba Autonomous Prefecture, a Tibetan region in Sichuan Province, which is much higher than in other provinces of the country (Shang et al. 2019).

### 5.3.2 *E. multilocularis* genotypes in China

Nakao et al. extracted the mitochondrial cob, nad2 and cox1 genes of *E. multilocularis* infected by different hosts from each continent and concatenated three mitochondrial genes, which were divided into 10 genotypes (A1–A10) in Asia, 5 genotypes (E1–E5) in Europe, 2 genotypes (N1–N2) in North America and 1 genotype (O1) in Inner Mongolia, China (Alvarez Rojas et al. 2020; Nakao et al. 2009). Six genotypes (A5–A10) were found in Sichuan, China, with more A5 genotypes, and the O1 genotype was found in Inner Mongolia, and the O1 genotype is currently found only in the Inner Mongolia region of China, where the genetic sequence differs significantly from other Asian, European and North American genotypes (Table 5.3). Guo et al. collected 127 AE samples from the Yili region of Xinjiang according to Nakao et al. and found 27 haplotypes, predominantly Asian genotypes and identical to A2, A5 and A10 genotypes (Fig. 5.2). The result showed 72.15% (57/79) of AE patients and 80.00% (4/5) of dogs were infected with A2 which was predominant and the most infected genotype in small rodents was A5, accounting

**Table 5.2** Estimated annual AE numbers of new cases in endemic provinces of China

| Province (Autonomous region) | Population at risk (/ 1, 000, 000) | Prevalence rate (%) | New cases per year (median) |
|---|---|---|---|
| Qinghai | 5.4 | 1.0 | 3766 |
| Tibet | 2.7 | 0.1 | 172 |
| Sichuan | 0.92 | 3.6 | 2390 |
| Gansu | 3.6 | 2.9 | 7676 |
| Xinjiang | 5.8 | 0.2 | 811 |
| Ningxia | 1.2 | 2.0 | 1770 |
| Inner Mongolia | 3 | 0.02 | 44 |

**Fig. 5.3** Habitat and trapping site for small rodents in Yili Prefecture, Xinjiang Autonomous Region

for 27.91% (12/43). On account of more isolate samples, proportion and distribution of A2 genotypes, A2 may be the most primitive sequence of all *E. multilocularis* according to the speculation of Alvarez Rojas et al. Similar results were found by Guo et al. in Yili, Xinjiang. Surprisingly, genotypes close to North American N1 and N2 sequences are also present in the Yili region of Xinjiang, respectively. In summary, these phenomena may be related to several ice ages in the Pleistocene (Aubry et al. 2010; Šnábel et al. 2020; Spotin et al. 2018), during which glaciation led to a drop in sea level and the formation of the Bering Sea–Land Bridge linking Asia and North America, known as the 'Bering Land Bridge' (Brunhoff et al. 2003; Rausch 2003), where the definitive hosts fox of *E. multilocularis* infected tapeworms could reach Japan, North America and Alaska. Eventually, as sea levels rose, the continents became 'isolated islands' and the *E. multilocularis* circulated through the hosts of each continent, with each continent having a different host species, eventually evolving into its own unique genotype (Fig. 5.4).

### 5.3.3 Transmission of *E. multilocularis*

The *E. multilocularis* usually maintains a circulating chain of domestic wildlife (Eckert and Deplazes 2004), with the pathogen circulating between foxes/dogs and

**Table 5.3** Confirmed occurrence of *E. multilocularis* haplotype assemblages in animal and human hosts in China

| Haplotype | Definitive host(s) | Intermediate host(s) | Human cases | Distribution |
|---|---|---|---|---|
| **Asia** | Not reported | Not reported | Not reported | Not reported |
| A1 | Not reported | Not reported | Not reported | Not reported |
| A2 | Microtus spp. | Dog | Reported | Xinjiang |
| A3 | Not reported | Not reported | Not reported | Not reported |
| A4 | Not reported | Not reported | Not reported | Not reported |
| A5 | Microtus spp. | Unknow | Reported | Sichuan and Xinjiang |
| A6 | Unknow | Unknow | Unknow | Sichuan |
| A7 | Unknow | Unknow | Unknow | Sichuan |
| A8 | Unknow | Unknow | Unknow | Sichuan |
| A9 | Unknow | Unknow | Unknow | Sichuan |
| A10 | Unknow | Dog | Reported | Sichuan and Xinjiang |
| **Europe** | | | | |
| E1 | Not reported | Not reported | Not reported | Not reported |
| E2 | Not reported | Not reported | Not reported | Not reported |
| E3 | Not reported | Not reported | Not reported | Not reported |
| E4 | Not reported | Not reported | Not reported | Not reported |
| E5 | Not reported | Not reported | Not reported | Not reported |
| **North American** | | | | |
| N1 | Not reported | Not reported | Not reported | Not reported |
| N2 | Not reported | Dog | Not reported | Xinjiang |
| **Inner Mongolia** | | | | |
| O1 | Not reported | Unknown[a] | Not reported | Inner Mongolia |

[a] Laboratory strain: no intermediate host is specified

rodents (Detry et al. 2005; Wnukowska et al. 2011). In areas close to human settlements, Microtus spp. mainly inhabit mountainous forested meadow grasslands, mountain grasslands and pre-mountain plains and valley bottoms below 2300 elevation (Guo et al. 2021a), in close proximity to pastoralist living conditions. In summer, pastoralists are highly dispersed in mountain pastures, often living in tents 10–50 m away from streams, and in these areas there are high densities of rodents, with a higher risk of infection in domestic dogs due to the proximity to the higher density intermediate hosts (Wang et al. 2021a; Guo et al. 2021a), combined with the fact that Microtus spp. are susceptible to AE and are highly susceptible to predation by foxes or domestic dogs, leading to a rodent-centred participation in two circular chains of domestic wild. The faeces excreted by infected definitive hosts canids carry large numbers of eggs (Guo et al. 2021b; Hou et al. 2021), contaminating pasture, water and food, leading to infection of Microtus spp., which become the main small rodent for AE transmission. In pastoral areas, where domestic dogs are

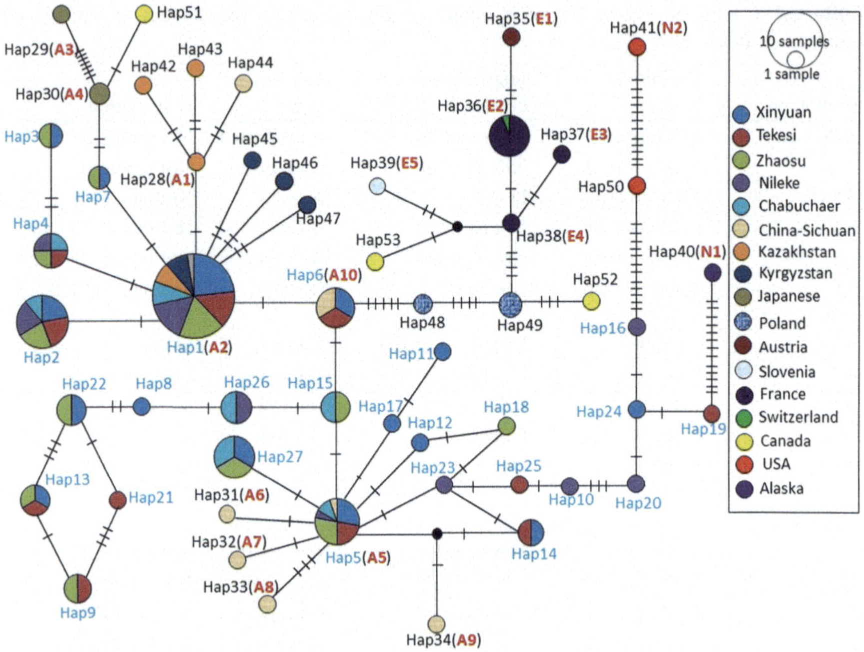


**Fig. 5.4** Map of Yili (China: Xinjiang) showing the region of origin of the haplotypes identified in samples of *E. multilocularis*

very common and infection rates are very high, large populations of red foxes (*Vulpes vulpes*) exist around villages, often with several red fox holes found less than 1–2 km in circumference (Guo et al. 2021a), creating a cycle between rodents and red foxes around villages and exacerbating the spread of AE, ultimately leading to Xinjiang being one of the most severely endemic areas in the whole China (Afonso et al. 2015; Guo et al. 2019).

### 5.3.4 The intermediate hosts of *E. multilocularis*

In China, a large number of studies have reported AE infections in small rodents, including the Xinjiang water vole (*Arvicola amphibius*): the long-clawed gerbil (*Meriones unguiculatus*) in Inner Mongolia (Tang et al. 2002), the *Microtus ilaeus* (Xu et al. 2002), the *Lasiopodomys brandti*, the plateau squirrel (*Neodon irene*) in Xinjiang, Inner Mongolia and Sichuan (*M. brandti, Microtus irene*) (Tang et al. 2004; Wang et al. 2008), Microtus spp. in Xinjiang, Chinese zokors (*Eospalax fontanierii*) in Ningxia and Gansu, plateau pika in Sichuan and Qinghai (Wang et al. 2009), and Daur in Gansu (*Ochotona daurica*), the Daur's yellow rat (*Spermophilus dauricus*) in Ningxia, and *Spermophilus erythrogenys* in Xinjiang (Wang et al. 2009), where the infection rate was 0.1% for rodents and nearly 10% for the

Ochotona (Wang et al. 2009). Lin et al. reported that *Mus musculus* in the Tacheng area of Xinjiang were infected with *E. multilocularis* (Lin and Hong 1991). Xu et al. reported that Achaillie's rodents in Yining County were infected with *E. multilocularis* and were found to have protoscoleces (PSCs) (Wang et al. 1989; Xu et al. 2002). Chai et al. reported *E. multilocularis* infection by *Neomys fodiens* (commonly known as water rats) from Nileke County in the Yili region and from Emin County in the Tacheng region (Chai et al. 2002). Yang et al. reported that vesicular infection was found in *Eolagurus przewalskii* in Hobokskar County, Tacheng Region, Xinjiang, which was the first report in China (Yang et al. 2011). In the high-elevation grasslands of Yili, Microtus spp. were found to be infected with vesicular along with *Ellobius trancrei* (Figs. 5.5 and 5.6), and in agricultural areas at lower elevations, *Mus musculus*, *Apodemus sylvaticus*, *Rattu norvegicus*, and *Mus musculus*, *Apodemus sylvaticus*, *Rattu Norvegicus*, *Meriones erythrourus* Pallas and *Rhombomys sopimus* Pallas, dominated by *Mus musculus* and *Apodemus sylvaticus*, were found for the first time in the semi-desert ecology in Chabuchaer County. AE infections were found for the first time in the semidesert ecosystem in Chabuchaer County (Guo et al. 2021a).

It is well documented that pigs are also susceptible to *E. multilocularis* infection, but that the development of *E. multilocularis* in the body is inhibited after infection, leading to calcification and necrosis, and that no PSCs are produced (Deplazes et al. 2017), so that infection in pigs has no transmission effect and doesn't participate in the circulatory chain. On open farms, pigs may have been infected by eating pasture contaminated with eggs (Sydler et al. 1998). In China, few studies have been reported on pigs being infected with *E. multilocularis* in other provinces and autonomous regions, with only Qiu et al. reporting a 3.13% (1/32) infection rate in pigs on the Qinghai-Tibet Plateau (Ji et al. 1995). On the Qinghai-Tibetan Plateau, domestic animals such as yaks, sheep and cattle are more common as intermediate

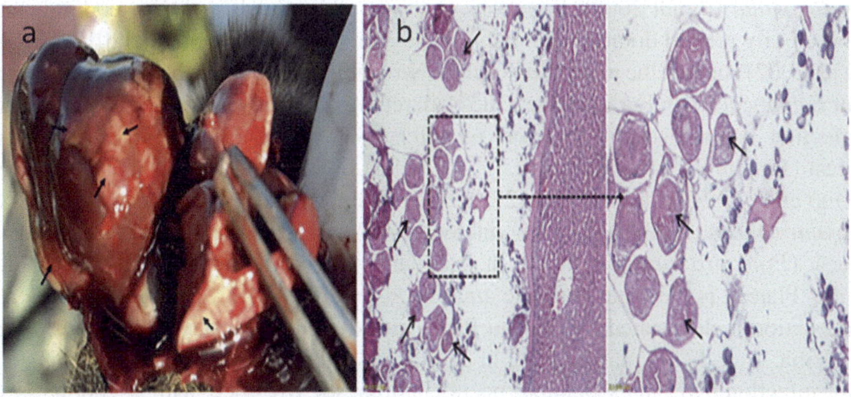


**Fig. 5.5** Infection with metacestode of *E. multilocularis* in Microtus spp. liver: (**a**) infection liver (arrow indicates metacestode of *E. multilocularis*); (**b**) the H&E staining of infection liver (arrow indicates PSCs)

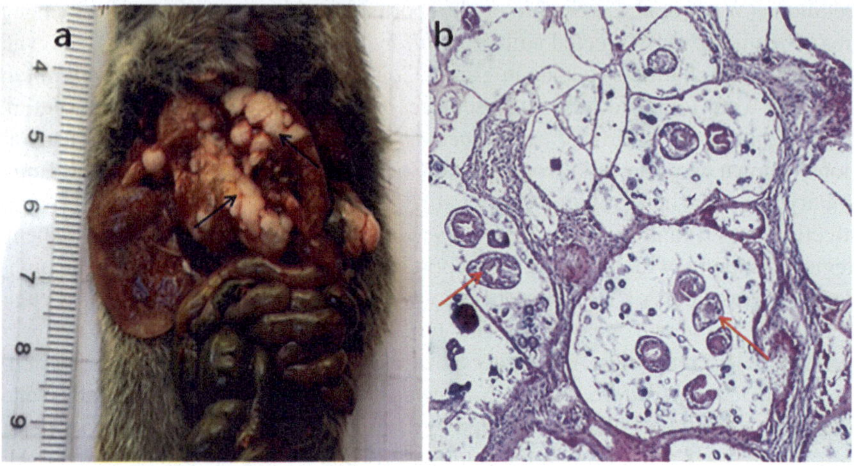


**Fig. 5.6** H&E showing the characteristic pathological response in alveolar echinococcosis (AE): (**a**) *E. multilocularis* metacestode; (**b**) liver section with protoscoleces (PSCs), 200× original magnification. Black arrow indicates cystic lesions; red arrow indicates PSCs

hosts for CE, and as intermediate hosts for *E. multilocularis* are mainly found in Qinghai, Sichuan and Tibet on the Qinghai-Tibetan Plateau (He and Wang 2001; Chiu et al. 2000; Ji et al. 1995), but AE infection in yaks and sheep has also been reported in Gannan Tibetan Autonomous Prefecture, Gansu Province, with infection rates of 0.14% (6/4309) and 0.14% (5/3645), respectively (Zhao et al. 2008). Compared to domestic animals, epidemiological data on wild large herbivores are limited due to sampling conditions and are mostly from rearing populations of individual species. Large herbivores known from the Tibetan Plateau include *Bos grunniens*, *Budorcas taxicolor*, *Pantholops hodgsonii*, *Pseudois nayaur* and *Procapra picticaudata* (Rendt 2000). Infections in these wild animals and their codominant and closely related domestic animals (cattle and sheep) (Ohiolei et al. 2019; Wang et al. 2002) suggest the possibility of these wild animals acting as intermediate hosts for *Echinococcosis*. For example, Guo et al. reported the prevalence of *Echinococcus* infection in blue sheep (*Pseudois nayaur*) (6.42%, 21/327) and Tibetan wildebeest (*Procapra picticaudata*) (6.57%, 13/198) in southwestern Qinghai Province (Guo et al. 1994). Cai et al. (2012) reported echinococcosis infections in *Pseudois nayaur* (6.02%, 21/349), Tibetan antelope (6.57%, 13/198) and Mongolian wildebeest (*Procapra gutturosa*) (1/1) in Qinghai Province. In summary, the Qinghai-Tibet Plateau is the main endemic area for *E. multilocularis*, and grassland rodent infestation is dominated by Plateau Pika, with the intermediate hosts found in all endemic areas being plateau pika, in addition to natural infection by *Lepus oiostolus*; infection by *E. multilocularis* occurs in domestic livestock, with infection rates significantly higher than those reported in foreign literature, and is the only area in China where *E. multilocularis* has been found to infest large livestock (Gongsang et al. 2017; Ji et al. 1995). This phenomenon suggests that pastures are heavily contaminated by definitive hosts. There are still many unknown problems that a natural source of epidemics circulating among wild animals and between domestic and wild animals, such as the identification of endemic areas and the presence of other pathogenic hosts. The prevalence of the disease in the population needs to be further investigated, in addition, the unique ecological environment of the Tibetan Plateau, the use of the *Rodentia* and *Lagomorpha* as intermediate hosts, the variability of adult worms in Tibetan isolates and the long maturation period of the worms also suggest the need for biochemical immunology and molecular biology to investigate whether there are genetic differences from other endemic areas (Weng et al. 2019; Wu et al. 2018).

### 5.3.5 The definitive hosts of *E. multilocularis*

The definitive hosts of *E. multilocularis* are mainly canids, including red fox, sand fox (*Vulpes corsac*) and Tibetan fox (*Vulpes ferrilata*) (Knapp et al. 2016), domestic dogs (*Canis familiaris*) (Raso et al. 2014) and raccoons (*Nyctereutes procyonoide*) (Raoul et al. 2010; Raso et al. 2014). Canids are infected through the capture of small mammals, and wild animals are also involved in the circulating chain of *E. multilocularis*, playing an important role in the wild cycle, as in Europe, North America, Central Asia and western China. In Tibetan areas of China, such as the Tibetan fox and the red fox (Jiang et al. 2012b), natural infection has been demonstrated (Budke et al. 2005a; Budke et al. 2005c; Varcasia et al. 2008). Han et al. detected lynx infected with *E. multilocularis* by faecal antigen in Banma County, Qinghai Province, with an infection rate of 33.33% (3/9) (Han et al. 2006), which is the first report in China, while Wang et al. also detected intestinal infection with *E. multilocularis* in cats in the Qilian Mountains and the Hehuang Valley, which is also the first report in China (Wang et al. 2002). Red foxes, sand foxes, Tibetan foxes and Tibetan dogs (*Tibetan mastiff*) are the main definitive hosts of *E. multilocularis* and are distributed in the main *E. multilocularis* endemic areas in western China. The detection rate varies due to various detection methods, such as faecal antigen, dissection of the intestine and pathological sections, fixation of worms, microscopic examination of the intestine, PCR amplification of adult worms, for example, Wang et al. detected *E. multilocularis* by intestinal dissection in Zhengduo County, Qinghai Province, Zeku County and Haibei Prefecture, Zeku County, Tibetan dog and Tibetan fox in Haibei Prefecture were 5.08% (3/59), 33.33% (4/12), respectively (Wang and Soha 2000). Qiu et al. detected 57.14% (12/21), 59.09% (13/22) and 11.54% (12/104) of red foxes, Tibetan foxes and dogs, respectively, in Shiqu County, Sichuan Province (Ji et al. 1995). Wang et al. reported that the infection rate of *E. multilocularis* in foxes was 30.56% (11/36) in the Tacheng area of northern Xinjiang, while wolves (*Canis lupus*) were found to be infected with *E. multilocularis* (Wang et al. 1989), which was the first report in China. Jiao et al. did faecal antigen testing on dogs in Fuhai, Emin and Yumin counties in northern Xinjiang (Jiao et al. 2003), and the detection rate was 18.02% (73/405), and by autopsy methods, they found that Tacheng, Fuhai, Tuoli and Yumin counties had an infection rate of 30.56% (11/36). Tang et al. conducted epidemiological surveys in Xinbaerhu Right Banner and Hulunbeier, Inner Mongolia, respectively, and the infection rate of red fox was 2.78% (1/36) and 33.33% (2/6), respectively (Tang et al. 2001). In Ningxia, Li et al. detected 27.27% (3/11) of red fox infections and Yang et al. detected 6.40% (48/750), 1.52% (10/650) and 3.37% (26/771) of infections in Xiji County, Yuanzhou district and Haiyuan County, respectively (Li et al. 1994).

## 5.4 *Echinococcus shiquicus*

Previously, the descriptions of the species of *Echinococcus* weren't scientific and the emergence of 'subspecies' of the same domain has caused much controversy. Although many species and subspecies have been described, most of them have been classified as *E. granulosus* (Xiao et al. 2005). However, the species genera of *E. granulosus* rely solely on the different species of intermediate hosts, the ungulates, as a criterion for classification (Brunetti et al. 2010; Brunhoff et al. 2003). At the 16th International Conference on Hydatid Disease held in Beijing, China, in 1993, Guo et al. (Budke et al. 2005a) reported pulmonary infections with *E. granulosus* in Plateau pika, of which 6 out of 21 species of plateau pika were found to be infected (Budke et al. 2006), and the Tibetan Plateau pika were the dominant intermediate host for *E. multilocularis* on the plateau (Fan et al. 2016; Wang et al. 2009). The lungs were found to be infected with 'single-chambered' cysts, similar to those found in the middle cestode stage of *E. granulosus*. In the adult stage, Qiu et al. first reported the *E. shiquicus* in 1995, but at that time they attributed the infection rate to the plateau Tibetan fox to an atypical *Echinococcus*, which was thought to be a variant of *E. multilocularis* (Ji et al. 1995). In an earlier phase of the study, Xiao et al. found that the mitochondrial DNA sequence of *E. multilocularis* infected by plateau pika was different from the published sequence of *Echinococcus* through genetic differentiation of *E. multilocularis* infected by isolated wild mammals from Shiqu County, and subsequent studies found that Tibetan foxes were infected with the same sequence of *E. shiquicus*. Finally, based on morphological features, host preference, molecular genetics and geographic distribution, Xiao et al. defined a new species- *E. shiquicus*. The body is approximately 1.2–1.8 mm long, with two body segments, no mature segments, only juvenile and gestational segments, and the reproductive pore is located anteriorly in the middle of the body segment (Xiao et al. 2005).

### 5.4.1 Transmission of *E. shiquicus*

Xiao et al. analysed mitochondria DNA sequences of fox and mouse specimens from the Qinghai-Tibetan Plateau in China and found significant base differences from all other strains of the genus *Echinococcus* (Xiao et al. 2005) which led to the identification of a new species of *E. shiquicus*. The epidemiology of *E. shiquicus* has been relatively poorly studied compared to that of *E. granulosus* and *E. multilocularis* (Craig et al. 2019; Wu et al. 2018). In endemic areas, the Tibetan fox is the definitive host of *E. shiquicus* (Jiang et al. 2012a; Xiao et al. 2005). Plateau pika has been confirmed at the parasitological level (morphology, histology and molecular biology) as its intermediate host (Xiao et al. 2005, 2006). In addition, several species of rodents have been infected (Wang et al. 2018; Wu et al. 2018). DNA from *E. shiquicus* tapeworms has also been detected in the faeces of sheepdogs (Boufana et al. 2013; Weng et al. 2020). This suggests the possibility of dogs acting as definitive hosts for *E. shiquicus* and the presence of *E. shiquicus* in the human environment.

### 5.4.2 Epidemiological characteristics of *E. shiquicus*

The main endemic areas of *E. shiquicus* are mainly concentrated in the eastern part of the Qinghai-Tibet Plateau. These endemic areas have the following characteristics: (1) they are all pure pastoral areas, with an altitude of 3000 m, and the climate is cold and dry; (2) the intermediate host is plateau pika but Rodentia are the main species, and a variety of rodents are also infected; (3) it highly overlaps with the endemic area of *E. multilocularis*, and the endemic area is often accompanied by the prevalence of *E. granulosus* and *E. multilocularis* (Weng et al. 2019). As natural foci that circulates among wild animals and between domestic animals and wild animals, there are still many unknown factors in the endemic area, such as the determination of the endemic area and whether there are other pathogenic hosts. The infection status of *E. shiquicus* in the population is still unclear and needs to be further investigated. There are three main reasons why humans aren't infected with *E. shiquicus*. First, humans aren't susceptible hosts for *E. shiquicus*. Second, due to the different ecological environments of humans and *E. shiquicus*, human and *E. shiquicus* seldom come into contact with each other. A large number of studies have shown that dogs play a vital role in the transmission of CE and AE (Xiao et al. 2012). However, in China, dogs haven't been found to be infected with *E. shiquicus* (Shang et al. 2014), and experiments have shown that dogs aren't susceptible to *E. shiquicus* (Xiao et al. 2005), and the small population of Tibetan foxes, the definitive host, also reduces human risk of infection with *E. shiquicus*. Finally, it may be due to the limitations of current diagnostic techniques. There may be cases of human infection with *E. shiquicus* in the Qinghai-Tibet Plateau. It is likely that the current diagnostic methods for CE and AE can't distinguish infection from CE and AE.

### 5.4.3 The intermediate Hosts of *E. shiquicus*

Compared with *E. granulosus* and *E. multilocularis*, there are fewer reports on *E. shiquicus*, and intermediate host infection is only reported in Shiqu County of Sichuan Province and Zema County and Dari County of Qinghai Province, all using

PCR Methods amplified mitochondrial identification (Wang et al. 2018; Han et al. 2006, 2009). In Qinghai Province, Han et al. reported that the infection rates of *Neodon fuscus* in Banma County and plateau pika in Dari County were 0.37% (1/224) and 11.30% (27/239), respectively (Han et al. 2006). Wang et al. reported that the infection rates of, *Cricetulus longicaudatus*, *Neodon fuscus*, *Phaiomys leucurus*, plateau pika, and *M. oeconomus limnophilus* in Shiqu County were 100% (1/1), 1.39% (2/144) 6.25% (1/16), 3.70% (5/135) and 11.36% (5/44), respectively, (Wang et al. 2018). Weng et al. also conducted an epidemiological investigation on plateau pikas, *Lasiopodomys fuscu* and Rodentia in Shiqu County in 2019 the infection rates were 10.00% (7/70), 3.66% (10/273) and 3.36% (10/298), respectively (Weng et al. 2020).

### 5.4.4 The definitive hosts of *E. shiquicus*

Compared with the intermediate hosts, there are only four reports on the definitive host, all of which are Tibetan fox infection (Jiang et al. 2012a; Xiao et al. 2005; Feng et al. 2013; Han et al. 2006). In Shiqu County, Sichuan Province, Xiao et al. also used the PCR method and found that the infection rate of Tibetan fox was 37.50% (6/16), which was discovered for the first time in China (Xiao et al. 2005); Jiang et al. collected Tibetan fox faeces in 2012 and identified the infection rate of Tibetan foxes by PCR-RFLP method as 25.68% (19/74) (Jiang et al. 2012a). In Qinghai Province, Han et al. used PCR method to amplify mitochondrial gene in Banma County, Qinghai Province, in 2006, and found that the infection rate was 17.39% (4/23) and 16.67% (1/6), respectively (Han et al. 2006).

## References


Afonso E, Knapp J, Tête N et al (2015) *Echinococcus multilocularis* in Kyrgyzstan: similarity in the Asian EmsB genotypic profiles from village populations of Eastern mole voles (*Ellobius tancrei*) and dogs in the Alay valley. J Helminthol 89:664–670

Agudelo Higuita NI, Brunetti E, Mccloskey C (2016) Cystic Echinococcosis. J Clin Microbiol 54:518–523

Al-Jawabreh A, Ereqat S, Dumaidi K et al (2017) The clinical burden of human cystic echinococcosis in Palestine, 2010-2015. PLoS Negl Trop Dis 11:e0005717

Alvarez Rojas CA, Kronenberg PA, Aitbaev S et al (2020) Genetic diversity of *Echinococcus multilocularis* and *Echinococcus granulosus* sensu lato in Kyrgyzstan: The A2 haplotype of *E. multilocularis* is the predominant variant infecting humans. PLoS Negl Trop Dis 14:e0008242

Alvi MA, Ali RMA, Khan S et al (2023) Past and present of diagnosis of echinococcosis: A review (1999-2021). Acta Trop 243:106925

An N, Shan B (2013) Investigation and analysis of porcine Echinococcosis in Bortala Mongolian Autonomous Prefecture, Xinjiang. J Anim Hus Vet med 32:126–126. (in Chinese)

Anvari D, Pourmalek N, Rezaei S et al (2021) The global status and genetic characterization of hydatidosis in camels (*Camelus dromedarius*): a systematic literature review with meta-analysis based on published papers. Parasitology 148:259–273

Aubry KB, Statham MJ, Sacks BN et al (2010) Phylogeography of the North American red fox: vicariance in Pleistocene forest refugia. Mol Ecol 18:2668–2686

Boufana B, Qiu J, Chen X et al (2013) First report of *Echinococcus shiquicus* in dogs from eastern Qinghai-Tibet plateau region, China. Acta Trop 127:21–24

Brunetti E, Kern P, Vuitton DA et al (2010) Expert consensus for the diagnosis and treatment of cystic and alveolar echinococcosis in humans. Acta Trop 114:1–16

Brunhoff C, Galbreath KE, Fedorov VB et al (2003) Holarctic phylogeography of the root vole (*Microtus oeconomus*): implications for late Quaternary biogeography of high latitudes. Mol Ecol 12:957–968

Budke CM, Campos-Ponce M, Qian W et al (2005a) A canine purgation study and risk factor analysis for echinococcosis in a high endemic region of the Tibetan plateau. Vet Parasitol 127:43–49

Budke CM, Deplazes P, Torgerson PR (2006) Global socioeconomic impact of cystic echinococcosis. Emerg Infect Dis 12:296–303

Budke CM, Jiamin Q, Craig PS et al (2005b) Modeling the transmission of *Echinococcus granulosus* and *Echinococcus multilocularis* in dogs for a high endemic region of the Tibetan plateau. Int J Parasitol 35:163–170

Budke CM, Jiamin Q, Qian W et al (2005c) Economic effects of echinococcosis in a disease-endemic region of the Tibetan Plateau. Am J Trop Med Hyg 73:2–10

Cai D, Li Y, Jiang Y et al (2019) The role of contrast-enhanced ultrasound in the diagnosis of hepatic alveolar echinococcosis. Medicine (Baltimore) 98:e14325

Cai H, Wang Y, Wang H et al (2012) Current situation of Echinococcus infection in different hosts in Qinghai plateau. Chinese J Endemiol 31:296–300 (in Chinese)

Cardona GA, Carmena D (2013) A review of the global prevalence, molecular epidemiology and economics of cystic echinococcosis in production animals. Vet Parasitol 192:10–32

Chai J, Jiang W, Islaine U et al (2002) Investigation on animal hosts of *Echinococcus multilocularis* in north Xinjiang, China. Chinese J Zoonoses 19:89–91 (in Chinese)

Chen C, Li Q, Liu J et al (2005) Study on the animal host of *Echinococcus multilocularis*. Chinese Qinghai J Anim Vet Sci 35:26–27 (in Chinese)

Chiu J, Liu F, Schantz P et al (2000) Pidemiological study on human hydatidosis in tibetan region of western sichuan. Chinese J Zoonoses 16:77–80 (in Chinese)

Chong S, Chen G, Dang Z et al (2022) *Echinococcus multilocularis* drives the polarization of macrophages by regulating the RhoA-MAPK signaling pathway and thus affects liver fibrosis. Bioengineered 13:8747–8758

Craig PS, Giraudoux P, Wang ZH et al (2019) Echinococcosis transmission on the Tibetan Plateau. Adv Parasitol 104:165–246

Cucher MA, Macchiaroli N, Baldi G et al (2016) Cystic echinococcosis in South America: systematic review of species and genotypes of Echinococcus granulosus sensu lato in humans and natural domestic hosts. Trop Med Int Health 21:166–175

Dán Á, Rónai Z, Széll Z et al (2018) Prevalence and genetic characterization of Echinococcus spp. in cattle, sheep, and swine in Hungary. Parasitol Res 117:3019–3022

Deplazes P, Rinaldi L, Alvarez Rojas CA et al (2017) Global distribution of alveolar and cystic echinococcosis. Adv Parasitol 95:315–493

Detry O, Honore C, Delwaide J et al (2005) Endemic alveolar echinococcosis in Southern Belgium? Acta Gastroenterol Belg 68:1–4

Eckert J, Deplazes P (2004) Biological, epidemiological, and clinical aspects of echinococcosis, a zoonosis of increasing concern. Clin Microbiol Rev 17:107–135

Eckert J, Gemmell MA, Meslin FX et al (2001) WHO/OIE manual on echinococcosis in humans and animals: a public health problem of global concern. Int J Parasitol 31:1717–1718

Eckert J, Thompson RC (2017) Historical aspects of echinococcosis. Adv Parasitol 95:1–64

Engelke C, Wiese AV, Schmudde I et al (2014) Distinct roles of the anaphylatoxins C3a and C5a in dendritic cell-mediated allergic asthma. J Immunol 193:5387–5401

Fan YL, Lou ZZ, Li L et al (2016) Genetic diversity in Echinococcus shiquicus from the plateau pika (Ochotona curzoniae) in Darlag County, Qinghai, China. Infect Genet Evol 45:408–414

Feng K, Huang F, Duo H et al (2013) Species identification of echinococcus isolates collected from canines and tibetan foxes in Chengduo County,Qinghai Province. Chinese J Parasitol Parasitic Dis 31:185–187 (in Chinese)

García-Méndez N, Manterola C, Totomoch-Serra A et al (2022) Proteomic profile of *Echinococcus granulosus*: a systematic review. J Parasitol 108:64–69

Giraudoux P, Pleydell D, Raoul F et al (2006) Transmission ecology of *Echinococcus multilocularis*: what are the ranges of parasite stability among various host communities in China? Parasitol Int 55(Suppl):S237–S246

Giraudoux P, Raoul F, Afonso E et al (2013b) Transmission ecosystems of *Echinococcus multilocularis* in China and Central Asia. Parasitology 140:1655–1666

Giraudoux P, Raoul F, Pleydell D et al (2013a) Drivers of *Echinococcus multilocularis* transmission in China: small mammal diversity, landscape or climate? PLoS Negl Trop Dis 7:e2045

Gongsang Q, Wu W, Yan X et al (2017) Investigation on distribution of alveolar echinococcosis in human and infection of intermediate host In Tibet Autonomous Region. J Pathogen Biol 12:1175–1179 (in Chinese)

Guo Y, Chu Y, Alxin K et al (2017) Emerging alveolar echinococcosis and gene fragments of Nad2 and Cob in Nileke County,Yili area of Xinjiang, China. Chinese J Zoonoses 12:1066–1070 (in Chinese)

Guo B, Guo G, Zhang L et al (2021a) Investigation on infection of *Echinococcus multilocularis* metacestode in small rodents in Chabchar County, Xinjiang. Chinese J Parasitol Parasitic Dis 39:327–332 (in Chinese)

Guo Z, He D, Li Y et al (1994) Infection of echinococcus species in wild animals in qinghai province. Chinese J Parasitol Parasitic Dis 12(S1):145–147 (in Chinese)

Guo L, Yang A, Hou W et al (2011) Prevention and control status and countermeasures of echinococcosis of livestock in Sichuan Province. Pruatacul Anim Husb 11:52–54(in Chinese)

Guo B, Zhang Z, Guo Y et al (2021b) High endemicity of alveolar echinococcosis in Yili Prefecture, Xinjiang Autonomous Region, the People's Republic of China: Infection status in different ethnic communities and in small mammals. PLoS Negl Trop Dis 15:e0008891

Guo B, Zhang Z, Zheng X et al (2019) Prevalence and molecular characterization of *echinococcus granulosus* Sensu Stricto in Northern Xinjiang, China. Korean J Parasitol 57:153–159

Gürler AT, Bölükbaş CS, Açıcı M et al (2019) Overview of *Echinococcus multilocularis* in Turkey and in the World. Turkiye Parazitol Derg 43:18–35

Han X, Wang H, Cai H et al (2009) Epidemiological survey on Echinococcosis in Darlag County of Qinghai Province. Chinese J Parasitol Parasitic Dis 27:22–26 (in Chinese)

Han X, Wang H, Qiu J et al (2006) Epidemiological investigation and analysis of vesicular and cystic echinococcosis in Banma County, Qinghai Province. Chinese J Zoonoses 22:189–190 (in Chinese)

He J, Qiu J, Liu F et al (2000) Epidemiological survey on hydatidosis in tibitan region of western sichuan ii. infection situation among domestic and wild animals. Chinese J Zoonoses 16:62–65 (in Chinese)

He D, Wang H (2001) A report on the epidemiological evaluation of hydatid disease in Zeku County, Qinghai Province. Endemic Dis Bulletin 16:36–38 (in Chinese)

Hong Z, Hong Q, Lin Z et al (2000) Porcine Echinococcosis. Fujian Anim Husb Vet Med 6:28 (in Chinese)

Hou J, Wen H, Wang M (2021) Changes of macrophage subsets and polarization in spleen of mice infected with *Echinococcus multilocularis* Chinese. J Parasitol Parasitic Dis 39:771–778 (in Chinese)

Hua R, Du X, He X et al (2022) Genetic diversity of *Echinococcus granulosus* sensu lato in China: epidemiological studies and systematic review. Transbound Emerg Dis 69:e1382–e1392

Huang D, Li R, Qiu J et al (2018) Geographical environment factors and risk mapping of human cystic Echinococcosis in Western China. Int J Environ Res Public Health 15:1729

Issued jointly by the National Health Planning Commission and twelve departments (2010) Action plan to control Echinococcosis (2010-2015) https://xueshu.baidu.com/usercenter/paper/show?paperid=1v4g0c50u93d0tm0271702a0ej550749&site=xueshu_se (in Chinese)

Jenkins DJ, Romig T, Thompson RC (2005) Emergence/re-emergence of Echinococcus spp.—a global update. Int J Parasitol 35:1205–1219

Ji Q, Chen X, Ren M (1995) Epidemiological study on Alveolar hydatid disease in Qinghai-xizang plateau. J Pract Parasitic Dis 3:106–109

Jiang W, Liu N, Zhang G et al (2012a) Specific detection of Echinococcus spp. from the Tibetan fox (*Vulpes ferrilata*) and the red fox (*V. vulpes*) using copro-DNA PCR analysis. Parasitol Res 111:1531–1539

Jiang L, Zhang Y, Liu M et al (2012b) Analysis on the reactivity of five subunits of antigen B family in serodiagnosis of echinococcosis. Exp Parasitol 131:85–91

Jiao W, Chai J, Fu C et al (2003) Field investigation of the control of *Echinococcus granulosus* infection in dogs bysubcutaneously plantation with a sustained slow release preparation of praziquantel. Chinese J Zoonoses 19:102–104 (in Chinese)

Karamon J, Stojecki K, Samorek-Pierog M et al (2017) Genetic diversity of Echinococcus multilocularis in red foxes in Poland: the first report of a haplotype of probable Asian origin. Folia Parasitol (Praha):64:2017.007

Khachatryan AS (2017) Analysis of lethality in echinococcal disease. Korean J Parasitol 55:549–553

Knapp J, Umhang G, Poulle ML et al (2016) Development of a real-time PCR for a sensitive one-step coprodiagnosis allowing both the identification of carnivore feces and the detection of Toxocara spp. and *Echinococcus multilocularis*. Appl Environ Microbiol 82:2950–2958

Li Y (2017) Diagnosis and prevention of a case of porcine Echinococcosis. Chinese J Trad Vet Sci 39

Li L, Chen X, Bai X et al (2008) Epidemiological survey of echinococcosis in Nanwan Group of Yumu Village, Xiji County, 2006. Ningxia Med J 30:834–844 (in Chinese)

Li T, Chen X, Zhen R et al (2010) Widespread co-endemicity of human cystic and alveolar echinococcosis on the eastern Tibetan Plateau, northwest Sichuan/southeast Qinghai, China. Acta Trop 113:248–256

Li T, Qiu J, Yang W et al (2005) Echinococcosis in Tibetan populations, western Sichuan Province, China. Emerg Infect Dis 11:1866–1873

Li L, Wang Z, Gao S et al (2006) Analysis of the results of echinococcosis control project in Jingyuan County, Gansu Province from 2001 to 2003. Endemic Dis Bulletin 21:43–44 (in Chinese)

Li W, Yang C, Zhang G et al (1994) Overwintering activity of ova of tapeworm polycephalus in nature. Chinese J Vet Med 30:16–17 (in Chinese)

Li K, Zhang L, Zhang H et al (2017a) Epidemiological investigation and risk factors of *Echinococcus granulosus* in yaks (*Bos grunniens*), Tibetan pigs and Tibetans on Qinghai Tibetan plateau. Acta Trop 173:147–152

Li K, Zhang L, Zhang H et al (2017b) Prevalence, associated risk factors, and phylogenetic analysis of *Echinococcus granulosus* isolated from freerange Tibetan pigs in Tibet, China. Trop Biomed 34:903–910

Lin Y, Hong L (1991) The biology and geographical distribution of *echinococus multilocularis* infection in China. Endemic Dis Bulletin 6(117-127):147–148 (in Chinese)

Liu F, Li X, Wei M et al (1992) Health education for prevention hydatidosis in Xinjiang I.Task, organization and planning of health education. Endemic Dis Bulletin 1992(S1):125–127 (in Chinese)

Loi F, Berchialla P, Masu G et al (2019) Prevalence estimation of Italian ovine cystic echinococcosis in slaughterhouses: a retrospective Bayesian data analysis, 2010-2015. PLoS One 14:e0214224

Luo Z, Xiao D, Ai J et al (2022) Epidemiological survey of echinococcosis in Gaize County, Tibet Autonomous Region. Chinese J Zoonoses 38:74–78 (in Chinese)

Mcmanus DP, Zhang W, Li J et al (2003) Echinococcosis. Lancet 362:1295–1304

Meng Q, Wang G et al (2014) Prevalence of hydatid cysts in livestock animals in Xinjiang, China. Korean J Parasitol 52(3):331–334

Nahorski WL, Knap JP, Pawlowski ZS et al (2013) Human alveolar echinococcosis in Poland: 1990-2011. PLoS Negl Trop Dis 7:e1986

Nakao M, Xiao N, Okamoto M et al (2009) Geographic pattern of genetic variation in the fox tapeworm *Echinococcus multilocularis*. Parasitol Int 58:384–389

Niu M, An Y (2012) The epidemiological investigation and prevention measures of echinococcosis in Sunan County, Gansu Province. Chinese J Clin Rational Drug Use 5:45–47 (in Chinese)

Noguera ZL, Charypkhan D, Hartnack S et al (2022) The dual burden of animal and human zoonoses: a systematic review. PLoS Negl Trop Dis 16:e0010540

Nuslaiti MA, Yan H et al (2016) Echinococcosis(hydatid disease)infection in livestock in 14 areas in Xinjiang, 2014. Chinese J Zoonoses 32:202–205 (in Chinese)

Ohiolei JA, Li L, Ebhodaghe F et al (2020) Prevalence and distribution of Echinococcus spp. in wild and domestic animals across Africa: a systematic review and meta-analysis. Transbound Emerg Dis 67:2345–2364

Ohiolei JA, Xia CY, Li L et al (2019) Genetic variation of Echinococcus spp. in yaks and sheep in the Tibet Autonomous Region of China based on mitochondrial DNA. Parasit Vectors 12:608

Otero-Abad B, Torgerson PR (2013) A systematic review of the epidemiology of echinococcosis in domestic and wild animals. PLoS Negl Trop Dis 7:e2249

Qian M, Abela-Ridder B, Wu W et al (2017) Combating echinococcosis in China: strengthening the research and development. Infect Dis Poverty 6:161

Qian Y, Ding W, Wu W et al (2019) A path to cooperation between China and Mongolia towards the control of echinococcosis under the belt and road initiative. Acta Trop 195:62–67

Raoul F, Deplazes P, Rieffel D et al (2010) Predator dietary response to prey density variation and consequences for cestode transmission. Oecologia 164:129–139

Raso FI, Otero-Abad B, Hegglin D et al (2014) Dynamics of the force of infection: insights from *echinococcus multilocularis* infection in foxes. PLoS Negl Trop Dis 8:e2731

Rausch RL (2003) Cystic echinococcosis in the Arctic and Sub-Arctic. Parasitology 127(Suppl):S73–S85

Rendt L (2000) BioOne Online Journals–Wildlife of the Tibetan Steppes. J Mammal 81:908–909

Romig T, Deplazes P, Jenkins D et al (2017) Ecology and life cycle patterns of echinococcus species. Adv Parasitol 95:213–314

Schantz PM, Wang H, Qiu J et al (2003) Echinococcosis on the Tibetan Plateau: prevalence and risk factors for cystic and alveolar echinococcosis in Tibetan populations in Qinghai Province, China. Parasitology 127:S107–S118

Schmidberger J, Uhlenbruck J, Schlingeloff P et al (2022) Dog ownership and risk for Alveolar Echinococcosis, Germany. Emerg Infect Dis 28:1597–1605

Shang Z, Gibb M, Leiber F et al (2014) The sustainable development of grassland-livestock systems on the Tibetan plateau: problems, strategies and prospects. Rangel J 36:267–296

Shang J, Zhang G, Yu W et al (2019) Molecular characterization of human echinococcosis in Sichuan, Western China. Acta Trop 190:45–51

Shen Y, Jiang W, Wang X (1995) 3 cases of hepatic echinococcosis. New J Gastroenterol (in Chinese)

Shumuye NA, Li L, Ohiolei JA et al (2023) Update on the genetic diversity and population structure of *Echinococcus granulosus* in Gansu Province, Tibet Autonomous Region, and Xinjiang Uygur Autonomous Region, Western China, inferred from mitochondrial cox1, nad1, and nad5 sequences. Parasitol Res 122:1107–1126

Sierra-Ramos R, Valderrama-Pome A (2017) Echinococcosis hyper-endemic and cystic fertility in pigs of the inter-andean valley of Huancarama, Peru. Rev Peru Med Exp Salud Publica 34:250–254

Smith A, Xie Y (2009) The Chinese veterinary field manual [M].

Šnábel V, Antolová D, Cavallero S et al (2020) On the geographic genetic variants of the cestode *Echinococcus multilocularis* with reference to the original descriptions from Bowles et al. (1992) and Bowles and McManus (1993), and their use. Parasitol Int 75:102039

Spotin A, Boufana B, Ahmadpour E et al (2018) Assessment of the global pattern of genetic diversity in *Echinococcus multilocularis* inferred by mitochondrial DNA sequences. Vet Parasitol 262:30–41

Stey C, Jost R, Ammann R (1993) Recurrent, life-threatening anaphylaxis as initial manifestation of cystic echinococcosis (granulosus) of the liver. Schweiz Med Wochenschr 123:1445–1447

Sun D (1992) Diagnosis and treatment of porcine Echinococcosis. Liaoning Animal Husbandry and Veterinary 4 (in Chinese)

Sydler T, Mathis A, Deplazes P, Eur J (1998) *Echinococcus multilocularis* lesions in the livers of pigs kept outdoors in Switzerland. Vet Pathol 4:43–46

Tang C, Chen J, Tang L et al (2002) Histologcal observation on Alveolaris hulunbeierensis in the liver of Microtus brandti in northeast Inner Mongolia, China. Chinese J Zoonoses 18:8-11T001,T002 (in Chinese)

Tang C, Quian Y, Kang Y et al (2004) Study on the ecological distribution of alveolar Echinococcus in Hulunbeier Pasture of Inner Mongolia. Parasitology 128:187–194

Tang C, Shu L, Tang L et al (2001) The species and bioepidemiology of alveolar echinococcus in xinbaerhu West County, Eastern Inner Mongolia. J Xiamen Univ: Nat Sci 40:503–511 (in Chinese)

Tian Y (2001) Diagnosis and control of porcine Echinococcosis. Anim Husb Vet Sci Inform:11–11 (in Chinese)

Tigre W, Deresa B, Haile A et al (2016) Molecular characterization of *Echinococcus granulosus* s.l. cysts from cattle, camels, goats and pigs in Ethiopia. Vet Parasitol 215:17–21

Torgerson PR (2003) Economic effects of echinococcosis. Acta Trop 85:113–118

Torgerson PR, Dowling PM (2001) Estimating the economic effects of cystic echinococcosis. Part 2: an endemic region in the United Kingdom, a wealthy, industrialized economy. Ann Trop Med Parasitol 95:177–185

Torgerson PR, Keller K, Magnotta M et al (2010) The global burden of alveolar echinococcosis. PLoS Negl Trop Dis 4:e722

Torgerson PR, Shaikenov BS, Baitursinov KK et al (2002) The emerging epidemic of echinococcosis in Kazakhstan. Trans R Soc Trop Med Hyg 96:124–128

Varcasia A, Garippa G, Pipia AP et al (2008) Cystic echinococcosis in equids in Italy. Parasitol Res 102:815–818

Vuitton DA, Mcmanus D, Rogan M et al (2020) International consensus on terminology to be used in the field of echinococcoses. Parasite 27:41

Wan J, Fu J, Niu Y et al (2001) A case of porcine Echinococcosis. Meat Ind 4:21 (in Chinese)

Wang H, Chai J, Liu F et al (2002) A study on ecological-epidemiology of two hydatid disease in Qinghai. Chinese J Parasitic Dis Control 15:284–287 (in Chinese)

Wang C, He H, Li M et al (2009) Parasite species associated with wild plateau pika (*Ochotona curzoniae*) in southeastern Qinghai Province, China. J Wildl Dis 45:288–294

Wang X, Liu J, Zuo Q et al (2018) *Echinococcus multilocularis* and *Echinococcus shiquicus* in a small mammal community on the eastern Tibetan Plateau: host species composition, molecular prevalence, and epidemiological implications. Parasit Vectors 11:302

Wang L, Qin M, Liu Z et al (2021b) Prevalence and spatial distribution characteristics of human echinococcosis in China. PLoS Negl Trop Dis 15:e0009996

Wang L, Quzhen G, Qin M et al (2022) Geographic distribution and prevalence of human echinococcosis at the township level in the Tibet Autonomous Region. Infect Dis Poverty 11:10

Wang H, Soha P (2000) Infections of larval and adult *echinococcus multilocularis* in human and animals in Qinghai province. Chinese J Parasitic Dis Control 13:120–123 (in Chinese)

Wang Z, Wang X, Liu X (2008) Echinococcosis in China, a review of the epidemiology of Echinococcus spp. EcoHealth 5:115–126

Wang W, Wu Y, Wu J (1989) The occurence of *echinococcus multilocularis* leukart, 1863 in fox and wolf in tacheng district, xinjiang. Endemic Dis Bulletin 4:8–1189 (in Chinese)

Wang H, Yin Y, Ma C et al (1991) Preliminary investigation of liver alveolar echinococcosis and liver cystic,echinococcosis in xiji county, Ningxia. Chinese J Parasitol Parasitic Dis 9:143–145 (in Chinese)

Wang H, Zhang C-S, Fang B-B et al (2021a) Dual role of hepatic macrophages in the establishment of the *Echinococcus multilocularis* Metacestode in mice. Front Immunol 11:600635

Wen H, Vuitton L, Tuxun T et al (2019) Echinococcosis: advances in the 21st century. Clin Microbiol Rev 32:e00075-18.

Weng X, Mu Z, Wei X et al (2020) The effects of dog management on Echinococcus spp. prevalence in villages on the eastern Tibetan Plateau, China. Parasit Vectors 13:207

Weng X, Mu Z, Wang X et al (2019) Prevalence of echinococcosis in small mammals in Shiqu County, Sichuan Province. J Pathogen Biol 14:1320–1323 (in Chinese)

Wnukowska N, Salamatin R, Golab E (2011) Human echinococcosis in Poland in 2003-2010 according to the serological tests results of NIPH-NIH. Przegl Epidemiol 65:455–458

Wu W (2017) Prevalence and distribution of two types of echinococcosis in China. China Animal Health 7:7–9 (in Chinese)

Wu W, Wang H, Wang Q et al (2018) A nationwide sampling survey on echinococcosis in China during 2012-2016. Chinese J Parasitol Parasitic Dis 36:1–14 (in Chinese)

Wu C, Zhang W, Ran B et al (2017) Genetic variation of mitochondrial genes among *Echinococcus multilocularis* isolates collected in western China. Parasit Vectors 10:265

Xiao N, Nakao M, Qiu J et al (2006) Dual infection of animal hosts with different Echinococcus species in the eastern Qinghai-Tibet plateau region of China. Am J Trop Med Hyg 75:292–294

Xiao N, Qiu J, Nakao M et al (2005) *Echinococcus shiquicus* n. sp., a taeniid cestode from Tibetan fox and plateau pika in China. Int J Parasitol 35:693–701

Xiao N, Zhou Z, Chen X et al (2012) Assessment of health education against echinococcosis in Tibetan region of Sichuan Province, China. Zhongguo Ji Sheng Chong Xue Yu Ji Sheng Chong Bing Za Zhi 30:6–11

Xu Q, Jiang W, Zheng Q et al (2002) *Microtus ilaeus* was found to be infected with *Echinococcus multilocularis* in Achale, Yining County. Endemic Dis Bulletin 17:83–83 (in Chinese)

Xu Z, Zhang C, Tan Y et al (2004) Prevalence of hydatidosis in Heilongjiang Province. Chinese J Parasitol Parasitic Dis 22:93–93 (in Chinese)

Yang Y, Craig PS, Sun T et al (2008) Echinococcosis in Ningxia Hui Autonomous Region, northwest China. Trans R Soc Trop Med Hyg 102:319–328

Yang Y, Gray DJ, Ellis MK et al (2009) Human cases of simultaneous echinococcosis and tuberculosis–significance and extent in China. Parasit Vectors 2:53

Yang Y, Rosenzvit MC, Zhang L et al (2005a) Molecular study of Echinococcus in west-central China. Parasitology 131:547–555

Yang Y, Sun T, Li Z et al (2005b) Echinococcosis, Ningxia, China. Emerg Infect Dis 11:1314–1316

Yang D, Zhao D, Guo H (2011) The first notes on the metacestode of *Echinococcus multilocularis* in Lagurus luteus Eversmann from the Hobukser Mongolian Autonomous County, Xinjiang Uygur Autonomous Region. Int J Med Parasitic Dis 38:186–188 (in Chinese)

Yu W, Li D, Chen X et al (2005) Epidemiological survey on human echinococcosis in four counties of Gaizi Tibetan autonomous prefecture, Sichuan. Parasitoses Infect Dis 3:170–172 (in Chinese)

Yu S, Wang H, Wu X et al (2008) Cystic and alveolar echinococcosis: an epidemiological survey in a Tibetan population in southeast Qinghai, China. Jpn J Infect Dis 61:242–246

Zait H, Kouidri M, Grenouillet FE et al (2016) Molecular characterization of *Echinococcus granulosus* sensu stricto and *Echinococcus canadensis* in humans and livestock from Algeria. Parasitol Res 115:2423–2431

Zhang T, Yang D, Zeng Z et al (2014) Genetic characterization of human-derived hydatid cysts of *echinococcus granulosus* Sensu Lato in Heilongjiang Province and the first report of G7 Genotype of *E. canadensis* in Humans in China. PLoS One 9:e109059

Zhang W, Zhang Z, Wu W et al (2015) Epidemiology and control of echinococcosis in central Asia, with particular reference to the People's Republic of China. Acta Trop 141:235–243

Zhang W, Zhang Z, Yimit T et al (2009) A pilot study for control of hyperendemic cystic hydatid disease in China. PLoS Negl Trop Dis 3:e534

Zhang G, Zuo B, Wu X et al (2001) Two cases of hepatic echinococcosis in Shanxi Province. Chinese J Parasitol Parasitic Dis 19:3–10 (in Chinese)

Zhao Y, Chong S, Cai X et al (2008) Epidemiological survey of echinococcosis in Tibetan areas of southern Gannan. Chinese J Epidemiol 29:1100–1100 (in Chinese)

# *Echinococcosis* Immune Response, Immunopathogenesis and Immune Evasion from the Human Host

6

Kalibixiati Aimulajiang and Baoping Guo

**Abstract**

*Echinococcosis* is a worldwide parasitic disease with great harm and complicated immune mechanisms. Immunoprophylaxis is an ideal way to prevent *Echinococcosis* in the future. *Echinococcosis* is an exogenous substance with antigenicity. After infection, it can induce the host to produce an immune response and undergo a series of cellular and molecular changes. In this chapter, we will have a broad overview of innate immunity, adaptive immunity, immune evasion, and host-*Echinococcosis* interactions resulting from *Echinococcosis* infection.

**Keywords**

*Echinococcosis* · Innate immunity · Adaptive immunity · Immune evasion

## 6.1 Introduction

The human body often has varying degrees of resistance to *Echinococcosis* infection, manifested as a series of immune responses. During the long-term parasitic process, *Echinococcus* tapeworms can find some strategies or mechanisms to enable

K. Aimulajiang (✉) · B. Guo
State Key Laboratory of Pathogenesis, Prevention and Treatment of High Incidence Diseases in Central Asia, Clinical Medicine Institute, The First Affiliated Hospital of Xinjiang Medical University, Urumqi, Xinjiang, PR China

Xinjiang Key Laboratory of Echinococcosis, Clinical Medicine Institute, The First Affiliated Hospital of Xinjiang Medical University, and WHO Collaborating Centre on Prevention and Case Management of Echinococcosis, Urumqi, Xinjiang, PR China





J. Li et al. (eds.), *Echinococcus: Control and Elimination of Echinococcosis with a Focus on China and Europe*, Parasitology Research Monographs 19,
https://doi.org/10.1007/978-3-031-54479-8_6

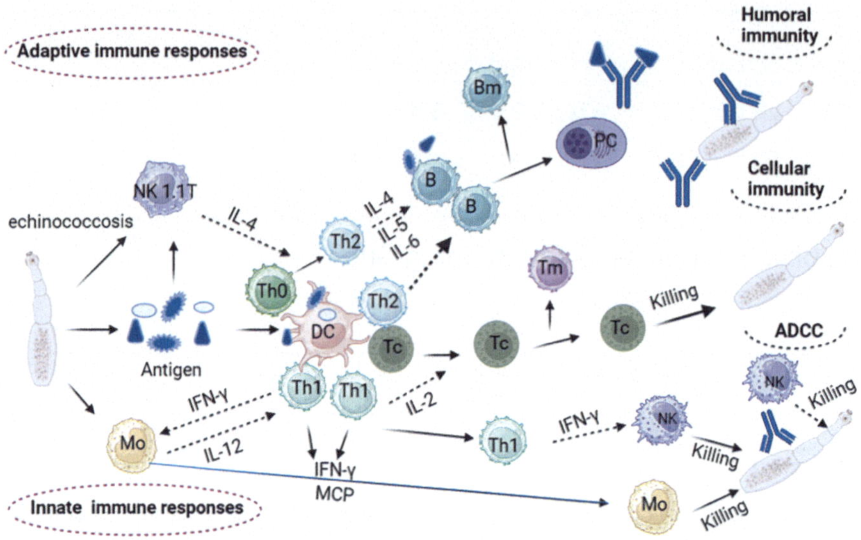


**Fig. 6.1** Immune response and association in *Echinococcosis* infection (Wen 2015)

them to continue to survive, reproduce, and resist the killing effects of the host's immune system and oxidative system in the host and have completed their life cycle. *Echinococcus*-infected organisms will have varying degrees of resistance, inducing the host to produce an innate immune response (non-specific immunity) and an acquired immune response (specific immunity). For hydatid infection, the body's specific immune and non-specific immune mechanisms both play an important role.

In the interaction between *Echinococcus* and host, *Echinococcus* mainly exhibits antigen specificity and immune evasion, while the host exhibits anti-insect immunity and immune tolerance. How *Echinococcosis* evades the immune recognition of the host has always been the subject of research by *Echinococcosis* scholars; some studies have reported that *Echinococcosis* can evade the host's immune system by means of antigen variation and antigen camouflage, and a large number of studies have followed *Echinococcus* can also actively change the host's immune response through a variety of mechanisms (Fig. 6.1).

## 6.2 The Role of Innate Immune Cells in *Echinococcus* Infection

### 6.2.1 Barrier Structure

The functions of the skin-mucosal barrier, blood-brain barrier and blood-placental barrier are: (1) Mechanically block the invasion of *Echinococcus*; (2) Secrete chemical substances to kill or inhibit the survival of *Echinococcus* eggs in vivo; (3) Normal The flora and its metabolites antagonized the growth of *Echinococcus*.

### 6.2.2 Immune Cells

1. Monocyte-macrophage

As one of the innate immune cells, macrophages not only mediate the body's endogenous and specific immune responses, but also affect immune tolerance. When the body is stimulated by the exogenous pathogen *E. multilocularis*, it triggers a complex immune response, leading to inflammation, and macrophages play an indispensable role in this process. Studies have confirmed that after liver resection in patients with *Alveolar Echinococcosis* (AE), the number of macrophages increases (Zhang et al. 2020); macrophages are involved in developing AE. Macrophages are extremely plastic, and they will trigger multiple functions after being stimulated by the outside world, and then play a dual role in AE, helping to clear *E. multilocularis* in the early stage and related to the immune escape mechanism of pathogens in the late stage (Weijian et al. 2021). Macrophages are divided into two types: one is M1 macrophages, also known as classically activated macrophages, which mediate the process of pro-inflammation, antigen presentation, and host immune clearance of pathogens and tumor cells; the other is M2 type Macrophages, also known as alternatively activated macrophages, play an anti-inflammatory role and promote tumor growth, invasion and metastasis (Xuan et al. 2015).

Wang et al (Wang and Fan 2018) showed that after mice were infected with *E. multilocularis*, the complex antigenic composition of *E. multilocularis* triggers the normal immune response of the body, polarizes macrophages to the M1 type, and releases white blood cells Interleukin-6 (IL-6), IL-12, tumor necrosis factor-α (TNF-α) and other inflammatory cytokines enhance the inflammatory response and assist the host to eliminate parasites; It can polarize macrophages to the M2 type, and suppress inflammatory responses by releasing transforming growth factor-β (TGF-β), IL-10 and other cytokines, thereby achieving immune evasion. M1 and M2 macrophages maintain a dynamic balance under normal conditions. Once the imbalance occurs due to external stimuli, macrophages cannot control the inflammatory response, which causes damage to the body. With the development of the disease, macrophages are mainly polarized to the M2 type, which destroys the polarization balance of macrophages, which may be an important mechanism of immune tolerance in the late stage of AE.

Monocytes complement liver macrophages, and both monocytes and Kupffer cells (KCs) may participate in Th1 and Th2 immune responses by differentiating into M1 or M2 at different stages of *E. multilocularis* infection (Zhang et al. 2020). KCs, the main immune cells in the liver, are normally responsible for clearing foreign substances. In addition, KCs protect the liver from damage by releasing cytokines, reactive oxygen species, etc., and play a key role in the acute and chronic responses of liver injury (Roberts et al. 2007). As *Echinococcosis* continues to expand and compress liver tissue, it can promote the activation and proliferation of a large number of KCs, and actively exert anti-inflammatory effects. In AE patients, anti-inflammatory cytokines secreted by KCs were higher than inflammatory cytokines (Liu et al. 2020). KCs play a key role in anti-*Echinococcosis* by maintaining liver homeostasis under normal and pathological conditions, thereby affecting *Echinococcosis* progression in the liver.

2. Dendritic cells

Dendritic cells (DCs), like macrophages, belong to mononuclear phagocytes, and there is significant heterogeneity in phenotype, tissue localization, proliferation potential and function. As a professional antigen-presenting cell, DCs possesses phagocytic properties and the function of presenting antigen to T cells. Studies have shown that DCs in mice infected with *E. multilocularis* are significantly higher than those in healthy control mice (Xu and Zhang 2021), and DCs in infected mice can specifically regulate CD4+ T and CD8+ T cell responses (Grubor et al. 2017), It was shown that DCs have immunosuppressive T cell regulatory functions in *Echinococcosis*. Jenne (Jenne et al. 2001) et al. proposed the special signs of DCs function in AE patients. *E. multilocularis* crude antigen failed to mature DCs but could induce the proliferation of autologous CD8+ T cells. The antigens of *E. multilocularis* directly stimulate and modulate the differentiation of DCs in mice infected with *E. multilocularis*, thereby impeding disease progression (Xu et al. 2016). DCs subpopulation and maturation may affect the differentiation of Th cells, one of the initial players in specific immune responses. Immature DCs show a higher level of antigen endocytosis, and the phagocytosis level of mature DCs is significantly lower than that of immature DCs, while the level of antigen presentation is significantly enhanced, which activates T helper cells to differentiate into Th1 or Th2 cells, thereby forming a link between innate immunity and adaptive immunity (Ma and Yin 2019). By detecting DCs in mice at different stages after *E. multilocularis* infection, it was found that mature DCs can effectively present antigens in the early stage of *E. multilocularis* infection, thereby inducing anti-parasitic immunity. The efficiency of antigen presentation to T cells is reduced, thus leading to immune escape of *E. multilocularis* (Wei et al. 2014). The maturation of DCs is related to the expression of phenotype molecules CD80, CD83 and CD86. Ma and Yin (2019) used flow cytometry to detect the expression level of DCs phenotype molecules and found that the DCs maturity of AE patients was significantly lower than that of healthy controls; that is, the immune response of T cells depends on the maturation of DC. In addition, mature DCs can secrete the Th1 cytokine IL-12. After infection with *E. multilocularis*, the DCs in the body have a maturation disorder, which affects the release of IL-12, which weakens the ability of Th cells to differentiate into Th1 cells and accelerates the growth rate of Th1 cells (Ma and Yin 2019). Previous studies have shown that indoleamine 2,3-dioxygenase (IDO) is an important molecular switch that regulates the immune tolerance of host DCs. Infection with *E. multilocularis* leading to immune evasion can induce a significant upregulation of indoleamine 2,3-dioxygenase (IDO) expression (Wang et al. 2015b), reduce the antigen presentation ability of DCs, and avoid the effective activation of T cells in the early stage of the immune response, thereby reducing the host's anti-infection ability (Fu and Jiang 2018). In summary, the interaction between the host and *Echinococcus* can be better elucidated at the DCs level.

3. Natural killer cells

Both innate and adaptive immunity are key to parasitic infection (Weatherhead et al. 2020). As active members of innate immunity, natural killer cells (NK) constitute approximately 25–50% of human hepatic lymphocytes and 10–15% of mouse hepatic lymphocytes. Most NK cells are cytotoxic, and they produce lytic granules containing a large number of molecules (perforin, granzyme, and granulysin in humans) that can induce stress cell death. They also express several members of the tumor necrosis factor (TNF) superfamily, such as FASL and TRAIL, which induce apoptosis of target cells by binding to the corresponding receptors FAS or TRAILR, respectively. NK cells produce a series of cytokines IFN-γ, TNF-α, IL-10, growth factor GM-CSF and chemokines CCL3, CCL4, CCL5, XCL1; NK cells can communicate with dendritic cells, macrophages and T cells interact to generate an immune response (Crinier et al. 2020). The significance of liver immunity has been increasingly substantiated by numerous studies, with some researchers even proposing that the liver functions as an innate immune organ (Kubes and Jenne 2018). Functional defects of NK cells are associated with the progression of pathogens and contribute to disease initiation (Peng et al. 2016). Many studies have shown the importance of NK cells in the parasite/host interaction at different stages (early and late) of *Echinococcosis (*Abulizi et al. 2019*;* Vuitton et al. 1989*)*. Impaired NK cell function may be a key factor in the immune evasion strategy of *Echinococcus*, and NK cells from the peripheral blood of AE patients have lower cytotoxicity (Vuitton et al. 1989). More NK cells were infiltrated in CLT compared with DLT in AE patients, impairing their function (Zhang et al. 2021). NK cell function is tightly regulated by a series of activating and inhibitory receptors (Long et al. 2013). Existing data have shown that *E. multilocularis* and its antigens inhibit the function of NK cells by up-regulating the expression of inhibitory receptors NKG2A, PD-1 and LAG-3 (Bellanger et al. 2020). Another study found that NK cells in the blood and CLT of AE patients expressed higher levels of the inhibitory receptor TIGIT, which can inhibit the activation of NK cells in tumors or chronic infections (Zhang et al. 2018; Stanietsky et al. 2013). In a mouse model of *E. multilocularis* infection, NK cell activity was shown to be reduced in liver and spleen of $TIGIT^{+}$ infected mice, as evidenced by decreased expression of CD69 and CD44, suggesting that NK cells aggregated in *E. multilocularis*-infected livers and depletion, which may be driven by continued worm development, during which TIGIT acts as a negative regulator of NK cells (Zhang et al. 2021). *E. multilocularis* infected mice and found that there were differences in the proportion of NK cells in the liver and spleen (Zhang et al. 2021). To sum up, it is better to elucidate the immune regulation effect of NK cells on the host after *Echinococcus* infection.

4. Polymorphonuclear cells (PMN)

Polymorphonuclear cells (i.e., neutrophils, eosinophils, and basophils) are key players in the innate immune machinery. They are often recruited to the site of infection by host/pathogen-derived chemotactic stimuli to defend against the first line of defense against pathogen invasion (Yang et al. 2017). PMN is the most abundant white blood cell in human circulation, and it is also the first responding cell to the site of infection or injury. It removes target cells through phagocytosis and confines them in phagosomes. The synergy between PMN-derived factors destroys the body's uptake. The integrity and vitality of the pathogenic microorganisms; with the initiation of the antibacterial response, the process of phagocytosis promotes the transcription of genes, accelerates apoptosis and finally terminates the inflammatory response (Aratani et al. 1999). Studies have found that PMN plays a key role in preventing the migration and development of *E.* multilocularis in rats primarily infected with Wistar (Joekel et al. 2021). In addition, macrophages play an important role in PMN depletion in rats. The hospital clinical data study found that the expression of neutrophil-to-lymphocyte ratio (NLR) and platelet-to-lymphocyte ratio (PLR) in the AE group was significantly higher than that in the CE group and the control group ($P < 0.05$). PLR was positively correlated with aspartate aminotransferase, alanine aminotransferase, total bilirubin, direct bilirubin, alkaline phosphatase, γ-glutamyl transpeptidase ($P < 0.05$), NLR is correlated with albumin content ($P < 0.05$) (Wang et al. 2021).

## 6.3 The Role of Adaptive Immune Response in *Echinococcus* Infection

After the *Echinococcosis* invades the host, the antigenic substance stimulates the host's immune system, and an immune response often occurs, producing a clearly targeted resistance, which can exert a clear or killing effect on the *Echinococcosis*, and also has a certain resistance to reinfection. Specific immunity and non-specific immunity are not completely opposite but an interrelated and complex dynamic process.

### 6.3.1 Changes and Significance of IgG Antibody Levels

Intact human immunoglobulin (IgG) antibodies consist of an antigen-binding fragment (Fab) and an Fcγ receptor-bound crystallizable fragment (Fc). Among the four subclasses of human IgG (IgG1, IgG2, IgG3, IgG4), their constant regions differ, especially in the hinge and CH2 domains; IgG1 has the highest Fc binding affinity, followed by IgG3, IgG2 and IgG4 (Yu et al. 2020). During the chronic phase of human *Cystic Echinococcosis* (CE), elevated serum antibody levels are common, especially IgG, IgM, and IgE, with the IgG subclasses IgG1 and IgG4 predominating. In humans, IgG4 antibody responses are associated with vesicle development, growth, and disease progression, whereas IgG1, IgG2, and IgG3 antibody production mainly occurs when vesicles are infiltrated or destroyed by the human host immune system (Riesle et al. 2014). When previous studies assessed IgG subclass production in human serum, the most important antibody subclasses in CE and AE were IgG1 and IgG4, respectively (Chandrasekhar and Parija 2009; Delunardo et al.

2010). Another study found that Th2 cells produced cytokines such as IL-4, IL-5, IL-6, IL-9, IL-10, and IL-13, which promoted the production of IgE, IgG1, and IgG4 subclasses (Riesle et al. 2014). In the diagnosis of *Echinococcosis* infection, detection of IgG subclasses can reveal changes in disease infection better than specific total IgG (Chirag et al. 2015).

### 6.3.2 Changes and Significance of IgE Antibody Levels

Host infection with parasites is usually dominated by Th2 immunity, including elevated levels of parasite antigen-specific and non-specific IgE, tissue and blood eosinophilia (sometimes basophilia), and intestinal pathological responses, including Crypt hyperplasia, goblet cell hyperplasia, and mucosal MC (MMC) hyperplasia (Finkelman et al. 1997; Anthony et al. 2007; Everts et al. 2010; Allen and Wynn 2011; Mukai et al. 2016). Data from epidemiological studies have shown that IgE antibodies are protective in humans infected with certain parasites, and parasite-specific IgE levels are positively correlated with host resistance to infection (Mukai et al. 2016; Hagan et al. 1991; Faulkner et al. 2002). Studies have shown that *E. granulosus*-specific IgE exists in 25% to 75% of patients with CE, and more than 75% of patients have specific IgE against AgB, Ag5 or recombinant antigens. Total IgE and specific IgE anti-*E. granulosus* antigen levels increased when the vesicles were numerous, centrally located in the liver, and contained asci. IgE and IgG4 levels correlated with IL-5 production; however, high eosinophil counts were not associated with high IL-5 expression in the same study (Vuitton 2004). IgE levels were positively correlated with disease severity and/or progression/regression after treatment. The decline of specific IgE after complete surgical resection of the cyst is slow, and serum-specific IgE can be detected in 55% of patients 1 year after surgery. In AE, total IgE and specific serum IgE (RAST technique using *E. granulosus* antigen) were elevated in 50% of patients, although allergic symptoms were rare. In patients treated with anti-parasitic drugs, disappearance of IgE and decline of IgG4-specific antibodies have been shown to correlate with regression of lesions (Vuitton 2004).

### 6.3.3 Synergy of Major Immune Cells

Adaptive immunity plays an important role in host defense against pathogens, in which CD4+ T helper cells play a major role. Naïve CD4 T cells, upon activation, differentiate into distinct Th subsets that produce lineage-specific cytokines (Fig. 6.2) (Kaplan et al. 2015). Th cell differentiation depends on the strength of T cell receptor (TCR) signaling and signaling triggered by activating polarizing cytokines and/or upregulating specific transcription factors. Several lineage-specific transcription factors determine Th cell fate and function. Although these master regulators cross-regulate each other, their expression may be dynamic. Sometimes, they are even co-expressed, resulting in substantial Th cell heterogeneity and

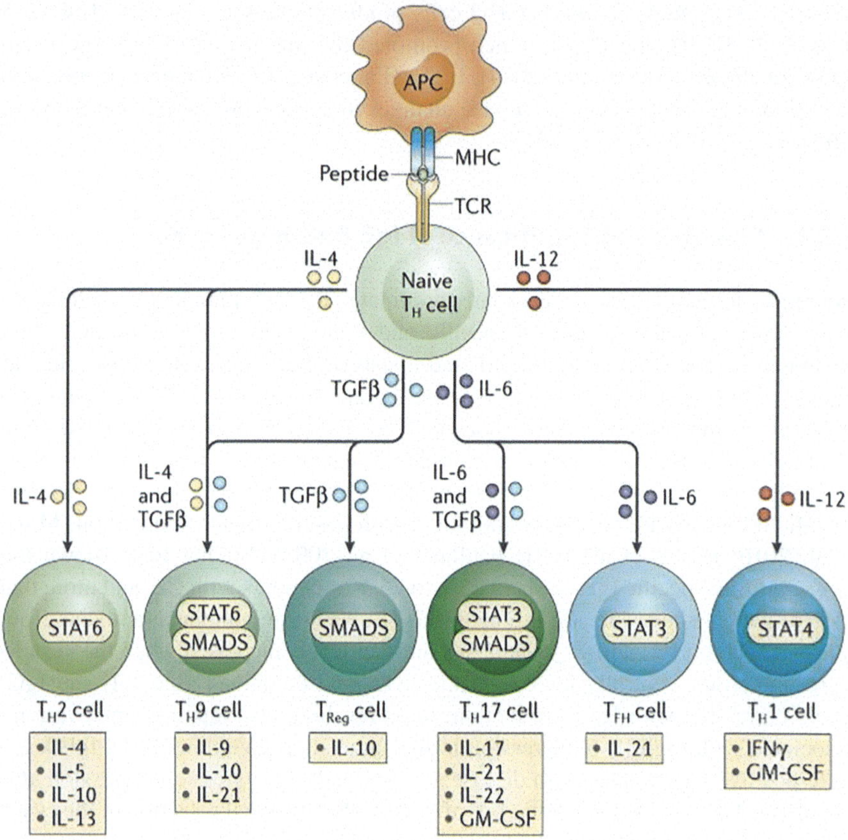


**Fig. 6.2** The influence of cytokine environment on Th cell differentiation (Kaplan et al. 2015)

plasticity. Similar regulation mediated by these master regulators has also been found in innate lymphoid cells (ILCs) (Zhu 2018). Cytokines are central mediators of the immune response, and CD4 T helper cells are professional cytokine-producing cells. Th cells play a key role in the adaptive immune response to infection through the production of effector cytokines; different Th subsets are involved in protective immunity against different pathogens (Murphy and Reiner 2002; Zhu and Paul 2008; Zhu et al. 2010).

Imbalances during parasitic infection are important immunological changes, and it is generally believed that Th1-type responses can enhance the host's immune defense against intracellular pathogen infection. In contrast, Th2-type responses are associated with infection progression, persistence, and chronicity (Bauder et al. 1999; Ortona et al. 2003). T cells gradually change to a certain direction with the course of the disease and participate in various pathological reactions, leading to tissue damage. It may be related to the selection of initial T cells to differentiate into Th1 or Th2 cells under the stimulation of parasite antigens. IL-2 and IL-4 are the main cytokines that induce the proliferation and differentiation of Th1 and Th2 cells, respectively, and the interaction of IL-2, IL-4 and IFN-γ is the key to maintaining the balance of Th1/Th2. Under normal conditions, the immune function of Th1/Th2 cells in the body is in a state of dynamic balance. Once this balance is broken, the body cannot maintain normal immune function and cause various diseases. During parasitic infection, the imbalance of Th1/Th2 is an important immunological change. Th cells gradually polarize in a certain direction with the development of the disease, and participate in various pathological reactions, leading to tissue damage, which may be stimulated by parasite antigens, the initial T selectivity is related to the differentiation of Th1 or Th2 cells (Ai Erken et al. 2012).

In the immune process of parasitic infection, Th17 cells have a two-way regulatory effect: in some cases, it can produce immune protection against parasitic infection in the body, while in other cases it can cause the host's immunopathological response. This function largely depends on shifts in the host's autoimmune state, the infection's severity, and the disease's treatment. Th17 secretes a large number of IL-17 cytokines, and immune response occurs in the early stage of *Echinococcosis* infection (Tuxun et al. 2012), affecting the immune response of the host, but Th1 and Th2 cells do not have these functions (Pang et al. 2012). Therefore, attention should be paid to it in future research, that is, in the early stage of parasitic infection (Zhao et al. 2012), try to effectively enhance the secretion of Th17 cells to improve the host's anti-infection immunity; in the late stage of infection, the host's pathology can be alleviated by inhibiting the secretion of Th17 cells Response (Zhao et al. 2012). Th1 cells mainly mediate cytotoxic effects by promoting the activation and proliferation of cytotoxic T cells (CTL), NK and macrophages (Mφ), and play an important role in anti-intracellular parasitic infection effect. The main function of Th2 cells is to assist B cells in proliferating, producing antibodies, and participating in the humoral immune response. In the late stage of an inflammatory response dominated by a Th1 response, there is often a gradually enhanced Th2 response (Vuitton 2003), which prevents the Th1 response from mediating damage. T cells secreted by Th1 and Th2 cytokines coexist in mice. Th1 and Th2 cell subsets secrete different cytokines and interact to form a complex regulatory network to regulate the immune response. In the early stage of infection, both types of *Echinococcosis* are dominated by Th1-type immune responses, which play a role in clearing *Echinococcosis* protoscoleces in the early stage of infection, and Th2-type immune responses are dominant in the late stage of infection (Brehm and Koziol 2017).

### 6.3.4 Th1/Th2 Balance of Host Infected with *Echinococcus*

The balance of Th1/Th2 plays an important role in the immune response to *Echinococcus* infection (Fig. 6.3). In the early stage of *Echinococcus* infection, the first manifests Th1 cell-mediated cellular immunity IL-2, IFN-γ and TNF-α showed advantages. IL-2 and IFN-γ promoted the proliferation and activity of CD8+ T cells, thereby inhibiting the exophytic growth and metastasis of *Echinococcus*. IFN-γ can inhibit the growth and metastasis of *Echinococcosis* in the defense mechanism

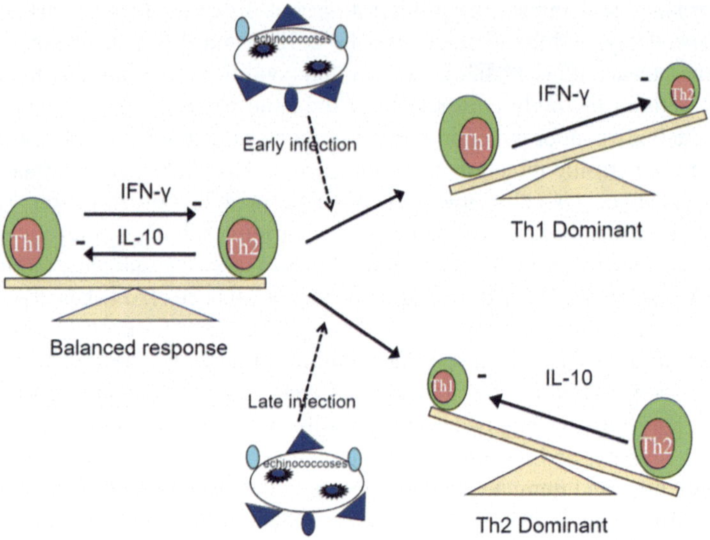


**Fig. 6.3** Th1/Th2 balance in *Echinococcus* infection (Wen 2015)

against parasitic infection (Brehm and Koziol 2017). The Th1-type and Th2-type immune responses exhibit reciprocal negative feedback in *Echinococcosis*, which is primarily mediated by cytokines. IFN-γ secreted by Th1 cells suppresses the production of Th2 cytokines, while IL-4 predominantly inhibits the secretion of Th1 cytokines by Th2 cells. Additionally, IL-10 can downregulate the function of Th1 cells. Similar to other parasitic infections, successful survival of *Echinococcosis* within the host necessitates active modulation of the host's immune system to evade detrimental immune reactions. Following infection with *Echinococcosis*, it establishes a long-term residence within the host. In advanced stages (beyond 3 months) of infection, there is a significant decline in lymphocyte transformation capacity and notable inhibition of cellular immune response. Therefore, in the host body, *Echinococcosis's* growth and development speed is accelerated, which indicates that the host lymphocytes shift from Th1 to Th2 (Brehm and Koziol 2017). Therefore, in *Echinococcosis* immunity, the Th1 and Th2 cytokine networks provide new clues for further exploring the mechanism of immune disorders during infection and using immune intervention methods to correct immune disorders and participate in immunotherapy.

### 6.3.5 The Role of Th9 in *Echinococcus* Infection

The Th9 immune response is a new type of immune response discovered in recent years. This type of immune response is mainly manifested by the secretion of IL-9 by Th9 cells, thereby inducing immune responses such as anti-parasitic infection, anti-tumor, autoimmunity and allergy (Licona-Limon et al. 2017; Jia and Wu 2014;

Neurath and Kaplan 2017; Rivera Vargas et al. 2017). Th9 cells are differentiated under the stimulation and special conditions of TGF β, IL-4 and IL-2 (Kaplan 2017). The expression of Th9 cell-related factors IL-9, PU.1, and IL-4 in patients with CE found that compared with healthy people, the expression of related factors in PBMC of patients infected with *E. granulosus* was significantly increased. In addition, the expressions of related cytokine IL-9 and transcription factor PU.1 in the liver tissue and PBMC of patients with hydatid were significantly increased compared with the control group (Chen et al. 2019). Another study found that activating the TGF-β/Smad signaling pathway may promote Th9 cell differentiation and IL-9 expression in patients with CE (Licona-Limon et al. 2017). The results preliminarily indicated that *Echinococcus* may have the ability to induce the host to produce a Th9 immune response, but the specific characteristics and mechanism are still unclear.

### 6.3.6 The Role of Th17 in *Echinococcus* Infection

Currently, an increasing body of evidence has demonstrated the pivotal role of Th17 cells in the host's immune response against infections and inflammatory damage (Wang et al. 2015a). Th17 cells are similar to Th1 and Th2 cells and are also differentiated from naive T cells (Tn). Tn is activated under the stimulation of antigens such as parasite eggs and excretion-secretion and differentiates into Th17 cells through the regulation of many cytokines. TGF-β and IL-6 are the initiators of Th17 cell differentiation, and the combination of TGF-β and IL-6 under the action of transcription factor STAT3 can promote the differentiation of primitive T cells into Th17 cells (Lechner et al. 2012). In the later stage of Th17 cell differentiation, the combination of IL-23 and its receptor stabilizes and promotes cytokine secretion. Th17 cells can secrete factors such as IL-17, IL-6, IL-22 and TNF-α, and these cytokines have the ability to collectively mobilize, recruit and activate neutrophils (Wang et al. 2015a; Wynn 2008; Bataller and Brenner 2005). IL-17 has a pro-inflammatory effect and can induce the expression of pro-inflammatory cytokines, such as IL-6, TNF-α, and chemokine metal matrix protease, causing high cell infiltration and tissue destruction, IL-17 as a pro-inflammatory factor that protects the host from eliminating extracellular pathogens, but at the same time causes autoimmune pathological damage (Pang et al. 2012; Zhao et al. 2012).

In the early stage of hydatid infection, the peripheral blood IL-17 of mice was significantly increased. In the middle and late stages of infection, the hydatid cysts in the peritoneal cavity of infected mice gradually increased, and a large number of parasites infected induced more Th17. The cells differentiated, and the level of IL-17 in the peripheral blood continued to increase. In addition, in the detection of *Echinococcus*-infected mice, it was found that the expressions of IL-17 and IL-6 in the liver of mice increased significantly as the infection progressed and remained relatively high. From the initial stage, it continued to increase. It shows that in the early and middle stages of infection, IL-17 recruits neutrophils to the infection site and plays a certain role in clearing the infection (Pang et al. 2012; Zhao et al. 2012). Three months after infection (Vuitton 2003), IL-17 still maintained a high level. It is speculated that IL-17 may be involved in the pathological damage of the host liver, but the specific mechanism of action needs further study. Th17 cells are an independent new subgroup with unique performance in immune cell differentiation, development and maintenance. For example, differentiation regulates Th1, which is similar to GATA-3 controlling Th2 differentiation. Th17 differentiation is also regulated by its transcription factor R0R-γT control. The expression level of ROR-γt mRNA began to increase in the early stage of *Echinococcus* infection, peaked at 1 month of infection, then decreased slowly, and returned to the level before infection at 10 months of infection. After the mice were infected with *Echinococcus*, Th1 cells and Th17 cells participated in the anti-infection immunity together one month after infection, and the expression level of the transcription factor R0R-γT of Th17 cells was higher than that of TGF-β. Cells may secrete it in large quantities and participate in pathogen clearance. One month after infection, the mRNA levels of transcription factors in Th1 cells and Th17 cells began to decrease, and the expression level of transcription factor GATA-3 in Th2 cells increased. At this time, the immune response showed a state of Th2 polarization, escaped the phenomenon and entered the chronic infection process. Therefore, the determination of Th1, Th2 and Th17 cells from the gene level or the cytokine level shows that after the host is infected with *Echinococcus*, Th1 and Th17 cells may be dominant at the same time in the early and middle stages and have the effect of anti-infection (Lechner et al. 2012). In the middle and late stages of infection, the dominant immune response dominated by Th2 cells gradually appeared, accompanied by the Th17 type cytokine IL-17. The chronic infection of aggravates the immunopathological damage of the body.

### 6.3.7 Treg/Th17 Balance of Host Infected with *Echinococcus*

CD4+ CD25+ FoxP3+ regulatory Treg cells and Th17 cells are as two distinct subsets from Th1 and Th2 cells. Treg cells exert anti-inflammatory effects by expressing forkhead/winged helix transcription factor (Foxp3), through contact-dependent inhibition or release Anti-inflammatory cytokines, IL-10 and TGF-b1 to maintain tolerance to self or external stimuli (Sakaguchi et al. 2006). They play a key role in the immune evasion of human parasitic infections such as *Echinococcosis*, *Leishmaniasis* and *Schistosomiasis* by attenuating the immune response to specific antigens or inhibiting the secretion of pro-inflammatory cytokines (Amri et al. 2009; Stäger et al. 2010; Tang et al. 2011). Treg cells and Th17 cells have opposing effects on inflammation, and the Th17/Treg balance controls inflammation and may play an important role in the pathogenesis of immune evasion. Th17/Treg has different levels of function in patients with Cystic *Echinococcosis* and healthy controls. Compared with the control group, the number of peripheral blood Treg, related cytokines (IL-10 and TGF- b1) and Foxp3 levels were significantly increased, and Th17 numbers, related cytokines (IL-17 and IL-23) and transcription factor (RORct) levels were decreased (Tuxun et al. 2012). The results showed that chronic CE patients had Th17/Treg functional imbalance, suggesting a potential role of Th17/Treg imbalance in the pathogenesis of immune evasion in *Echinococcosis*.

## 6.4 *Echinococcus* Infection and Host Immune Evasion

*Echinococcus* cysts have been reported to survive up to 53 years in humans (Spruance 1974) and up to 16 years in horses (Roneus et al. 1982). The parasite's ability to survive in the host for such a long time and have the potential to resist infection suggests that the parasite possesses strategies to subvert or avoid protective immune responses. Theoretically, there are two mechanisms by which the host immune response can be disrupted: passive evasion, whereby the parasite avoids the damaging effects of the immune response, and immunomodulation, whereby the active interaction with the immune system reduces the impact on the parasite response (Riffkin et al. 1996). Some immune evasion mechanisms used by *Echinococcus* (Wakelin 1997) are listed in Table 6.1. A distinctive feature of *E. granulosus* is the formation of two cysts. One is the cyst-derived acellular laminated capsule. The other is the host fibrous capsule, which typically surrounds fully

**Table 6.1** Proposed mechanisms of avoidance from host-protective responses by the metacestode stage of *Echinococcus*

| Mechanism | Reference(s) |
|---|---|
| Sequestration and antigenic disguise | (Bresson-Hadni et al. 1990; D'Amelio et al. 1989; Dixon 1997; Lightowlers et al. 1984) |
| Molecular mimicry | (Maizels and McSorley 2016; Frosch et al. 1991) |
| Antigen or DNA polymorphism | (Bretagne et al. 1996; Chemale et al. 2001; Chow et al. 2001; Gottstein 1991; Rosenzvit et al. 1999; Monteiro et al. 2008; Zhang et al. 1998, 1999, 2000) |
| Production of proteases | (Diaz et al. 2000; Ferreira et al. 1995; Marco and Nieto 1991; Tort et al. 1999; Willis et al. 1997) |
| Protease inhibitor | (Xu et al. 2022; Shepherd et al. 1991) |
| Interference with complement activity | (Ferreira et al. 1995; Cicardi et al. 1985) |
| Alteration of macrophage and leukocyte function | (Joekel et al. 2021; Annen et al. 1981; Playford and Kamiya 1992; Silva-Alvarez et al. 2016) |
| Alteration of lymphoid organ architecture | (Fotiadis et al. 1999; Ali-Khan 1978; Jiang et al. 2022; Riley and Dixon 1987; Riley et al. 1985) |
| Depletion of T lymphocytes | (Playford and Kamiya 1992; Jenne et al. 1997; Kizaki et al. 1993a, b; Wang et al. 2017) |
| Induction of suppressor cells | (Kizaki et al. 1991, 1993b; Taratuto and Venturiello 1997) |
| Alteration of antibody responses | (Lu et al. 2018; Baz et al. 1995) |
| Alteration of lymphocyte proliferative responses | (Bauder et al. 1999; Emery et al. 1996; Playford and Kamiya 1992; Kizaki et al. 1991; Baz et al. 1995) |
| Inhibition of effector cell chemotaxis | (Playford and Kamiya 1992; Riganò et al. 2001) |

developed viable cysts of *E. granulosus* and is formed probably by filtration of eosinophils (Slais and Vanek 1980) and of fibroblasts and mesothelial cells (Richards et al. 1983). These tissue structures shield the parasite from immune attack. The host's immune response wanes over time, nadir 6 months after infection (Fotiadis et al. 1999). The expression of IL-2 and IL-4 could not be detected in diseased cells infected early (Bauder et al. 1999), possibly due to host immunosuppression. Furthermore, about 30% of CE patients had undetectable antibody levels in serum (Chi et al. 1990; Craig et al. 1996; Riganò et al. 1998), a feature that also appears to occur in sheep infections (Judson et al. 1985; Lightowlers et al. 1986). Parasite circulating antigens can clear specific antibodies, and circulating antigens and immune complexes can be detected in some negative sera (Craig 1986). This suggests that antibody production may be regulated during infection, possibly through periodic antigen release from vesicles and/or through Th cell activation and down-regulation of B cells. C57BL/6 mice were infected with *Echinococcus granulosus*. In the early and late stages of infection, it was found that the proliferation of splenocytes was inhibited, and the proliferation was reversed by adding anti-IFN-γ (Dai and Gottstein 1999). Splenocytes from late-infected mice expressed only basal levels of IL-10, but levels of inducible nitric oxide synthase were greatly increased. The immunosuppression observed in chronic AE is mainly dependent on nitric oxide produced by host macrophages rather than IL-10 (Dai and Gottstein 1999; Emery et al. 1996).

## 6.5 Interaction Between *Echinococcus* and Parasitic Host

It is a gradual process that *Echinococcus* parasitizes the liver and forms space-occupying lesions. At present, in terms of immunology research, the host immune response caused by *Echinococcus* infection in the early stage is mainly Th1, and it turns to Th2 after 3 months. Mainly, *Echinococcus* achieves the purpose of parasitic host through immune evasion. At the same time, the granuloma formed by *Echinococcus* locally in the host body makes the infiltrating cells around the lesion unable to effectively participate in the effector stage of cellular immune response, because the central part of the granuloma Cells secretes more IL-10 and less IFN-γ than peripheral cells. IFN-γ is necessary for effector macrophages to kill *Echinococcus*, and IL-10 can inhibit the activation of macrophages, but at the same time, it will also limit the growth and transfer of AE and protect the body. With persistent infection, *Echinococcus* can cause up-regulation of immune cell-related apoptosis genes C-myc and TGF-β mRNA expression in the parasitic site and down-regulation of anti-apoptosis-related gene Bcl-2 mRNA expression, thereby inducing host mature CD4+ T cells to significantly CD8+ T cells increase, the ratio of CD4+/CD8+ cells decreases rapidly, and finally an immunosuppressive state dominated by CD8+ T cells is formed, resulting in the continued growth and development of *Echinococcus* and maintenance of infection (Zhang et al. 2008; Wang et al. 2013; Tuxun et al. 2015).

In view of the invasive growth of *E. multilocularis* in the host liver, pathological analysis of clinical liver samples found that there was an "active infiltration and proliferation zone" between the liver lesions of AE patients and normal liver tissue, which was considered to be a mixture of a large number of increasing inflammatory cells and liver cells intertwined results. At the same time, there are expressions of pro-inflammatory factors (such as TNF-α, IL-6) and anti-inflammatory factors (such as IL-10, TGF-β) in the "infiltration and proliferation zone" between liver lesions and normal liver tissue in AE patients, in turn, it causes damage to liver cells in or around the "proliferation zone," eventually leading to "swelling" of the host liver, irreversible fibrosis, damage to liver function, and even liver failure (Margos et al. 2011).

## References

Abulizi A, Shao Y, Aji T et al (2019) *Echinococcus multilocularis* inoculation induces NK cell functional decrease through high expression of NKG2A in C57BL/6 mice. BMC Infect Dis 19(1):792

Ai Erken MH, Zhao JM, Guan XY, Wen H, Wang YH (2012) Immune responses on allograft heart transplantation in inbred rats infected with *Echinococcosis multilocularis*. Chin Med J (Engl) 125(24):4412–4417

Ali-Khan Z (1978) Cellular changes in the lymphoreticular tissues of C57L/J mice infected with *Echinococcus multilocularis* cysts. Immunology 34(5):831–839

Allen JE, Wynn TA (2011) Evolution of Th2 immunity: a rapid repair response to tissue destructive pathogens. PLoS Pathog 7(5):e1002003

Amri M, Mezioug D, Touil-Boukoffa C (2009) Involvement of IL-10 and IL-4 in evasion strategies of *Echinococcus granulosus* to host immune response. Eur Cytokine Netw 20(2):63–68

Annen JM, Köhler P, Eckert J (1981) Cytotoxicity of *Echinococcus granulosus* cyst fluid in vitro. Zeitschrift fur Parasitenkunde (Berlin, Germany) 65(1):79–88

Anthony RM, Rutitzky LI, Urban JF Jr, Stadecker MJ, Gause WC (2007) Protective immune mechanisms in helminth infection. Nat Rev Immunol 7(12):975–987

Aratani Y, Koyama H, Nyui S, Suzuki K, Kura F, Maeda N (1999) Severe impairment in early host defense against Candida albicans in mice deficient in myeloperoxidase. Infect Immun 67(4):1828–1836

Bataller R, Brenner DA (2005) Liver fibrosis. J Clin Invest 115(2):209–218

Bauder B, Auer H, Schilcher F et al (1999) Experimental investigations on the B and T cell immune response in primary alveolar echinococcosis. Parasite Immunol 21(8):409–421

Baz A, Hernández A, Dematteis S, Carol H, Nieto A (1995) Idiotypic modulation of the antibody response of mice to *Echinococcus granulosus* antigens. Immunology 84(3):350–354

Bellanger AP, Courquet S, Pallandre JR, Godet Y, Millon L (2020) *Echinococcus multilocularis* vesicular fluid induces the expression of immune checkpoint proteins in vitro. Parasite Immunol 42(6):e12711

Brehm K, Koziol U (2017) Echinococcus-host interactions at cellular and molecular levels. Adv Parasitol 95:147–212

Bresson-Hadni S, Liance M, Meyer JP, Houin R, Bresson JL, Vuitton DA (1990) Cellular immunity in experimental *Echinococcus multilocularis* infection. II. Sequential and comparative phenotypic study of the periparasitic mononuclear cells in resistant and sensitive mice. Clin Exp Immunol 82(2):378–383

Bretagne S, Assouline B, Vidaud D, Houin R, Vidaud M (1996) *Echinococcus multilocularis*: microsatellite polymorphism in U1 snRNA genes. Exp Parasitol 82(3):324–328

Chandrasekhar S, Parija SC (2009) Serum antibody & Th2 cytokine profiles in patients with cystic echinococcosis. Indian J Med Res 130(6):731–735
Chemale G, Haag KL, Ferreira HB, Zaha A (2001) *Echinococcus granulosus* antigen B is encoded by a gene family. Mol Biochem Parasitol 116(2):233–237
Chen YPN, Zhao X, Zhao H, An MT, Yan F, Zhang FB, Ding JB (2019) Detection of IL-9、PU.1、IRF-4 and IL-4 mRNA in patient with cystic echinococcus. Int J Lab Med 40(8):908–912. (in chinese)
Chi P, Zhang W, Zhang Z et al (1990) Cystic echinococcosis in the Xinjiang/Uygur Autonomous Region, People's Republic of China. I. Demographic and epidemiologic data. Trop Med Parasitol 41(2):157–162
Chirag S, Fomda BA, Khan A et al (2015) Detection of hydatid-specific antibodies in the serum and urine for the diagnosis of cystic echinococcosis in patients from the Kashmir Valley, India. J Helminthol 89(2):232–237
Chow C, Gauci CG, Cowman AF, Lightowlers MW (2001) A gene family expressing a host-protective antigen of *Echinococcus granulosus*. Mol Biochem Parasitol 118(1):83–88
Cicardi M, Frangi D, Bergamaschini L, Gardinali M, Sacchi G, Agostoni A (1985) Acquired C1 inhibitor deficiency with angioedema symptoms in a patient infected with *Echinococcus granulosus*. Complement (Basel, Switzerland) 2(2–3):133–139
Craig PS (1986) Detection of specific circulating antigen, immune complexes and antibodies in human hydatidosis from Turkana (Kenya) and Great Britain, by enzyme-immunoassay. Parasite Immunol 8(2):171–188
Craig PS, Rogan MT, Allan JC (1996) Detection, screening and community epidemiology of taeniid cestode zoonoses: cystic echinococcosis, alveolar echinococcosis and neurocysticercosis. Adv Parasitol 38:169–250
Crinier A, Narni-Mancinelli E, Ugolini S, Vivier E (2020) SnapShot: natural killer cells. Cell 180(6):1280–1280.e1281
D'Amelio R, De Rosa F, Pontesilli O et al (1989) Hydatid disease: analysis of parasite antigens in circulating immune complexes and in preformed hydatid antigen-antibody complexes. Med Microbiol Immunol 178(3):177–186
Dai WJ, Gottstein B (1999) Nitric oxide-mediated immunosuppression following murine *Echinococcus multilocularis* infection. Immunology 97(1):107–116
Delunardo F, Ortona E, Margutti P et al (2010) Identification of a novel 19 kDa Echinococcus granulosus antigen. Acta Trop 113(1):42–47
Diaz A, Willis AC, Sim RB (2000) Expression of the proteinase specialized in bone resorption, cathepsin K, in granulomatous inflammation. Mol Med 6(8):648–659
Dixon JB (1997) Echinococcosis. Comp Immunol Microbiol Infect Dis 20(1):87–94
Emery I, Liance M, Deriaud E, Vuitton DA, Houin R, Leclerc C (1996) Characterization of T-cell immune responses of *Echinococcus multilocularis*-infected C57BL/6 J mice. Parasite Immunol 18(9):463–472
Everts B, Smits HH, Hokke CH, Yazdanbakhsh M (2010) Helminths and dendritic cells: sensing and regulating via pattern recognition receptors, Th2 and Treg responses. European J Immunol 40(6):1525–1537
Faulkner H, Turner J, Kamgno J, Pion SD, Boussinesq M, Bradley JE (2002) Age-and infection intensity-dependent cytokine and antibody production in human trichuriasis: the importance of IgE. J Infect Dis 185(5):665–672
Ferreira AM, Wurzner R, Hobart MJ, Lachmann PJ (1995) Study of the in vitro activation of the complement alternative pathway by *Echinococcus granulosus* hydatid cyst fluid. Parasite Immunol 17(5):245–251
Finkelman FD, Shea-Donohue T, Goldhill J et al (1997) Cytokine regulation of host defense against parasitic gastrointestinal nematodes: lessons from studies with rodent models. Ann Rev Immunol 15:505–533
Fotiadis C, Sergiou C, Kyrou I et al (1999) Experimental echinococcus infection in the mouse model: pericystic cellular immunity reaction and effects on the lymphoid organs of immunocompetent and thymectomized mice. In vivo (Athens, Greece) 13(6):541–546

Frosch PM, Frosch M, Pfister T, Schaad V, Bitter-Suermann D (1991) Cloning and characterisation of an immunodominant major surface antigen of *Echinococcus multilocularis*. Mol Biochem Parasitol 48(2):121–130

Fu YMR, Jiang T (2018) Effect of multilocular cyst fluid on IDO expression in mouse bone marrow derived dendritic cells. Chin J Zoonoses 34(11):1001–1005. (in Chinese)

Gottstein B (1991) *Echinococcus multilocularis*: antigenic variance between different parasite isolates. Parasitol Res 77(4):359–361

Grubor NM, Jovanova-Nesic KD, Shoenfeld Y (2017) Liver cystic echinococcosis and human host immune and autoimmune follow-up: a review. World J Hepatol 9(30):1176–1189

Hagan P, Blumenthal UJ, Dunn D, Simpson AJ, Wilkins HA (1991) Human IgE, IgG4 and resistance to reinfection with *Schistosoma haematobium*. Nature 349(6306):243–245

Jenne L, Arrighi JF, Sauter B, Kern P (2001) Dendritic cells pulsed with unfractionated helminthic proteins to generate anti-parasitic cytotoxic T lymphocyte. Parasite Immunol 23(4):195–201

Jenne L, Kilwinski J, Scheffold W, Kern P (1997) IL-5 expressed by CD4+ lymphocytes from *Echinococcus multilocularis*-infected patients. Clin Exp Immunol 109(1):90–97

Jia L, Wu C (2014) Differentiation, regulation and function of Th9 cells. Adv Exp Med Biol 841:181–207

Jiang T, Sun W, Aji T et al (2022) Single-cell heterogeneity of the liver-infiltrating lymphocytes in individuals with chronic *Echinococcus multilocularis* infection. Infect Immun 90(11):e0017722

Joekel DE, Nur S, Monné Rodriguez J et al (2021) Agranulocytosis leads to intestinal *Echinococcus multilocularis* oncosphere invasion and hepatic metacestode development in naturally resistant Wistar rats. Parasitology 148(1):53–62

Judson DG, Dixon JB, Clarkson MJ, Pritchard J (1985) Ovine hydatidosis: some immunological characteristics of the seronegative host. Parasitology 91(Pt 2):349–357

Kaplan MH (2017) The transcription factor network in Th9 cells. Semin Immunopathol 39(1):11–20

Kaplan MH, Hufford MM, Olson MR (2015) The development and in vivo function of T helper 9 cells. Nat Rev Immunol 15(5):295–307

Kizaki T, Ishige M, Bingyan W, Day NK, Good RA, Onoé K (1993b) Generation of CD8+ suppressor T cells by protoscoleces of *Echinococcus multilocularis* in vitro. Immunology 79(3):412–417

Kizaki T, Ishige M, Kobayashi S et al (1993a) Suppression of T-cell proliferation by CD8+ T cells induced in the presence of protoscolices of *Echinococcus multilocularis* in vitro. Infect Immun 61(2):525–533

Kizaki T, Kobayashi S, Ogasawara K, Day NK, Good RA, Onoé K (1991) Immune suppression induced by protoscoleces of *Echinococcus multilocularis* in mice. Evidence for the presence of CD8dull suppressor cells in spleens of mice intraperitoneally infected with *E. multilocularis*. J Immunol (Baltimore, Md: 1950) 147(5):1659–1666

Kubes P, Jenne C (2018) Immune Responses in the Liver. Ann Rev Immunol 36:247–277

Lechner CJ, Gruner B, Huang X, Hoffmann WH, Kern P, Soboslay PT (2012) Parasite-specific IL-17-type cytokine responses and soluble IL-17 receptor levels in Alveolar Echinococcosis patients. Clin Dev Immunol 2012:735342

Licona-Limon P, Arias-Rojas A, Olguin-Martinez E (2017) IL-9 and Th9 in parasite immunity. Semin Immunopathol 39(1):29–38

Lightowlers MW, Rickard MD, Honey RD (1986) Serum antibody response following parenteral immunization with hydatid cyst fluid in sheep infected with *Echinococcus granulosus*. Am J Trop Med Hyg 35(4):818–823

Lightowlers MW, Rickard MD, Honey RD, Obendorf DL, Mitchell GF (1984) Serological diagnosis of *Echinococcus granulosus* infection in sheep using cyst fluid antigen processed by antibody affinity chromatography. Aust Vet J 61(4):101–108

Liu Y, Tian F, Shan J et al (2020) Kupffer cells: important participant of hepatic alveolar Echinococcosis. Front Cell Infect Microbiol 10:8

Long EO, Kim HS, Liu D, Peterson ME, Rajagopalan S (2013) Controlling natural killer cell responses: integration of signals for activation and inhibition. Ann Rev Immunol 31:227–258

Lu LL, Suscovich TJ, Fortune SM, Alter G (2018) Beyond binding: antibody effector functions in infectious diseases. Nat Rev Immunol 18(1):46–61

Ma XJSM, Yin QC (2019) Maturation of dendritic cells in peripheral blood of patients with alveolar echinococcosis. Chin Trop Med 19(2):111–115. (in Chinese)

Maizels RM, McSorley HJ (2016) Regulation of the host immune system by helminth parasites. J Allergy Clin Immunol 138(3):666–675

Marco M, Nieto A (1991) Metalloproteinases in the larvae of *Echinococcus granulosus*. Int J Parasitol 21(6):743–746

Margos MC, Grandgirard D, Leib S, Gottstein B (2011) In vitro induction of lymph node cell proliferation by mouse bone marrow dendritic cells following stimulation with different *Echinococcus multilocularis* antigens. J Helminthol 85(2):128–137

Monteiro KM, Zaha A, Ferreira HB (2008) Recombinant subunits as tools for the structural and functional characterization of *Echinococcus granulosus* antigen B. Exp Parasitol 119(4):490–498

Mukai K, Tsai M, Starkl P, Marichal T, Galli SJ (2016) IgE and mast cells in host defense against parasites and venoms. Semin Immunopathol 38(5):581–603

Murphy KM, Reiner SL (2002) The lineage decisions of helper T cells. Nat Rev Immunol 2(12):933–944

Neurath MF, Kaplan MH (2017) Th9 cells in immunity and immunopathological diseases. Semin Immunopathol 39(1):1–4

Ortona E, Rigano R, Buttari B et al (2003) An update on immunodiagnosis of cystic echinococcosis. Acta Trop 85(2):165–171

Pang NNDJ, Zhao H, Li YJ, Li L, Ma HM (2012) The change characteristics and function of interleukin-17 and interleukin-6 in *echinococcus multilocularis* infection mice. Chinese J Exp Surg 28(12):2131–2133. (in Chinese)

Peng H, Wisse E, Tian Z (2016) Liver natural killer cells: subsets and roles in liver immunity. Cellular Mol Immunol 13(3):328–336

Playford MC, Kamiya M (1992) Immune response to Echinococcus multilocularis infection in the mouse model: a review. Jpn J Vet Res 40(2–3):113–130

Richards KS, Arme C, Bridges JF (1983) *Echinococcus granulosus* equinus: an ultrastructural study of murine tissue response to hydatid cysts. Parasitology 86(Pt 3):407–417

Riesle S, García MP, Hidalgo C, Galanti N, Saenz L, Paredes R (2014) Bovine IgG subclasses and fertility of *Echinococcus granulosus* hydatid cysts. Veterinary Parasitol 205(1–2):125–133

Riffkin M, Seow HF, Jackson D, Brown L, Wood P (1996) Defence against the immune barrage: helminth survival strategies. Immunol Cell Biol 74(6):564–574

Riganò R, Profumo E, Bruschi F et al (2001) Modulation of human immune response by *Echinococcus granulosus* antigen B and its possible role in evading host defenses. Infect Immun 69(1):288–296

Riganò R, Profumo E, Ioppolo S, Notargiacomo S, Teggi A, Siracusano A (1998) Cytokine patterns in seropositive and seronegative patients with *Echinococcus granulosus* infection. Immunol Lett 64(1):5–8

Riley EM, Dixon JB (1987) Experimental *Echinococcus granulosus* infection in mice: immunocytochemical analysis of lymphocyte populations in local lymphoid infections during early infection. Parasitology 94(Pt 3):523–532

Riley EM, Dixon JB, Kelly DF, Cox DA (1985) The immune response to *Echinococcus granulosus*: sequential histological observations of lymphoreticular and connective tissues during early murine infection. J Comp Pathol 95(1):93–104

Rivera Vargas T, Humblin E, Vegran F, Ghiringhelli F, Apetoh L (2017) T(H)9 cells in anti-tumor immunity. Semin Immunopathol 39(1):39–46

Roberts RA, Ganey PE, Ju C, Kamendulis LM, Rusyn I, Klaunig JE (2007) Role of the Kupffer cell in mediating hepatic toxicity and carcinogenesis. Toxicol Sci 96(1):2–15

Roneus O, Christensson D, Nilsson NG (1982) The longevity of hydatid cysts in horses. Vet Parasitol 11(2–3):149–154

Rosenzvit MC, Zhang LH, Kamenetzky L, Canova SG, Guarnera EA, McManus DP (1999) Genetic variation and epidemiology of *Echinococcus granulosus* in Argentina. Parasitology 118(Pt 5):523–530

Sakaguchi S, Ono M, Setoguchi R et al (2006) Foxp3+ CD25+ CD4+ natural regulatory T cells in dominant self-tolerance and autoimmune disease. Immunol Rev 212:8–27

Shepherd JC, Aitken A, McManus DP (1991) A protein secreted in vivo by *Echinococcus granulosus* inhibits elastase activity and neutrophil chemotaxis. Mol Biochem Parasitol 44(1):81–90

Silva-Alvarez V, Folle AM, Ramos AL et al (2016) *Echinococcus granulosus* Antigen B binds to monocytes and macrophages modulating cell response to inflammation. Parasit Vectors 9:69

Slais J, Vanek M (1980) Tissue reaction to spherical and lobular hydatid cysts of *Echinococcus granulosus* (Batsch, 1786). Folia Parasitol (Praha) 27(2):135–143

Spruance SL (1974) Latent period of 53 years in a case of hydatid cyst disease. Arch Intern Med 134(4):741–742

Stäger S, Joshi T, Bankoti R (2010) Immune evasive mechanisms contributing to persistent *Leishmania donovani* infection. Immunol Res 47(1–3):14–24

Stanietsky N, Rovis TL, Glasner A et al (2013) Mouse TIGIT inhibits NK-cell cytotoxicity upon interaction with PVR. European J Immunol 43(8):2138–2150

Tang CL, Lei JH, Wang T et al (2011) Effect of CD4+ CD25+ regulatory T cells on the immune evasion of *Schistosoma japonicum*. Parasitol Res 108(2):477–480

Taratuto AL, Venturiello SM (1997) Echinococcosis. Brain Pathol (Zurich, Switzerland) 7(1):673–679

Tort J, Brindley PJ, Knox D, Wolfe KH, Dalton JP (1999) Proteinases and associated genes of parasitic helminths. Adv Parasitol 43:161–266

Tuxun T, Ma HZ, Apaer S et al (2015) Expression of toll-like receptors 2 and 4 and related cytokines in patients with hepatic cystic and alveolar Echinococcosis. Mediators Inflamm 2015:632760

Tuxun T, Wang JH, Lin RY et al (2012) Th17/Treg imbalance in patients with liver cystic echinococcosis. Parasite Immunol 34(11):520–527

Vuitton DA (2003) The ambiguous role of immunity in echinococcosis: protection of the host or of the parasite? Acta Trop 85(2):119–132

Vuitton DA (2004) Echinococcosis and allergy. Clin Rev Allergy Immunol 26(2):93–104

Vuitton DA, Bresson-Hadni S, Laroche L et al (1989) Cellular immune response in *Echinococcus multilocularis* infection in humans. II. Natural killer cell activity and cell subpopulations in the blood and in the periparasitic granuloma of patients with alveolar echinococcosis. Clin Exp Immunol 78(1):67–74

Wakelin D (1997) Immune response to Echinococcus infection: parasite avoidance and host protection. Parassitologia 39(4):355–358

Wang DXWH, Fan HN (2018) Study on the role of macrophage polarization during *E. multilocularis*-infection in mice. Chin High Alt Med Biol 39(2):118–122. (in Chinese)

Wang B, Liang S, Wang Y et al (2015a) Th17 down-regulation is involved in reduced progression of schistosomiasis fibrosis in ICOSL KO mice. PLoS Negl Trop Dis 9(1):e0003434

Wang J, Müller S, Lin R et al (2017) Depletion of FoxP3(+) Tregs improves control of larval *Echinococcus multilocularis* infection by promoting co-stimulation and Th1/17 immunity. Immun Inflamm Dis 5(4):435–447

Weijian E, Wang Z, Pang M, Lu Y, Fan H (2021) The Correlation Between Platelet-to-Lymphocyte Ratio and Neutrophil-to-Lymphocyte Ratio with Hepatic Echinococcosis. J Inflamm Res 14:2403–2409

Wang J, Zhang C, Wei X et al (2013) TGF-β and TGF-β/Smad signaling in the interactions between *Echinococcus multilocularis* and its hosts. PloS One 8(2):e55379

Wang Y, Zhou H, Shen Y et al (2015b) Impairment of dendritic cell function and induction of CD4(+)CD25(+)Foxp3(+) T cells by excretory-secretory products: a potential mechanism of immune evasion adopted by *Echinococcus granulosus*. BMC Immunol 16:44

Weatherhead JE, Gazzinelli-Guimaraes P, Knight JM et al (2020) Host immunity and inflammation to pulmonary helminth infections. Front Immunol 11:594520

Wei XL, Xu Q, Rexiti FL, Zhu M, Lin RY, Wen H (2014) Dynamic changes of DC and T cell subsets in mice during *Echinococcus multilocularis* infection. Cent Eur J Immunol 39(1):19–24

Weijian E, Yongliang L, Bingmin Q (2021) Association between serum macrophage polarization-related factors and liver fibrosis in echinococcosis multilocularis. J Clin Hepatol 37(12):2813–2818

Wen H (2015) Hydatid disease. People's Medical Publishing House, Beijing (in Chinese)

Willis AC, Díaz A, Nieto A, Sim RB (1997) *Echinococcus granulosus* antigen 5 may be a serine proteinase. Parasite Immunol 19(8):385

Wynn TA (2008) Cellular and molecular mechanisms of fibrosis. J Pathol 214(2):199–210

Xu Y, Pang NN, Guo ZS (2016) The dynamic changes of CD11C+ CD45RA- myeloid dendritic cells in the mice infected with *Echinococcus multilocularis*. J Xinjiang Med Univ 39(5):565–568. (in Chinese)

Xu X, Qian X, Gao C et al (2022) Advances in the pharmacological treatment of hepatic alveolar echinococcosis: From laboratory to clinic. Front Microbiol 13:953846

Xu KWH, Zhang L (2021) Research progress on the mechanisms underlying the impairment of host hepatocytes by *Echinococcus multilocularis*. Chin J Parasitol Parasit Dis 39(2):256–260. (in Chinese)

Xuan W, Qu Q, Zheng B, Xiong S, Fan GH (2015) The chemotaxis of M1 and M2 macrophages is regulated by different chemokines. J Leukocyte Biol 97(1):61–69

Yang F, Feng C, Zhang X, Lu J, Zhao Y (2017) The diverse biological functions of neutrophils, beyond the defense against infections. Inflammation 40(1):311–323

Yu J, Song Y, Tian W (2020) How to select IgG subclasses in developing anti-tumor therapeutic antibodies. J Hematol Oncol 13(1):45

Zhang Q, Bi J, Zheng X et al (2018) Blockade of the checkpoint receptor TIGIT prevents NK cell exhaustion and elicits potent anti-tumor immunity. Nat Immunol 19(7):723–732

Zhang L, Eslami A, Hosseini SH, McManus DP (1998) Indication of the presence of two distinct strains of *Echinococcus granulosus* in Iran by mitochondrial DNA markers. Am J Trop Med Hyg 59(1):171–174

Zhang L, Gasser RB, Zhu X, McManus DP (1999) Screening for different genotypes of *Echinococcus granulosus* within China and Argentina by single-strand conformation polymorphism (SSCP) analysis. Trans R Soc Trop Med Hyg 93(3):329–334

Zhang S, Hue S, Sene D et al (2008) Expression of major histocompatibility complex class I chain-related molecule A, NKG2D, and transforming growth factor-beta in the liver of humans with alveolar echinococcosis: new actors in the tolerance to parasites? J Infect Dis 197(9):1341–1349

Zhang LH, Joshi DD, McManus DP (2000) Three genotypes of *Echinococcus granulosus* identified in Nepal using mitochondrial DNA markers. Trans R Soc Trop Med Hyg 94(3):258–260

Zhang C, Lin R, Li Z et al (2020) Immune Exhaustion of T Cells in Alveolar Echinococcosis Patients and Its Reversal by Blocking Checkpoint Receptor TIGIT in a Murine Model. Hepatology (Baltimore, Md) 71(4):1297–1315

Zhang C, Wang H, Li J et al (2021) Involvement of TIGIT in Natural Killer Cell Exhaustion and Immune Escape in Patients and Mouse Model With Liver *Echinococcus multilocularis* Infection. Hepatology (Baltimore, Md) 74(6):3376–3393

Zhao HPN, Ma HM, Li YJ, Zhang L, Liu XX (2012) Balance changes of Tregs and Th17 cell-related cytokines in mice infected with *echinococcus multilocularis*. Chinese J Pathog Biol 7(2):129–131. (in Chinese)

Zhu J (2018) T Helper Cell Differentiation, Heterogeneity, and Plasticity, vol 10. Cold Spring Harb Perspect Biol, p 10

Zhu J, Paul WE (2008) CD4 T cells: fates, functions, and faults. Blood 112(5):1557–1569

Zhu J, Yamane H, Paul WE (2010) Differentiation of effector CD4 T cell populations. Ann Rev Immunol 28:445–489

# Pathology and Clinical Features of Echinococcosis

# 7

Haining Fan

**Abstract**

Echinococcosis includes cystic echinococcosis (CE) and alvealar echinococcosis (AE), and the clinical and pathological features are completely different. The diagnostic techniques of two types of echinococcosis mainly include serological detection and imaging detection. AE is invasive and metastasis to other organs, such as lung, brain, and bone. The clinical presentation includes abdominal pain, jaundice, and portal hypertension. The prognosis of patients without treatment is poor. Germinal cells are contained in the cyst wall of CE, which is the main site of parasite development. CE grows continuously and causes mechanical compression for surrounding tissues. Cyst fluid of CE can cause allergic reactions, and some cysts are associated with bile leakage. The prognosis of CE is better than AE.

**Keywords**

Cystic echinococcosis · Alveolar echinococcosis · Pathology · Clinical feature

The signs and symptoms of echinococcosis depend on the organ involved, the pathogen and the size of the cyst. Cyst echinococcosis (CE) and alveolar echinococcosis (AE) are caused by the species *Echinococcus granulosus* and *Echinococcus multilocularis*. However, clinically, in terms of the clinical manifestations, the course of the disease, and the prognosis of the disease, AE and CE are completely different.

H. Fan (✉)
Affiliated Hospital of Qinghai University, Xining, China
e-mail: fanhaining@medmail.com.cn





J. Li et al. (eds.), *Echinococcus: Control and Elimination of Echinococcosis with a Focus on China and Europe*, Parasitology Research Monographs 19,
https://doi.org/10.1007/978-3-031-54479-8_7

## 7.1 Pathology and Clinical Features of Alveolar Echinococcosis

### 7.1.1 Pathological Feature of Alveolar Echinococcosis

A typical lesion appears as macroscopical fibrous tissue with a conglomerate of scattered vesicles. The necrosis of tissues surrounds alveolar structures with a layer of inflammation as it reacts to the foreign body, which does not have defined limits with the adjacent liver tissues (Coyle and Junghanss 2020). Active parasitic tissues can be observed around the lesion. Older lesions often show scattered calcification, especially in hepatic lesions. The microscopic appearance of AE is typically seen with irregular cysts, such as those with diameters of less than 30 mm. Most of the alveolar lesions show a PAS-positive laminated layer and an inner germinal layer of cyst walls. The germinal layer may generate new lesions and then exhibit a tumor-like growth, with the potential for invasion and distant metastasis (Yimingjiang et al. 2023). After an infection of AE, without clinical intervention, the prognosis of most patients is not ideal (Koch et al. 2003).

### 7.1.2 Clinical Feature of Alveolar Echinococcosis

The liver is the primary organ in humans and natural intermediate hosts that is affected by AE (Kern et al. 2003); the right lobe of the liver is most commonly invaded (Stojkovic et al. 2014). Calcification can be seen in the lesion and in the proliferating vesicles. In the early stages, symptoms such as upper-quadrant pain or discomfort are not significant. In patients with a hepatic AE lesion, most patients become symptomatic when the disease advances. The initial symptoms include jaundice, epigastric pain, fever, weight loss, and fatigue. Then, the dysfunction of liver and portal hypertension are often associated with the progressive growth of the lesion. Without treatment, the main cause of death is hepatic failure, biliary obstruction, AE metastasis, or complications (Bresson-Hadni et al. 2000). In cases also featuring biliary fistula, patients are at risk of developing bacterial cholangitis (Hosch et al. 2008). What is more, Budd–Chiari syndrome, inferior vena cava thrombosis, biliary leakage, spontaneous or traumatic rupture, and other serious adverse events can also be found in patients with AE (Wang et al. 2021; Yang et al. 2017; Yamamoto et al. 2022). In spinal AE, most cases have presented with repetitive neck and back pain, and surgeons often attribute this to spinal tuberculosis first (Song et al. 2022). Turgut et al. wrote in a summary that patients with spinal echinococcosis experienced muscle weakness of the limbs, back pain, bowel and bladder dysfunction, and variable sensory disturbances (Turgut 1997). As AE may mimic tumor-like growth, in cases with cerebral invasion, patients may present with headaches and dizziness. If the patient has a history of AE and CT or MRI of the

**Table 7.1** Imaging indications of AE

| | US | CEUS | CT | MRI | PET-CT |
|---|---|---|---|---|---|
| Signs and utilization | (i) Lesions with clear or unclear boundaries<br>(ii) Biliary dilatation or vascular involvement | Poor blood-supply lesion | (i) The number, size, and location of and the change in vascular and biliary structures<br>(ii) Lesions with clear or unclear boundaries<br>(iii) Lesions may invade local organs | (i) MRI can be used for evaluating parasitic activity<br>(ii) Higher sensitivity and specificity than CT when determining the number, size, location, and boundaries of lesions | Evaluating and monitoring disease activity |
| Note | (i) Lack of large-sample clinical assessment and diagnostic studies<br>(ii) Operator dependent | Same as US | May be misdiagnosed as liver hemangioma, liver cancer, or another disease | (−) | (−) |

*US* ultrasound; *CEUS* contrast-enhanced ultrasound; *MRI* magnetic resonance imaging; *PET-CT* positron emission computed tomography

brain shows solid or multilocular cystic masses with clear boundaries, accompanied by calcification and peripheral edema, cerebral invasion should be highly suspected (Cakir et al. 2022).

In the B-scan sonography, AE lesions show highly diverse sonomorphological characteristics. Echinococcosis multilocularis Ulm classification-ultrasound (EMUC-US) is used to illuminate five B-scan sonographic patterns: hailstorm-like, pseudocystic, ossification-like, hemangioma-like, and metastasis-like (Kratzer et al. 2015). Magnetic resonance imaging (MRI) and computed tomography (CT) also play important roles in evaluating parasitic activity and diagnosing diseases. Table 7.1 lays out the imaging indications of AE.

The primary infection is only rarely found in organs other than the liver. Lung metastasis, brain metastasis, and cerebral metastasis have been reported in some cases (Schweizer et al. 2022; Kern 2010; Liu et al. 2021; Xu et al. 2022; Ju and Liu 2023). However, because a limited number of cases have been reported, the symptoms of AE metastasis exhibit significant differences, and their clinical manifestations require more research to investigate. When metastasis occurs, the symptoms caused by the primary site are often serious.

## 7.2 Pathology and Clinical Features of Cyst Echinococcosis

### 7.2.1 Pathological Feature of Cyst Echinococcosis

The cyst is fluid filled and spherical, which consists of an inner germinal layer of cells supported by an acellular, laminated membrane of variable thickness. Histologically, the thin spherical germinal layer is the primary site of parasite development. A layer of granulomatous reaction appears outside the cyst, which contains mucins and calcium inositol hexakisphosphate deposits (Díaz et al. 2011). The content of the cyst depends on the stage of the cyst's development, and the calcification may increase as the cyst grows. Similar to AE, most of the time, a CE lesion exhibits a PAS-positive laminated layer.

### 7.2.2 Clinical Feature of Cyst Echinococcosis

The symptoms vary depending on the primarily infected organ and the size of the cyst. The most common site is the liver, followed by the lungs. In patients who have liver cysts, abdominal pain is the most common symptom, and the right lobe of the liver is most commonly affected (Stojkovic et al. 2014). As the lesion grows, it compresses the liver tissue and causes portal hypertension and biliary obstruction, among other problems. In some cases, cystobiliary fistula, biliary leakage, spontaneous or traumatic rupture, and Budd–Chiari syndrome can be observed (Said-Ali et al. 2013; Bruzinskaite et al. 2007). The cyst can be infected by various bacteria, and therefore, some patients will be hospitalized because of an infection with cystic lesions. Generally, the prognosis of CE is much better than that of AE. But if the disease is not treated in time, it may also lead to the death of the patient. In patients with lung cysts, the symptoms are not usually significant, and the lesion is always found under routine chest radiography. As the cyst grows, chest pain with cough, dyspnea, and hemoptysis are typical in some patients (Díaz et al. 2011; Nono et al. 2012). When the lesion compresses the bronchus, pneumonia can be observed. The rupture of the lesions may lead to a lung abscess and bacterial overgrowth. Bone CE is a challenge because of the insidious course of the disease. Some researchers have reported cases of scapula and clavicular CE. In these cases, patients may present with the swelling of their bones and sinuses, causing pain. X-ray and CT images may reveal cystic masses and destruction from the swelling bone (Wang et al. 2023; Verghese et al. 2022). Moreover, CE can also be seen in the muscles and the uterine system, and those cases should be kept in mind as possible locations of this disease (Yağmur et al. 2022; Arian and Kazerani 2022). CE has also been reported in organs such as the spleen, brain, and almost every other organ (Thakar and Sunil 2023; Kılınç et al. 2022). The imaging indications of CE can be found in Table 7.2.

**Table 7.2** Imaging indications of CE

| | US | CEUS | CT | MRI | PET-CT |
|---|---|---|---|---|---|
| Signs and utilization | Evaluation of WHO-IWG classification and preliminarily evaluating disease activity | No indication | Evaluating the number, size, location, and boundary of and any changes in the vascular and biliary structures | Higher sensitivity and specificity than those of CT | No indication |
| Note | (i) Lack of large-sample clinical assessment and diagnostic study (ii) Operator dependent | (−) | Not suitable for staging the disease | (−) | (−) |

*US* ultrasound; *CEUS* contrast-enhanced ultrasound; *MRI* magnetic resonance imaging; *PET-CT* positron emission computed tomography

## References

Arian M, Kazerani M (2022) Primary hydatid cyst in the adductor muscles of thigh: a case report. Clin Case Rep 10(12):e6664

Bresson-Hadni S, Vuitton DA, Bartholomot B et al (2000) A twenty-year history of alveolar echinococcosis: analysis of a series of 117 patients from eastern France. Eur J Gastroenterol Hepatol 12(3):327–336

Bruzinskaite R, Marcinkute A, Strupas K et al (2007) Alveolar echinococcosis, Lithuania. Emerg Infect Dis 13:1618–1619

Cakir IM, Bekci T, Eryuruk U (2022) A case of disseminated alveolar echinococcosis mimicking metastatic malignant disease with cerebral involvement. Rev Soc Bras Med Trop 55:e02842022

Coyle CM, Junghanss T (2020) 133–alveolar echinococcosis (alveolar hydatid disease). In: Hunter's tropical medicine and emerging infectious diseases, 10th edn. Elsevier, pp 954–956

Díaz A, Casaravilla C, Allen JE et al (2011) Understanding the laminated layer of larval Echinococcus II: immunology. Trends Parasitol 27(6):264–273

Hosch W, Stojkovic M, Jänisch T et al (2008) MR imaging for diagnosing cysto-biliary fistulas in cystic echinococcosis. Eur J Radiol 66(2):262–267

Ju H, Liu C (2023) Cerebral Alveolar Echinococcosis. N Engl J Med 388(5):453

Kern P (2010) Clinical features and treatment of alveolar echinococcosis. Curr Opin Infect Dis 23:505–512

Kern P, Bardonnet K, Renner E et al (2003) European echinococcosis registry: human alveolar echinococcosis, Europe, 1982-2000. Emerg Infect Dis 9(3):343–349

Kılınç F, Çay Ü, Gündeşlioğlu ÖÖ, Alabaz D, Oktay K, OPehlivanktay UA (2022) Recurrence from the spinal region of the patient whose treatment was completed with liver and lung cystic echinococcosis: a rare pediatric case of spinal cystic echinococcosis. Turkiye Parazitol Derg 46(3):246–248

Koch S, Bresson-Hadni S, Miguet JP et al (2003) Experience of liver transplantation for incurable alveolar echinococcosis: a 45-case European collaborative report. Transplantation 75:856–863

Kratzer W, Gruener B, Kaltenbach TE et al (2015) Proposal of an ultrasonographic classification for hepatic alveolar echinococcosis: echinococcosis multilocularis Ulm classification-ultrasound. World J Gastroenterol 21(43):12392–12402

Liu C, Fan H, Ge RL (2021) A case of human hepatic alveolar echinococcosis accompanied by lung and brain metastases. Korean J Parasitol 59(3):291–296

Nono JK, Pletinckx K, Lutz MB et al (2012) Excretory/ secretory-products of Echinococcus multilocularis larvae induce apoptosis and tolerogenic properties in dendritic cells in vitro. PLoS Negl Trop Dis 6(2):e1516

Said-Ali Z, Grenouillet F, Knapp J et al (2013) FrancEchino network: detecting nested clusters of human alveolar echinococcosis. Parasitology 140(13):1693–1700

Schweizer M, Schmidberger J, Schlingeloff P, Kratzer W (2022) Contrast-enhanced ultrasound (CEUS) in patients with metastasis-like hepatic alveolar echinococcosis: a cohort study [published online ahead of print, 2022 May 21]. J Ultrasound:111

Song T, Peng S, Zhou X, Jiang L, Zhang J (2022) Case report: diagnosis of vertebral alveolar echinococcosis upon next-generation sequencing in a suspected tuberculosis. Front Surg 9:984640

Stojkovic M, Gottstein B, Junghanss T (2014) Echinococcosis. In: Manson's tropical diseases, 23rd edn. W.B. Saunders, London, pp 795–819.e3

Thakar S, Sunil A (2023) Cerebral cystic echinococcosis. N Engl J Med 388(5):e10

Turgut M (1997) Hydatid disease of the spine: a survey study from Turkey. Infection 25:221–226

Verghese SC, George ST, Duraisamy P, Karthikeyan TM, Kumar SA, Arasu S (2022) A rare case report of primary hydatid disease of the scapula–an unforeseen diagnosis. J Orthop Case Rep 12(8):98–101

Wang X, Huang J, Su L, Ma Q, Ma C, Xie Z (2023) Complete excision of giant clavicular hydatid cyst: a case report. BMC Infect Dis 23(1):178

Wang C, Qiu Y, Wang W (2021) Application of ex vivo liver resection and autotransplantation in treating Budd-Chiari syndrome secondary to end-stage hepatic alveolar echinococcosis: a case series. Medicine (Baltimore) 100(34):e27075

Xu X, Gao C, Qian X et al (2022) Treatment of complicated hepatic alveolar echinococcosis disease with suspicious lymph node remote metastasis near the inferior vena cava-abdominal aorta: a case report and literature review. Front Oncol 12:849047

Yağmur I, Kocaman OH, Dere O, Demir M, Katı B, Boleken ME (2022) Multiorgan echinococcosis with uterine involvement causing bilateral Hydronephrosis in a child: case report. Iran J Parasitol 17(3):425–430

Yamamoto Y, Sakamoto Y, Kamiyama T et al (2022) A case of alveolar echinococcosis in the liver that ruptured into the pericardium treated by a combination of hepatectomy and albendazole. Surg Case Rep 8(1):63

Yang X, Qiu Y, Wang W et al (2017) Risk factors and a simple model for predicting bile leakage after radical hepatectomy in patients with hepatic alveolar echinococcosis. Medicine (Baltimore) 96(46):e8774

Yimingjiang M, Aini A, Tuergan T, Zhang W (2023) Differential gene expression profiling in alveolar echinococcosis identifies potential biomarkers associated with angiogenesis. Open forum Infect Dis 10(2):ofad031

# Diagnosis of Echinococciasis: Parasitological, Immunological, and Molecular Tools

**8**

Jian Li

**Abstract**

Echinococcosis is a serious zoonotic disease caused by the larvae of several *Echinococcus* species forming a hydatid cyst, which can parasitize humans and animals. Diagnosis is a prerequisite for disease prevention and treatment. Common diagnostic methods include epidemiological investigation, imaging diagnosis, parasitological detection, immunological testing, and molecular diagnosis. In the current chapter, the diagnostic techniques, including parasitological detection, immunological testing, and molecular diagnosis, were reviewed and discussed.

**Keywords**

*Echinococcus* · Laboratory diagnosis · Morphology · Immunodiagnosis · Molecular diagnosis

## 8.1 Introduction

Human echinococcosis (hydatidosis, or hydatid disease) is a zoonotic infection caused by larval stages (metacestodes) of cestode species of the genus *Echinococcus*. The disease is widely distributed worldwide, and it continuously spreads with the development of animal husbandry, becoming the most important global public health and economic problem (Agudelo Higuita et al. 2016). In China, echinococcosis is regarded as the most vital zoonotic parasitic disease, and the annual direct pecuniary loss is more than 3 billion yuan (Guan et al. 2019).

J. Li (✉)
Department of Human Parasitology, School of Basic Medical Sciences, Hubei University of Medicine, Shiyan, People's Republic of China





J. Li et al. (eds.), *Echinococcus: Control and Elimination of Echinococcosis with a Focus on China and Europe*, Parasitology Research Monographs 19,
https://doi.org/10.1007/978-3-031-54479-8_8

Currently, five *Echinococcus* spp. can cause echinococcosis. It includes *Echinococcus granulosus*, *Echinococcus multilocularis*, *Echinococcus vogeli*, *Echinococcus oligarthrus*, and *Echinococcus shiquicus*. Of these, cystic echinococcosis (CE) is produced by the larval stage of *E. granulosus* (sensu lato). It is the form most frequently encountered and recognized as a serious public health problem in China. For *E. granulosus*, individuals are aberrant intermediate hosts and become infected by ingesting infective eggs. Six-hooked oncospheres are released in the intestine, penetrate the intestinal wall, and migrate through the circulatory system into various organs. Subsequently, hydatid cysts develop in a variety of organs, especially organs including the liver and double lungs. If the cysts rupture, the liberated protoscolices may generate secondary cysts in tissues and organs of the host (secondary echinococcosis). *E. granulosus* is distributed globally, predominantly in pastoral and rural areas of Africa, the Middle East, the Mediterranean, Europe, Central Asia, South America and Western China (Deplazes et al. 2017). Alveolar echinococcosis (AE) is caused by the larval stage of *E. multilocularis,* which is becoming increasingly common. In untreated cases, AE has a high mortality rate. For *E. multilocularis*, humans are aberrant intermediate hosts and become infected by ingesting eggs. Oncospheres are released in the intestine, and cysts develop within the liver. Metastasis or dissemination to other organs (e.g., lungs, brain, heart, bone) may occur if protoscolices are released from cysts, occasionally also called "secondary echinococcosis." *E. multilocularis* is mainly distributed in the northern hemisphere, and the main epidemic area is the Tibetan Plateau in China. It is worth mentioning that *E. shiquicus* was recently identified as a new species discovered only in Shiqu County on the Tibetan Plateau of China (Jin et al. 2021; Zhu et al. 2019; Li et al. 2018). *E. multilocularis* and *E. shiquicus* are infrequently detected in China (https://www.cdc.gov/dpdx/echinococcosis/).

Diagnostic methods for echinococcosis include epidemiological surveys, clinical sign examinations, pathogen detection, molecular detection, serological examinations, and imageological examinations, including B-scan ultrasonography, magnetic resonance imaging, and computerized tomography. In this chapter, the progress and application of parasitological detection, immunological testing, and molecular detection for *Echinococcus* have been reviewed and discussed.

## 8.2 Parasitological Tools

For the diagnosis of parasitic diseases, including echinococcosis, etiological examination methods are regarded as the gold standard and can be used as important evidence for confirmation. The diagnosis of echinococciasis should be based on an etiological examination. In particular, if stages of *Echinococcus* or protocephalococcus were detected from surgical extract, sputum, pleural effusion or ascites, the disease can be confirmed as echinococcosis.

For patients with pulmonary echinococcosis, the wall of hydatid cysts, daughter cysts, even granddaughter cysts, protoscoleces or partial apical hooks of protoscoleces may be found after cyst rupture in the sputum. Although the cystic wall and daughter cyst can be detected easily by the naked eye of humans, they still need to be identified by histological examination to reconfirm. Patients with hepatic echinococcosis can be examined by fine needle puncture guided by B-ultrasound or by surgical removal of echinococcosis. It is important to note that the fluid should be prevented from oozing to prevent secondary echinococcosis or allergic reactions during diagnostic puncture. Although the liver puncture technique can effectively diagnose the disease, patient compliance is not high, and it requires professional and experienced medical personnel to operate. It also has a risk of a secondary infection, so these problems limit its application.

As definitive hosts, adult worms parasitize in the small intestine. The gravid proglottid and eggs rupturing from gravid proglottid can be detected in feces by microscopy. It is worth noting that there is a risk of infection during surgery. For egg detection, eggs of *Echinococcus* spp. are very similar to other tapeworm eggs and are not easily distinguished based on information on morphology, size, color, contents, and eggshell. In practice, gravid proglottids are not easily detected in excreted feces.

For etiological examination, there are several difficulties in the diagnosis process for echinococcosis, especially in the early stages of infection when there are no obvious clinical symptoms. Furthermore, the complicated examination process limits its use in the field and clinical diagnosis. It is challenging to apply for the detection of early infection as well as animal infection for time-consuming, complicated operation and high requirements for operators (Grond et al. 2017). Therefore, other convenient methods, tools and techniques are needed to replace or complement this method.

## 8.3 Immunological Tools

For *Echinococcus* spp. parasites and echinococciasis, immunological tests are often used to aid diagnosis and epidemiological investigation. It mainly detects antigens, which include crude antigen, purified antigen, recombinant antigen, and circulating antigen, as well as antibodies, including IgE and IgG, from human serum. The detailed information is as follows.

### 8.3.1 Diagnostic Antigen

High-quality antigens can guarantee and improve the sensitivity and specificity of diagnosis. At present, the commonly used antigens for detection include crude antigen and purified antigen of hydatid cystic fluid, specific antigen, crude antigen and purified antigen of protoscolex. Ag5 and AgB are the main antigens in both hydatid cystic fluid and protoscolex. These antigens, including EgCF, Acr5 and EgB, have potential diagnostic value for CE. For AE, Em2, Em18, Em2plus, Em10, Em13, EmII/3, and RecEm18 are considered candidates for serological diagnosis (Feng et al. 2010).

Crude antigen and purified antigen from cystic fluid have the disadvantages of cross-reaction with other tapeworms, causing teniasis, and are not suitable for large-scale production. These deficiencies limit the scope of its use. Because recombinant proteins and synthetic peptides have the advantages of high sensitivity and specificity, low cost for scale production, and high yield, they have been widely used in clinical diagnosis and field investigation in recent years.

### 8.3.2 Diagnostic Methods

For serodiagnosis of *Echinococcus* spp. parasites and echinococciasis, the common methods are described as follows.

1. Enzyme-linked immunosorbent assay (ELISA)

   Serological testing is an important diagnostic method for *E. granulosus* infection in domestic animals. ELISA is the most effective, sensitive, and specific commercial kit available for serological diagnosis. In a previous study, the authors used the fusion protein *E. granulosus* mefAg-1 to evaluate CE in sheep by indirect ELISA. The findings suggest that the fusion protein could be a potential antigen candidate for CE serodiagnosis in sheep (Tianli et al. 2019). In another study, they characterized *E. granulosus* glutaredoxin 1 (Eg-Grx1) and developed an indirect ELISA (iELISA) for CE serodiagnosis. The results showed that the specificity and sensitivity were 64.3% and 1:3200, respectively. This demonstrated that the established iELISA method may contribute to the surveillance of sheep CE in epidemic areas (Song et al. 2016).

   ELISA is the most recommended method for the detection of antigens or antibodies. However, it is still cumbersome and requires professional personnel and professional equipment. Therefore, there is an urgent need for simple and fast methods that do not require complex procedures. This would satisfy the requirement of POCT.
2. Lateral chromatographic strips (LFS)

   Strips are based on colloidal gold as the color medium, and the use of immunological antigen antibodies can be a specific binding principle in the process of chromatography to complete this reaction and achieve the purpose of detection. Compared with other methods, it can avoid the interference of endogenous enzymes, the destruction of antigens by hydrogen peroxide and the influence of carcinogens in enzymatic methods on the human body (Gao et al. 2012).

   Gao Chunhua et al. established a colloidal gold immunochromatographic test strip method for convenient and rapid CE diagnosis, and ELISA conducted parallel detection studies. The results showed that the test strip method detected serum sensitivity of CE patients was 85.8%. The total specificity was 94%. Compared to ELISA, there was no significant difference.
3. Dot immunogold filtration assay (DIGFA)

   Clinical diagnosis and postsurgical assessment of CE depend on laboratory serodiagnosis and ultrasound examinations. A previous study used DIGFA and recombinant antigen (rAgB) to compare their diagnostic effects with those of natural antigens (crude fluid antigen, protoscolex antigen). This demonstrated that rAgB can improve the specificity for serodiagnosis of CE (Chen et al. 2015). In another study, they developed a platform with DIGFA and four native antigens (EgCF, AgB, EgP, and Em2). The overall sensitivity was 80.7% for CE and 92.9% for AE. The developed rapid test can be used for both clinical diagnostic support and large-scale screening in endemic CE and AE areas (Feng et al. 2010).

4. Latex agglutination test (LAT)

   A previous study used LAT with native 8-kDa antigen to detect the CE in buffaloes. The diagnostic sensitivity was 98.67%. This demonstrated that LAT should be convenient for hydatid cyst infection (Sheeba et al. 2016).

Immunological detection is common for screening parasites. Cross-reaction is one of its disadvantages. Therefore, it can only be used as an auxiliary method. The disease needs to be diagnosed in conjunction with other methods, including molecular diagnosis and etiological examination.

## 8.4 Molecular Tools

For infectious diseases, including parasitosis, molecular diagnosis can also be an important basis for confirmation. For diagnosis, it has the same important status as etiological examination. Furthermore, it has become increasingly used in the diagnosis and detection of parasites and parasitic diseases due to its unique advantages, including small sample size, high sensitivity and specificity, and fast response. Molecular diagnosis can be roughly divided into the following categories: (1) polymerase chain reaction (PCR)-based methods, including conventional PCR, nested PCR, multiple-PCR, restriction fragment length polymorphism (RFLP-PCR), quantitative real-time PCR, and droplet digital PCR (ddPCR); (2) isothermal amplification methods, including loop-mediated isothermal amplification (LAMP), recombinase aided isothermal amplification assay (RAA) or recombinase polymerase amplification (RPA). The detailed information is as follows.

1. PCR-based methods

   PCR is widely used in nucleic acid detection because of its high sensitivity, specificity, and timeliness (Li and Makrigiorgos 2009; Garibyan and Avashia 2013).

### PCR

Zan et al. used the ND6 gene as the target gene to develop a molecular method for *E. granulosus* in dogs by PCR. The analytical sensitivity was 4 pg of DNA, and only *E. granulosus* could be successfully amplified (Jia-fei et al. 2019).

**Real-Time PCR**
Wang et al. developed an efficient and specific detection method for *E. granulosus* by TaqMan real-time PCR. The assay showed a good linear relationship, and the correlation coefficient reached 0.998. The sensitivity reached $10^2$ copies/μL (Jia-fei et al. 2019).

**Degenerate Primer q-PCR**
Wang et al. used the cytochrome b gene as the target gene to establish a molecular method of detecting *E. granulosus* by degenerate primer q-PCR. The PCR had good specificity and no cross-amplifications of the host. This result indicated that PCR offers technical support for hydatid disease detection (Bing-jie et al. 2021).

2. Isothermal Amplification Methods

   Isothermal amplification can be used for nucleic acid amplification at constant temperature and simple conditions. For hydatid disease, the methods are displayed as follows.

**Loop-Mediated Isothermal Amplification (LAMP)**
Xiangzhen Xu et al. used LAMP to evaluate the sensitivity of the detection of *E. granulosus* parasites with 12S rRNA. The detection limit was 1 egg/reaction. This illustrates that the LAMP assay is a simple method for hydatid disease pathogen detection (Xu et al. 2011).

**Recombinase Aided Isothermal Amplification Assay (RAA)**
Yang Dai et al. established a nucleic acid assay for the detection of *E. granulosus* based on a recombinase-aided isothermal amplification assay (RAA) with 12S rRNA as the target. The lowest detection limit was 10 copies/μL for plasmid and 0.1 μg/μL for genomic DNA. There was no cross-reaction with other parasites (Ding et al. 2020). Ning Xiao et al. developed a sensitive, specific and rapid tool for identifying *E. granulosus* G1 based on RAA. The established method can amplify genomic DNA within 40 min, and the minimum detection limit was 10 pg, which is consistent with that of conventional PCR (Zhou et al. 2020).

For molecular diagnosis, it can accurately diagnose specific diseases but still relies on equipment and specialized technicians to operate it. Thus, it is suitable for areas with good health and medical conditions.

## 8.5 Conclusions

Over the past few decades, the development of multiple technologies has spawned many new detection techniques and supported diagnostic development. For the diagnosis of parasites (*Echinococcus* spp.) and parasitic diseases (Echinococcosis), there are many tests, each of which has advantages but also disadvantages. In practical applications, reasonable selection should be made according to actual application scenarios, and sometimes, the combination of several methods, including etiological diagnosis, immunological diagnosis and molecular diagnosis, must be considered to improve the diagnostic timeliness. In China, the prevention and control of echinococcosis has a long way to go, and strong technical support is necessary. Successful experiences in controlling and eliminating filariasis, kala-azar, and malaria also need to be learned and used for reference. Only in this way can the goal of early elimination be achieved.

## References


Agudelo Higuita NI, Brunetti E, McCloskey C (2016) Cystic echinococcosis. J Clin Microbiol 54(3):518–523. https://doi.org/10.1128/jcm.02420-15. From NLM

Bing-jie W, Li Z, Wan-li B, Lan-li D, Yun-ying C, Zhuang-zhi Z (2021) Establishment and application of TaqMan real-time PCR asays for detection of Echinococus granulosus. Chinese J Zoonoses 37(08):678–682+687. From Cnki

Chen X, Chen X, Lu X, Feng X, Wen H (2015) The production and comparative evaluation of native and recombinant antigens for the fast serodiagnosis of cystic echinococcosis with dot immunogold filtration assay. Parasite Immunol 37(1):10–15. https://doi.org/10.1111/pim.12151. From NLM

Deplazes P, Rinaldi L, Alvarez Rojas CA, Torgerson PR, Harandi MF, Romig T, Antolova D, Schurer JM, Lahmar S, Cringoli G et al (2017) Global distribution of alveolar and cystic echinococcosis. Adv Parasitol 95:315–493. https://doi.org/10.1016/bs.apar.2016.11.001. From NLM

Ding X, Liu YH, Ni B-X, Wang X-T, Xu X-Z, Ying Q-J, Dai Y, Cao J (2020) Establishment of a nucleic acid assay for detection of Echinococcus granulosus based on recombinase-aided isothermal amplification assay. Chin J Schisto Control 32(04):340–344. https://doi.org/10.16250/j.32.1374.2020071. From Cnki

Feng X, Wen H, Zhang Z, Chen X, Ma X, Zhang J, Qi X, Bradshaw H, Vuitton D, Craig PS (2010) Dot immunogold filtration assay (DIGFA) with multiple native antigens for rapid serodiagnosis of human cystic and alveolar echinococcosis. Acta Trop 113(2):114–120. https://doi.org/10.1016/j.actatropica.2009.10.003. From NLM

Gao CH, Shi F, Wang JY, Yang YT, Zhu HH (2012) Establishment and evaluation of colloid gold labeled immunochromatographic strip test for rapid diagnosis of alveolar echinococcosis. Zhongguo Ji Sheng Chong Xue Yu Ji Sheng Chong Bing Za Zhi 30(2):90–94. From NLM

Garibyan L, Avashia N (2013) Polymerase chain reaction. J Invest Dermatol 133(3):1–4. https://doi.org/10.1038/jid.2013.1. From NLM

Grond S, Eichmann TO, Dubrac S, Kolb D, Schmuth M, Fischer J, Crumrine D, Elias PM, Haemmerle G, Zechner R et al (2017) PNPLA1 deficiency in mice and humans leads to a defect in the synthesis of omega-O-Acylceramides. J Invest Dermatol 137(2):394–402. https://doi.org/10.1016/j.jid.2016.08.036. From NLM

Guan W, Zhang X, Wang X, Lu S, Yin J, Zhang J (2019) Employing parasite against cancer: a lesson from the canine tapeworm Echinococcus Granulocus. Front Pharmacol 10:1137. https://doi.org/10.3389/fphar.2019.01137. From NLM

https://www.cdc.gov/dpdx/echinococcosis/

Jia-fei Z, Gyu SH, Ning W, Cheng G, Chun-yan L, Yue X, Xiao-Bin G, Guang-You Y (2019) Detection of Echinococcus granulosus infection in dogs on the mitochondrial ND6 gene by Copro-PCR assay. Chinese J Zoonoses 35(07):626–632. From Cnki

Jin X, Li Y, Yang X, Zheng Y (2021) Modulatory effects of Echinococcus multilocularis emu-let-7-5p on the immunological functions of RAW264.7 macrophages. Front Vet Sci 8:663497. https://doi.org/10.3389/fvets.2021.663497. From NLM

Li J, Makrigiorgos GM (2009) COLD-PCR: a new platform for highly improved mutation detection in cancer and genetic testing. Biochem Soc Trans 37(Pt 2):427–432. https://doi.org/10.1042/bst0370427. From NLM

Li JQ, Li L, Fan YL, Fu BQ, Zhu XQ, Yan HB, Jia WZ (2018) Genetic diversity in Echinococcus multilocularis from the plateau vole and plateau Pika in Jiuzhi County, Qinghai Province, China. Front Microbiol 9:2632. https://doi.org/10.3389/fmicb.2018.02632. From NLM

Sheeba A, Sangaran A, Latha BR (2016) Diagnosis of cystic echinococcosis in buffaloes by native 8 kDa antigen using latex agglutination test (LAT). J Parasit Dis 40(4):1401–1405. https://doi.org/10.1007/s12639-015-0700-2. From NLM

Song X, Yan M, Hu D, Wang Y, Wang N, Gu X, Peng X, Yang G (2016) Molecular characterization and serodiagnostic potential of a novel dithiol glutaredoxin 1 from Echinococcus granulosus. Parasit Vectors 9:456. https://doi.org/10.1186/s13071-016-1741-9. From NLM

Tianli L, Xifeng W, Zhenzhong T, Lixia W, Xingxing Z, Jun Q, Qingling M, Shasha G, Ying C, Xuepeng C (2019) Multi-epitope fusion protein Eg mefAg-1 as a serodiagnostic candidate for cystic echinococcosis in sheep. Korean J Parasitol 57(1):61–67. https://doi.org/10.3347/kjp.2019.57.1.61. From NLM

Xu X-Z, Jin X-L, Li J, Jiang W-C, Jiang G (2011) Preliminary study on detection of Echinococcus granulosus DNA by loop-mediated isothermal amplification. Chin J Schisto Control 23(05):558-560+565. https://doi.org/10.16250/j.32.1374.2011.05.016. From Cnki

Zhou H-R, Xu Q-L, Ai L, Chen M-X, Yu Q, Xiao N (2020) Establishment of a method of recombinase aided isothermal nucleic acid amplification and preliminary evaluation of its use to detect and identify Echinococcus granulosus G1. J Pathogen Biol 15(06):655–660. https://doi.org/10.13350/j.cjpb.200607. From Cnki

Zhu GQ, Li L, Yan HB, Wu YT, Li WH, Fu BQ, Jia WZ (2019) Advances in research on echinococcus shiquicus tapeworm. Zhonghua Yu Fang Yi Xue Za Zhi 53(1):112–117. https://doi.org/10.3760/cma.j.issn.0253-9624.2019.01.017. From NLM

# Light and Electron Microscopical Aspects of Stages of *Echinococcus* Species Around the World

# 9

Heinz Mehlhorn

**Abstract**

The relationships between the different species of the Echinococcus group are still under intense discussion today. Many of them occur worldwide in different variations, endangering the health of humans and their domestic animals. However, there are also other strains that are only of local importance.

**Keywords**

Worms endangering human health · Pathways of infection · Means of diagnosis · Treatment

## 9.1 Important *Echinococcus* Species Endangering Human and Animal Health (Figs. 9.1, 9.2, 9.3 and 9.4)

1. *Echinococcus shiquicus*

   This species is found as a new species in Tibetan foxes (*Vulpes ferrilata*) and is considered a variant of *Echinococcus multilocularis* (Xiao et al. 2006). The final hosts are foxes, while pikas (*Ochontona curzoni*) are intermediate hosts. *Length*: 1.3–1.7 mm.

   This is a new species from the Qinghai-Tibet plateau region of China, and it may spread over the continents due to tourism and the transportation of food and local animals.

H. Mehlhorn (✉)
Emeritus of Department of Parasitology, Heinrich Heine University Düsseldorf, Düsseldorf, Germany
e-mail: mehlhorn@uni-duesseldorf.de



J. Li et al. (eds.), *Echinococcus: Control and Elimination of Echinococcosis with a Focus on China and Europe*, Parasitology Research Monographs 19,
https://doi.org/10.1007/978-3-031-54479-8_9

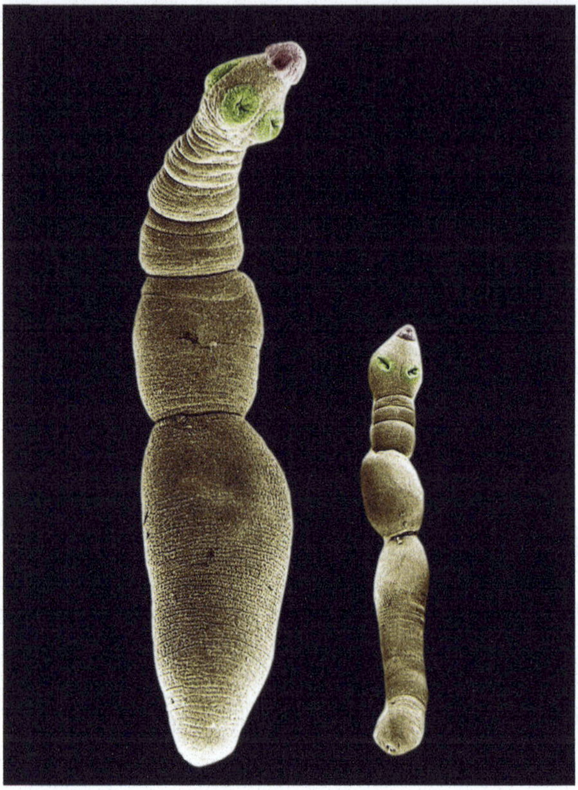

**Fig. 9.1** (**a**) Scanning electron micrograph of adult worms of *Echinococcus granulosus* (left; 2.5–6 mm) and *E. multilocularis* (1–3 mm) Note the different sizes. (**b**) Light micrograph of an adult (stained) worm of *Echinococcus granulosus* showing the large egg-filled proglottis, which becomes pushed off and thus sets free the infectious eggs

2. *Echinococcus vogeli*

   This species induces the so-called *polycystic echinococcosis* in Central and South America. The adult cestodes live in the liver, lung and other organs of the *bush (forest) dogs* (*Speothos venatius*) and their metacestodes in *pacas* in neotropical forests. Domestic dogs may become infected by the uptake of protoscolices, which may excrete infectious eggs and infect humans after oral uptake. Intermediate hosts are pacas. *Length*: 3.9–5.6 mm.

   This species is considered the causative agent of polycystic echinococcosis and the most pathogenic species of the *Echinococcus* species.

3. *Echinococcus oligarthrus*

   This species is based in the subtropical regions of Central and South America such as Argentina and occurs in two different populations, which had been considered as inducer of a neglected disease by the WHO. The final (definitive) hosts are wild felids such as *Puma concolor*, jaguar. Intermediate hosts are humans, rodents, pacas. Parasitized organs are muscles and internal organs and the disease is described as unicystic echinococcosis.

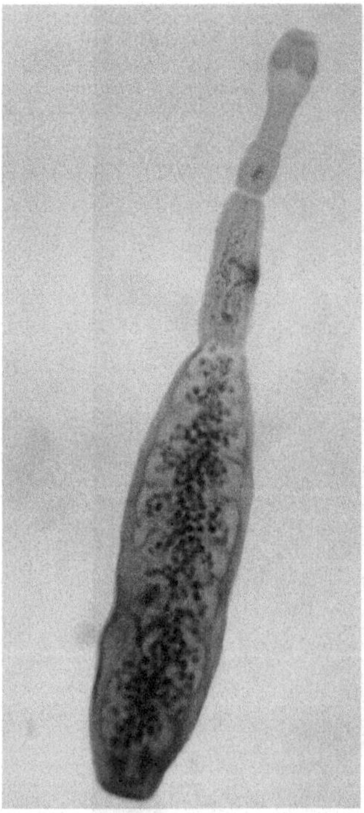

**Fig. 9.1** (continued)

*Echinococcus granulosus* has developed different strains, all of which can be transmitted by dogs. The strains of this worldwide (due to animal transport) occurring species have been spread to many countries around the globe, forming variations.

- Sheep strain/pig strain/cattle strain/camel strain
- Cervid strain/wolf strain
- Cattle strain
- Horse strain/donkey strain
- Pig strain/fox strain
- Camel strain

4. *Echinococcus granulosus* and *E. felidis*

   Both species are sister taxa and occur in the same locations in Southern and Eastern Africa as well as in many countries worldwide. Their lifecycles with respect to hosts are rather similar. They also spread in Asia to at least 270 million people.

5. *Echinococcus equinus*

   This species is closely related (sympatry) to *E. ortleppi*, *E. canadensis* and *E. multilocularis*. Domestic dogs are the definitive hosts, while horses, donkeys and zebras are intermediate hosts, where the liver is parasitized. They are common in Eurasia and Africa.

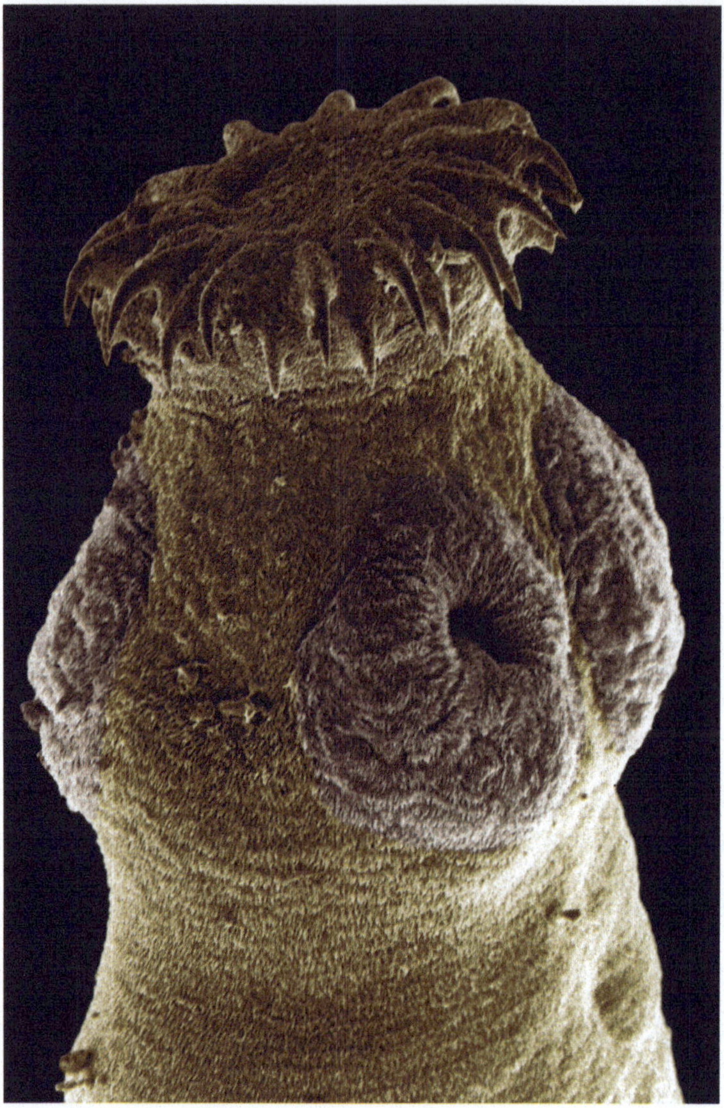

**Fig. 9.2** Scanning electron micrograph of the scolex region of *Echinococcus multilocularis*. Note the presence of a coronary of hooks and 4 suckers, which fix the worm inside the host's intestine

6. *Echinococcus multilocularis*

   This species occurs in Alaska, USA, Asia and Europe. Adults occur in foxes, dogs and cats; intermediate hosts are rodents, foxes, humans, horses, pigs, dogs etc.—parasitized organs are liver and several other organs (alveolar echinococcosis).

7. *Echinococcus canadensis*

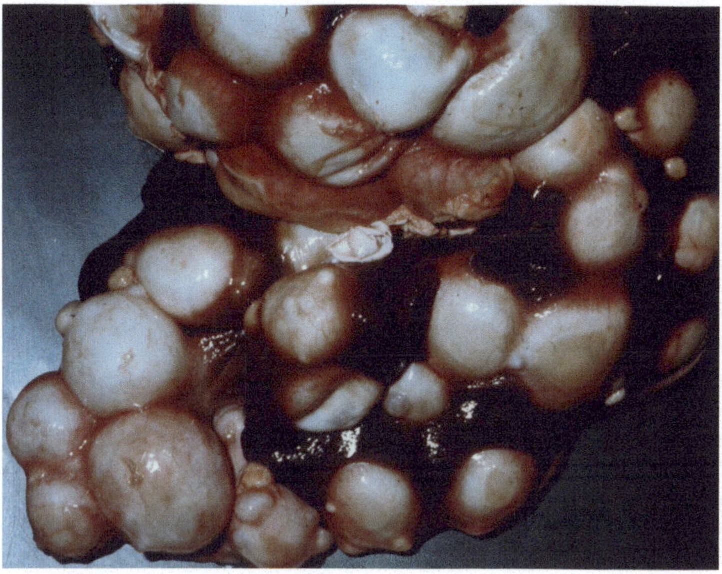

**Fig. 9.3** (**a**) Macrophoto of a sheep liver being infected with numerous hydatids of the tapeworm *Echinococcus granulosus*. Note the enormous destruction. (**b**) Macrophoto of an infected human liver containing the strands of a so-called alveococcus cyst of the fox tapeworm *Echinococcus multilocularis*, which is life-threatening

Some scientists consider this species to have a monophyletic origin, while others claim a sistership to *E. ortleppi*. Dogs and wolves are definitive hosts. Pigs, camels, cervids and humans are intermediate hosts, where lungs and liver are infected.

8. *Echinococcus ortleppi*

   This species infects mainly cattle as intermediate hosts and only rarely humans. Dogs are definitive hosts in regions of Eurasia and Africa.

9. *Echinococcus felidis*

   *E. felidis* was described in 1939 in the fur of African lions and hyenas, but was later included in the *E. granulosus* type (as subspecies or strain).

10. *Echinococcus intermedius* sensu lato

    This species occurs, e.g., in pigs and has been successfully controlled with the help of praziquantel but is still spreading.

11. *Further species*

    Since many *Echinococcus* species occur worldwide and look rather similar, it can be presumed that many more species exist than described today.

12. *Therapy*

    *Final hosts* **(dogs, cats): Application of praziquantel (1 × 5 mg/kg body weight, orally or subcutaneously).** *Importance*: **Hot cleaning of sleeping places of infected domestic dogs/cats, washing their fur after treatment, since eggs may stick herein. Treatment should be repeated every 2 months in regions with many infected mice. Intermediate hosts:** *Humans*: **Cysts due to** *E. granulosus* **can be surgically removed;**

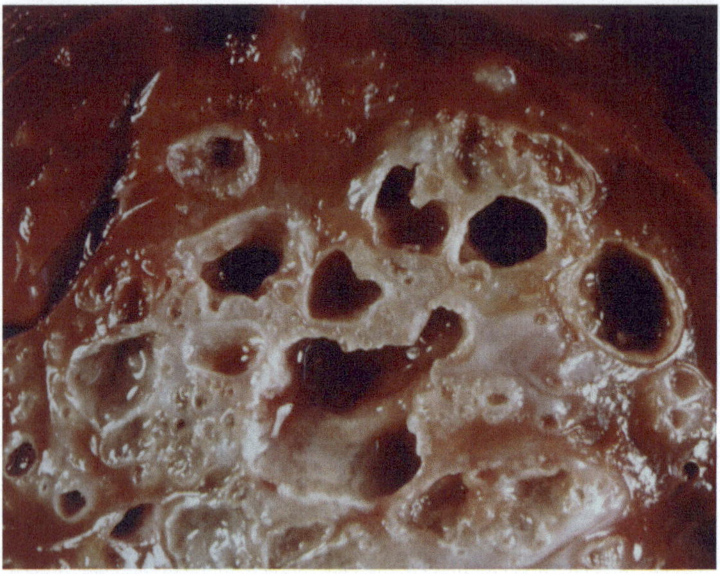

**Fig. 9.3** (continued)

**however, cysts due to** *E. multilocularis* **must be treated practically lifelong with albendazole or mebendazole, since these products do not kill the growing cysts but only stop their enlargement.** *Note*: **Recently Swiss veterinarians showed that dogs can also become intermediate hosts of** *E. multilocularis* **(Table 9.1).**

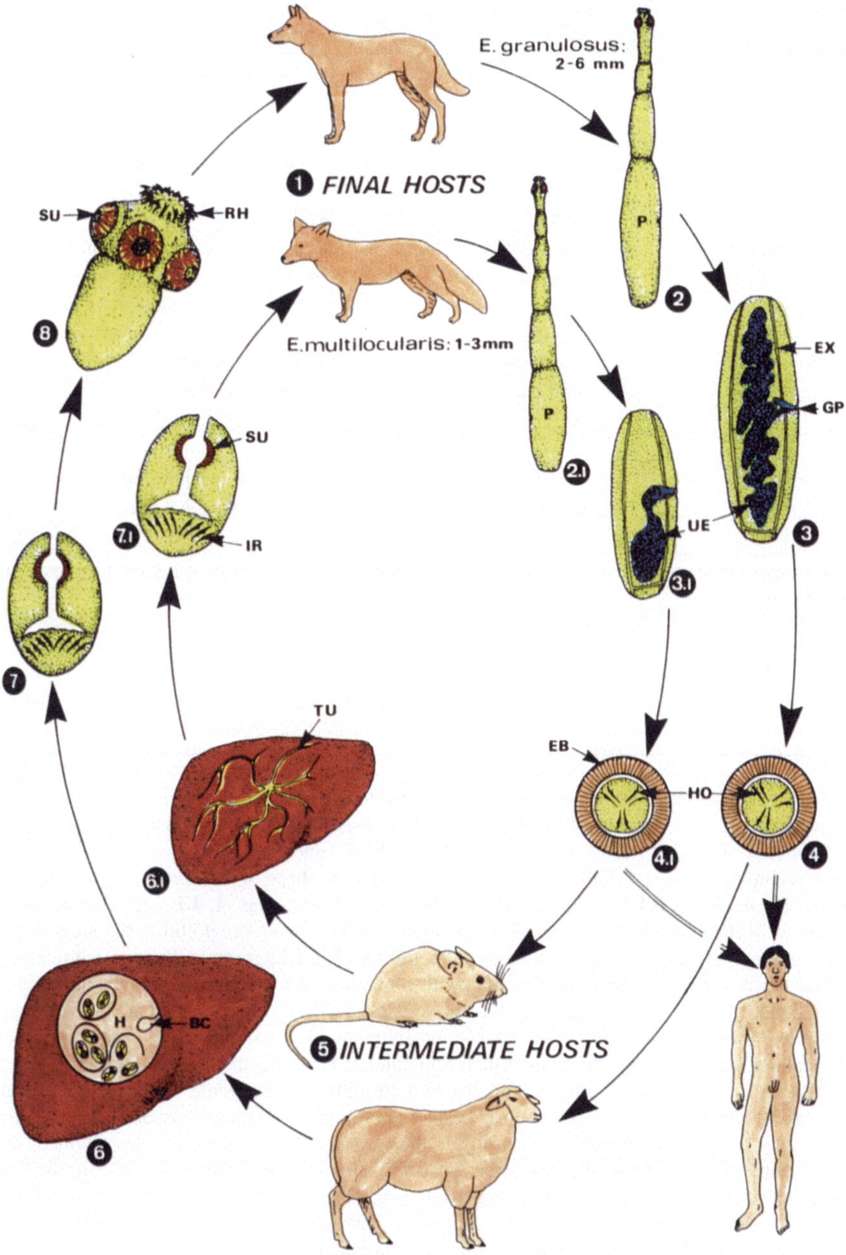
E. granulosus:
2-6 mm
1 FINAL HOSTS
E.multilocularis: 1-3mm
SU
RH
P
EX
GP
UE
IR
TU
EB
HO
H
BC
5 INTERMEDIATE HOSTS

**Table 9.1** Species of the genus *Echinococcus*

| Species, strain, genotypes (g) | Final hosts | Intermediate hosts |
|---|---|---|
| *E. canadensis*<br>*G 8, G 10* | Dogs, wolves | Cervids, *humans* |
| *E. equinus*<br>*G 4* | Dogs | Horses, donkeys, zebras |
| *E. felidis* | Lions | Wildlife |
| *E. granulosus*<br>*G 1, G 2, G 3* | Dogs, canids | Sheep, cattle, goat, camel, pig, *humans* |
| *E. intermedius*<br>*G 6, G 7, G 9* | Dogs | Pigs, goats, camels, *humans* |
| *E. multilocularis* | Foxes, dogs, cats | Rodents, pigs, horses, *humans* |
| *E. oligoarthus* | Wild cats | Rodents, *humans* |
| *E. ortleppi*<br>*G 5* | Dogs | Cattle, sheep, goats, *humans* |
| *E. shiquicus* | Tibet fox | Pika (Family Ochotonidae) |
| *E. vogeli* | South American forest dog | Rodents, pakas, *humans* |

*G* genotypes (as far as known); results collected from results of the groups of Eckert, Deplazes and Thompson

**Fig. 9.4** Diagrammatic representation of the life cycles of *Echinococcus granulosus* (1–8) and *E. multilocularis* (1.1–7.1). **1, 1.1** Final hosts may be dogs, cats or foxes with clear, species-specific preference. **2–3.1** Adult worms, which live in the small intestine of the final host, may be differentiated according to the size of the terminal proglottids (P), the shape of the uterus (UE) and the size of rostellar hooks. **3, 3.1** Excreted proglottids filled with fertile eggs. **4, 4.1** Eggs containing an infectious oncosphaera larva are released from the detached drying proglottid in the feces of the host; eggs are indistinguishable from those of *Taenia* spp. **5, 5.1** Eggs are orally ingested by intermediate hosts (e.g., mice, sheep etc.) or man with contaminated food. **6, 6.1** Inside the intestine of the intermediate hosts (including man), the oncosphaera hatches, enters the wall and may migrate (via blood) to many organs. Cysts are formed mostly in the liver and lung; in *E. granulosus* large unilocular hydatids occur, which are filled with fluid (containing thousands of protoscolices), whereas in *E. multilocularis* stages a tubular system infiltrates the whole organ (giving rise to alveolar aspects in sections). **7–8.1** In brood capsules of both cyst types protoscolices are formed, which may become evaginated (8) even inside their cysts. Evaginated or not, protoscolices are fully capable of infecting final hosts when these animals ingest infected organs of intermediate hosts. *BC* brood capsule; *EB* embryophore of the egg; *EX* excretory vessels; *GP* genital pore; *H* hydatid; *HO* hooks of oncosphaera; *IR* invaginated rostellar hooks; *P* proglottid; *RH* rostellar hooks; *SU* sucker; *TU* tubular system; *UE* uterus containing eggs

## Further Reading


Abdybekova AM, Zhang Z, Sultanov AA, Abdibayeva AA, Zhaksylykova AA, Junisbayeva SM, Aubakirov MZ, Akhmetova GD, Torgerson PR (2019) Genotypes of *Echinococcus* isolated from domestic livestock in Kazakhstan. J Helminthol 94:e69

Arrabal JP, Avila HG, Rivero MR, Camicia F, Salas MM, Costa SA, Nocera CG, Rosenzvit MC, Kamenetzky L (2017) *Echinococcus oligarthrus* in the subtropical region of Argentina: first integration of morphological and molecular analyses determines two distinct populations. Vet Parasitol 240:60–67

Avcioglu H, Guven E, Balkaya I, Kirman R, Akyuz M, Bia MM, Gulbeyen H, Yaya S (2021) *Echinococcus multilocularis* in red foxes in Turkey: increasing risk in urban. Acta Trop 216:105826

Banda F, Sikasunge C, Addy F, Wassermann M, Mackenstedt U, Kern P, Daugschies A, Mulinge E, Romig T (2020) *Echinococcus ortleppi* and *Echinococcus canadensis* G6/7 affect domestic animals in western Zambia. Acta Trop 211:105648

Bonelli P, Dei Giudici S, Peruzzu A, Piseddu T, Santucciu C, Masu G, Mastrandrea S, Delogu ML, Masala G (2020) Genetic diversity of *Echinococcus granulosus* sensu stricto in Sardinia (Italy). Parasitol Int 77:102120

Boufana B, Lett WS, Lahmar S, Buishi I, Bodell AJ, Varcasia A, Casulli A, Beeching NJ, Campbell F, Terlizzo M, McManus DP, Craig PS (2015) *Echinococcus equinus* and *Echinococcus granulosus* sensu stricto from the United Kingdom: genetic diversity and haplotypic variation. Int J Parasitol 45(2–3):161–166

Brehm K, Koziol U (2016) *Echinococcus*-host interactions at cellular and molecular levels. Adv Parasitol 95:147–212

Citterio CV, Obber F, Trevisiol K, Dellamaria D, Celva R, Bregoli M, Ormelli S, Sgubin S, Bonato P, Da Rold G, Danesi P, Ravagnan S, Vendrami S, Righetti D, Agreiter A, Asson D, Cadamuro A, Ianniello M, Capelli G (2021) *Echinococcus multilocularis* and other cestodes in red foxes (*Vulpes vulpes*) of Northeast Italy, 2012-2018. Parasit Vectors 14(1):29

D'Alessandro A, Rausch RL (2008) New aspects of neotropical polycystic (*Echinococcus vogeli*) and unicystic (*Echinococcus oligarthrus*) echinococcosis. Clin Microbiol Rev 21(2):380–401

Daipert-Garcia D, Pavan MG, Neves LBD, Almeida FB, Siqueira NG, Santos GBD, Dias-Correia TP, Ferreira HB, Rodrigues-Silva R (2019) Genetic diversity of *Echinococcus vogeli* in the western Brazilian Amazon. Mem Inst Oswaldo Cruz 114:e190149

Deplazes P, von Samson-Himmelstjerna G, Zahner H, Joachim A, Mathis A, Taubert A, Strube C (2020) Lehrbuch der Parasitologie für die Tiermedizin, 4th edn. Thieme, Stuttgart

Díaz Á, Fernández C, Pittini Á, Seoane PI, Allen JE, Casaravilla C (2015) The laminated layer: recent advances and insights into *Echinococcus* biology and evolution. Exp Parasitol 158:23–30

Fan J, Wu H, Li K, Liu X, Tan Q, Cao W, Liang B, Ye B (2020) Transcriptomic features of *Echinococcus granulosus* protoscolex during the encystation process. Korean J Parasitol 58(3):287–299

Halajian A, Luus-Powell WJ, Roux F, Nakao M, Sasaki M, Lavikainen A (2017) *Echinococcus felidis* in hippopotamus, South Africa. Vet Parasitol 243:24–28

Hua R, Xie Y, Song H, Shi Y, Zhan J, Wu M, Gu X, Peng X, Yang G (2019) *Echinococcus canadensis* G8 tapeworm infection in a sheep, China, 2018. Emerg Infect Dis 25(7):1420–1422

Hüttner M, Nakao M, Wassermann T, Siefert L, Boomker JD, Dinkel A, Sako Y, Mackenstedt U, Romig T, Ito A (2008) Genetic characterization and phylogenetic position of *Echinococcus felidis* (Cestoda: Taeniidae) from the African lion. Int J Parasitol 38(7):861–868

Hüttner M, Siefert L, Mackenstedt U, Romig T (2009) A survey of *Echinococcus* species in wild carnivores and livestock in East Africa. Int J Parasitol 39(11):1269–1276

Jansen F, Claes M, Bakkers E, Aryal A, Madimba KC, Gabriël S, Dermauw V, Van Hul A, Vervaeke M, Dorny P (2020) *Echinococcus multilocularis* in red foxes in North Belgium: prevalence and trends in distribution. Vet Parasitol Reg Stud Rep 22:100470

Knapp J, Meyer A, Courquet S, Millon L, Raoul F, Gottstein B, Frey CF (2021) *Echinococcus multilocularis* genetic diversity in Swiss domestic pigs assessed by EmsB microsatellite analyzes. Vet Parasitol 293:109429

Lymbery AJ (2017) Phylogenetic pattern, evolutionary processes and species delimitation in the genus *Echinococcus*. Adv Parasitol 95:111–145

Lymbery AJ, Jenkins EJ, Schurer JM, Thompson RC (2015) *Echinococcus canadensis*, *E. borealis*, and *E. intermedius*. What's in a name? Trends Parasitol 31(1):23–29

Maldonado LL, Arrabal JP, Rosenzvit MC, Oliveira GC, Kamenetzky L (2019) Revisiting the phylogenetic history of helminths through genomics, the case of the new *Echinococcus oligarthrus* genome. Front Genet 10:708

Mehlhorn H (ed) (2016) Encyclopedia of parasitology, 3 volumes, 4th edn. Springer, Berlin

Mehlhorn H (2016) Human parasites, 8th edn. Springer, Heidelberg

Mehlhorn H (2016) Animal parasites, 8th edn. Springer, Heidelberg

Mehlhorn H, Heukelbach J (eds) (2022) Infectious tropical diseases and One Health in Latin America. Parasitology research monographs, vol 16. Springer International

Mehlhorn H, Klimpel S (eds) (2019) Parasite and disease spread by major rivers on earth - past and future perspectives. Parasitology research monographs, vol 12. Springer, Heidelberg

Miles S, Magnone J, García-Luna J, Ancarola ME, Cucher M, Dematteis S, Frischknecht F, Cyrklaff M, Mourglia-Ettlin G (2021) Ultrastructural characterization of the tegument in protoscoleces of *Echinococcus ortleppi*. Int J Parasitol 51(12):989–997

Santa MA, Musiani M, Ruckstuhl KE, Massolo A (2021) A review on invasions by parasites with complex life cycles: the European strain of *Echinococcus multilocularis* in North America as a model. Parasitology 148(13):1532–1544

Šarkūnas M, Vienažindienė Ž, Rojas CAA, Radziulis K, Deplazes P (2019) Praziquantel treatment of dogs for four consecutive years decreased the transmission of *Echinococcus intermedius* G7 to pigs in villages in Lithuania. Food Waterborne Parasitol 15:e00043

Tamarozzi F, Deplazes P, Casulli A (2020) Reinventing the wheel of *Echinococcus granulosus* sensu lato transmission to humans. Trends Parasitol 36(5):427–434

Toews E, Musiani M, Checkley S, Visscher D, Massolo A (2021) A global assessment of *Echinococcus multilocularis* infections in domestic dogs: proposing a framework to overcome past methodological heterogeneity. Int J Parasitol 51(5):379–392

Wen H, Vuitton L, Tuxun T, Li J, Vuitton DA, Zhang W, McManus DP (2019) Echinococcosis: advances in the twenty-first century. Clin Microbiol Rev 32(2):e00075–e00018

Woolsey ID, Miller AL (2021) *Echinococcus granulosus* sensu lato and *Echinococcus multilocularis*: a review. Res Vet Sci 135:517–522

Xiao N, Qiu J, Nakao M, Li T, Yang W, Chen X, Schantz PM, Craig PS, Ito A (2006) *Echinococcus shiquicus*, a new species from the Qinghai-Tibet plateau region of China: discovery and epidemiological implications. Parasitol Int 55(Suppl):S233–S236

Yan HB, Li L, Li W, Zhu G, Li JQ, Wu Y, Zhang N, Wu Y, Li M, Zhang L, Yao G, Tian W, Li L, Li W, Guo A, Dai G, Fu B, Ohiolei JA, Jia WZ (2021) *Echinococcus shiquicus* in Qinghai-Tibet plateau: population structure and confirmation of additional endemic areas. Parasitology 148(7):879–886

Zaffarano GP, de Klerk-Lorist LM, Junker K, Mitchell E, Bhoora RV, Poglayen G, Govender D (2021) First report of cystic echinococcosis in rhinos: a fertile infection of *Echinococcus equinus* in a Southern white rhinoceros (*Ceratotherium simum simum*) of Kruger National Park, South Africa. Int J Parasitol Parasites Wildl 14:260–266

# CT and MR Features of Echinococcosis

# 10

Yanqiu Sun, Chenhong Zhang, and Dengfeng Tian

**Abstract**

Although the incidence rate of echinococcosis has decreased with the improvement of living standards, echinococcosis is still a common and frequently occurring disease in pastoral areas. Different parts of the disease cause different harm to the human body. Early diagnosis and surgical treatment are of great significance to improve the survival of patients. We retrospectively analyzed the CT and MRI features of 15 cases of hydatid disease confirmed by surgery and pathological examination. Among the 15 patients, there were 10 cases of cystic echinococcosis and 6 cases of alveolar echinococcosis, including 6 cases of hepatic cystic echinococcosis, 2 cases of hepatic alveolar echinococcosis, 1 case of cerebral alveolar echinococcosis, 1 case of bone cystic echinococcosis, 1 case of bone alveolar echinococcosis, 1 case of neck cystic echinococcosis, 1 case of adrenal alveolar echinococcosis, 1 case of intraspinal cystic echinococcosis, and 1 case of mediastinal alveolar echinococcosis. The diagnosis of echinococcosis occurring outside the liver is relatively difficult; here is a summary of CT and MRI findings of different types of hydatid diseases at different sites of onset.

**Keywords**

Cystic echinococcosis · Alveolar echinococcosis · Tomography · Magnetic resonance imaging · Imaging features

Y. Sun (✉) · C. Zhang · D. Tian
Department of Radiology, Qinghai Provincial People's Hospital, Xining, Qinghai, China





J. Li et al. (eds.), *Echinococcus: Control and Elimination of Echinococcosis with a Focus on China and Europe*, Parasitology Research Monographs 19,
https://doi.org/10.1007/978-3-031-54479-8_10

Hydatid disease, also known as echinococcosis, is a zoonotic disease caused by the larvae of the echinococcus cestode parasitizing mammals. Approximately four million people worldwide have been infected by the encysted worms, and 60 million people are at risk of infection, posing a serious risk to public health security and economic development worldwide (Infectious Disease Imaging Group, Radiology Branch, Chinese Medical Association, Infection Imaging Special Committee, Radiology Branch, Chinese Medical Doctor Association 2021).

The main types of human infections are cystic echinococcosis (CE) and alveolar echinococcosis (AE), which are caused by two species of cestodes of the genus Echinococcus, Echinococcus granulosus and Echinococcus multilocularis (Qian and Zhou 2018; McManus et al. 2012). The northwestern part of Cna is one of the most prevalent areas of encapsulation worldwide, with a population prevalence of 0.5–5.0% (Qian and Zhou 2018; Filippou 2013). Encapsulated echinococcosis can infect all body organs, with the liver predominating, followed by the lungs, and other organs being relatively uncommon. Some studies have reported that 65–80% of CE occurs in the liver, while AE occurs in up to 98% of the liver (Wen et al. 2019; Bhutani and Kajal 2018).

The onset of echinococcosis is insidious, and there are no clinical symptoms when the early lesions are small. When the lesions are large (>10 cm), they may cause symptoms of pressure on the surrounding organs and vascular system, such as abdominal distension and vague pain. Therefore, the clinical symptoms may vary depending on the size and location of the lesion, but they are not specific. Life history in the endemic area can be one of the diagnostic clues. Serological findings are helpful, but there are false negatives and positives, while imaging is an important method for diagnosing encapsulated echinococcosis (Li 2016). In recent years, with the rapid development of imaging technology, a variety of new indicators have been provided for the diagnosis of encapsulated echinococcosis, which play a crucial role in the early diagnosis, preoperative evaluation and efficacy monitoring of encapsulated echinococcosis, bringing the imaging diagnosis of encapsulated echinococcosis to a brand new level. In this paper, we will review the CT and MR features of common sites of encapsulation and some rare sites.

## 10.1 Hepatic Cystic Echinococcosis

### 10.1.1 Growth Characteristics and Pathological Process of Hepatic Cystic Echinococcosis

Hepatic cystic echinococcosis refers to the eggs of Taenia are unshelled and separated by the effect of gastric acid in the human gastric and duodenal circulation. When the eggs circulate into the portal system through small intestinal wall veins, they will be blocked by hepatic sinuses and grow near the portal. The different courses of disease are divided into early, middle, and late stages.

In the early stages of hepatic cystic echinococcosis infection, the CE cyst wall has two layers: an outer layer of fibrous tissue composed of epithelioid cells, eosinophils and foreign-body giant cells and an inner layer composed of the germinal layer and corneum. The germinal layer comprises the brood capsules, scolex and daughter cysts, and each scolex in the cyst may develop into a pericyst. With the prolongation of the disease course and the infection, damage and aging of the cyst wall, the endocyst and the ectocyst can be separated. In the late stage, extensive vitreous, organized and fibrotic cyst walls gradually appear necrosis and calcification. The scolex and exact pericyst could not be seen inside the cyst, and atrophic hepatocytes and bile ducts could be seen outside the cyst wall.

### 10.1.2 Classification of Hepatic Cystic Echinococcosis

#### 10.1.2.1 Clinical Classification

Currently, the standard classification of CE is mainly used in clinical practice, formulated by WHO Informal Working Group on Echinococcosis (WHO/IWGE) on the basis of Gharbi ultrasonic classification (Gharbi et al. 1981). CE is classified into six types: cystic lesion (CL), single cystic type (CE1), multiple daughter cysts type (CE2), internal capsule collapse type (CE3), solid type (CE4), and calcification type(CE5). According to the size of cysts, CE can be divided into three categories: less than 5 cm in diameter is called small echinococcosis and denoted as S; between 5 and 10 cm in diameter is called medium echinococcosis and denoted as M; more than 10 cm is called large echinococcosis and denoted as L. Biological characteristics: active (group 1: CE1 and CE2 types), Transitional (group 2: CE3a and CE3b types) and inactive (group 3: CE4 and CE5 types).

The WHO/IWGE consensus classification of alveolar echinococcosis (AE) is the PNM system (P = parasitic mass in the liver, N = involvement of neighboring organs, and M = metastasis) proposed by Kern (Kern et al. 2006) et al. It differs from CE classification based on imaging, mainly for clinical treatment. Wen Hao (Wen et al. 2002) et al. proposed the PJVM system (P = parasitic mass in the liver, J = jaundice, V = vessels, M = metastasis) according to clinical needs in 2002. This classification comprehensively describes the extent of lesions, the degree of direct invasion, and the distant metastasis of hepatic AE, which is of great significance for surgeons who want to make treatment plans.

#### 10.1.2.2 Imaging Classification

The imaging classification of CE is consistent with the clinical classification, but there is no consensus on the classification of AE. The currently recognized classification method of AE is the MRI classification method proposed by Kodama et al. (Kodama et al. 2003), based on the features of hepatic AE with microcysts. According to the composition of solid and cystic components and the distribution pattern of microcysts in AE lesions, the imaging features of AE lesions can be divided into the following five types: multiple small round cysts without a solid component (type 1); multiple small round cysts with a solid component (type 2); a solid component surrounding large and/or irregular cysts with multiple small round cysts (type 3); a solid component without cysts (type 4); a large cyst without a solid component (type 5). It reflects the biological evolution of AE to some extent.

### 10.1.3 CT and MR Image Findings of Hepatic Echinococcosis

#### 10.1.3.1 CT and MR Image Findings of CE

CT image Findings of CE:CE1 type showed round or round-like cystic lesions with smooth and sharp edges, showed water-like density inner the lesion, the cyst wall could be visualized when it is thicker, and the lesion is not enhanced on contrast-enhanced scan (Fig. 10.1a); CE2 type showed daughter cyst sign, according to the content of cyst fluid of the mother cyst and the arrangement of daughter cysts, it showed "cyst in the cyst," "honeycomb sign," "wheel-like sign" (Fig. 10.1b), the density of cyst fluid in the mother cyst is higher than that in the daughter cyst ; CE3 type manifested the cyst wall separates, showed "double wall sign," the internal capsule collapsed and floated in the cyst fluid, it showed "drift belt sign" (Fig. 10.1c), if the cyst is complicated with biliary fistula or infection, gas appeared in the cyst, it showed "water-lily sign"; CE4 type showed solid soft tissue mass usually with thicker cyst wall, the density in the cyst is inhomogeneous, and the lesion is not enhanced on contrast-enhanced scan (Fig. 10.1d); CE5 type showed thick-shelled calcified cyst wall, the density of the contents in cyst is increased with partial or total calcification (Fig. 10.1e).

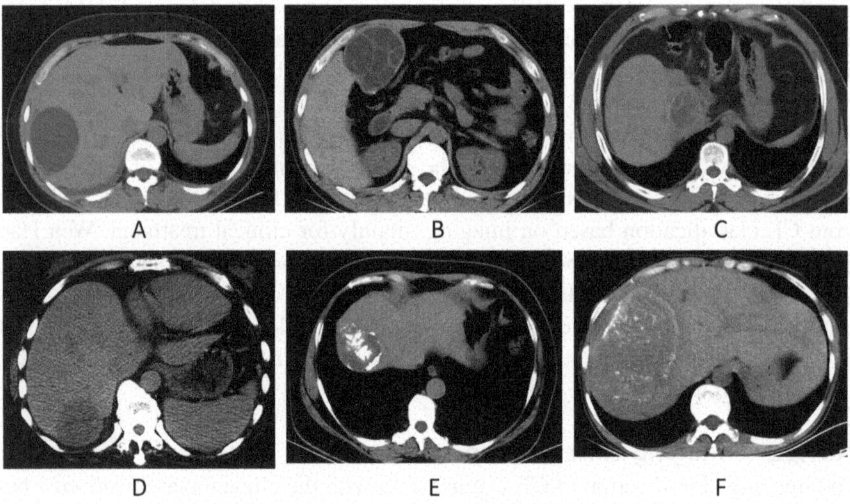


**Fig. 10.1** CT imaging of hepatic echinococcosis. (**a**) shows hepatic cystic echinococcosis CE1 type, and the CT scan showed a round watery density lesion in the right lobe of the liver. (**b**) shows CE2 type, and the non-contrast CT scan shows a multilocular cystic mass shadow in the S4 segment of the liver, showing a "intracapsular cyst." (**c**) shows CE3 type, and the non-contrast CT scan shows that the cyst wall in the hydatid cyst collapses and floats in the cyst fluid in a "streamer sign." (**d**) shows CE4 type, and the non-contrast CT scan shows a solid soft tissue mass in the right posterior lobe of the liver (S7 segment), with uneven internal density and clear borders, and some protruding from the liver surface. (**e**) is CE5 type, and non-contrast CT scan shows calcification of the cyst wall and cyst contents. (**f**) shows hepatic alveolar echinococcosis (AE), and noncontrast CT scan shows an inhomogeneous solid mass in the right lobe of the liver, with small vesicles and calcifications in the lesion

MR image Findings of CE:CE1 type showed round or round-like cystic lesions with smooth and sharp edges, had long T1 and long T2 signals, the signal is uniform, the cyst wall appeared hypointense on T2WI with uniform thickness (Fig. 10.2a). CE2 type showed "cyst in the cyst" imaging, it showed "rose petals sign" and "spoke wheel sign" when the mother cyst contained multiple daughter cysts (Fig. 10.2b). The signal of the daughter cyst is lower than that of the mother cyst on T1WI and higher than that of the mother cyst on T2WI. CE3 type is characterized by the internal capsule collapsing and floating in the cyst fluid, forming a "double wall sign" (Fig. 10.2c). CE4 type showed mixed signal inner lesion and cerebriform slightly short T2 signal; the cyst wall always appeared hypointense on T2WI.MRI is superior to CT in showing the collapsed and shrunken internal capsule.

The cyst wall calcification of CE5 type appears hypointense on T1WI and T2WI (Fig. 10.2d).

#### 10.1.3.2 CT and MR Image Findings of AE

The CT imaging showed an irregular parenchymal mass. The lesion itself is not enhanced on contrast-enhanced scan, but its boundary is clearly shown because of the enhancement of the surrounding liver parenchyma; various amounts of calcification are often found in the contents of the lesion, and small cysts and small circles

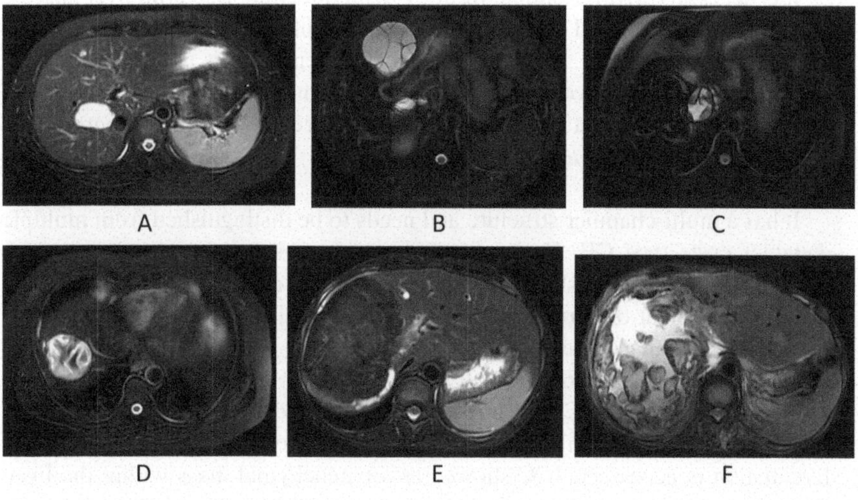


**Fig. 10.2** MRI imaging of hepatic echinococcosis. (**a**) shows hepatic cystic echinococcosis CE1 type, and T2WI shows a cystic lesion with a smooth edge of the right lobe of the liver with uniform signal. (**b**) shows CE2 type, and T2WI shows that polycystic hydatid cysts show "intracystic" and "rose petal sign." (**c**) shows CE3 type, and T2WI shows that the collapsed inner capsule is suspended in the capsule fluid to form a "streamer sign." (**d**) shows CE5 type, T2WI shows a round lesion in the right lobe of the liver, and the calcified part of the cyst wall or capsule contents shows hypointensity. (**e**–**f**) showed hepatic alveolar echinococcosis (AE), and T1WI and T2WI showed that the right lobe lesion was mainly hypointense, and the liquefaction necrosis area in the lesion showed a "cave sign" or "map sign"

of calcification are the most characteristic, liquefaction necrosis appeared in the center of the lesion, forming a "map sign" appearance (Fig. 10.1f); the edge of the liver parenchyma adjacent to the lesion is constricted and depressed, and the contralateral liver lobe or segment showed compensatory enlargement.

Hepatic AE showed irregular solid lesions with infiltrative growth and unclear margins. The lesions are mainly hypointense on T1WI and T2WI, especially hypointense on T2WI; small cysts appeared slightly hypointense on T2WI; liquefaction necrosis might appear in the center of the lesion, which is manifested as a "melting hole sign" or "map sign" (Fig. 10.2e, f); the inner of the lesion is not enhanced on contrast-enhanced scan. MRCP is the best method for showing small cysts.

### 10.1.4 Differential Diagnosis

1. Simple hepatic cysts

   It is necessary to differentiate CE from single cyst type. Most patients with simple hepatic cysts have no epidemiological history, no cystic wall, no calcification, homogeneous cystic fluid, no typical imaging features such as "cyst sand sign," "double wall," and "curved calcification." Immunological tests often show negative reactions.
2. Bacterial liver abscess

   It is necessary to differentiate from CE with infection. Bacterial liver abscess appeared as thick-walled cystic lesions with a visible gas-liquid plane inside the cyst. An enhanced CT scan showed enhancement of the abscess wall with peripheral low-density edema zone. The symptoms of systemic poisoning are severe, with a significant increase in white blood cell count, and the serological tests for echinococcosis are mostly negative.
3. Hepatic cystadenoma

   It has a multi-chamber structure and needs to be distinguished from multiple daughter cysts type CE. Hepatic cystadenoma often had intraluminal growth solid mural nodules and showed slight enhancement on contrast-enhanced scan; there is no circular or curved calcification of the cycle wall, and the intra-cystic septal showed enhancement.
4. Intrahepatic cholangiocarcinoma (ICC)

   Both liver AE and ICC can show contraction and depression of the liver margin near the lesion and dilation of the bile duct at the edge of the lesion, so differentiation is necessary. ICC showed as a parenchymal mass within the liver. The edges showed lace-like enhancement and the inside showed late enhancement. However, the hepatic AE lesion is not enhanced, so the small cysts and calcifications are important imaging features to distinguish them.

## 10.2 Pulmonary Echinococcosis

Pulmonary cystic echinococcosis (CE), also known as pulmonary hydatidosis, is caused by infection with Echinococcus granulosus. It is endemic in the Qinghai-Tibetan plateau region of China. Common in the hepatic, pulmonary hydatidosis accounts for about 20% of echinococcosis (Lupia et al. 2021). Some studies suggest that it is common in women aged 40–49 years (Ma et al. 2020). The clinical presentation of pulmonary CE is characterized by cough, hemoptysis, and wasting as the first symptoms, most commonly in the middle and lower lobes of the right pulmonary (Mao et al. 2019). The consensus staging scheme developed by the WHO Informal Working Group on Echinococcosis (IWGE) is based on Gharbi ultrasound staging. CE is classified into six types: cystic lesion (CE), single-cyst (CE1), multi-cysts (CE2), internal cyst rupture (CE3), consolidation (CE4) and calcified lesion (CE5). The CT presentation of pulmonary CE is a single or multiple round or round-like cystic lesions in the pulmonary (Fig. 10.3) with a CT value of 3–20 HU, a smooth cyst wall, and a "daughter cyst sign" as a specific image, with pleural effusion seen when the cyst ruptures, enhanced scan of the hydatid does not intensify. The MRI lesion may show "vesicular sign," "double contour sign," "rosette sign," " spoke-wheel sign," and "cortical ribbon sign." T1WI shows a low signal and T2WI shows a high signal, which can clearly show the cystic mass with pneumonia, pulmonary cancer-like changes and the displacement of surrounding bronchi and blood vessels by compression (Cao 2018). Li Ting et al. (Li and Bao 2020) mentioned that the CT value of CE lesions tended to increase gradually with the change of CE1 to CE5 typing, and the ADC value of each type tended to decrease, which could reflect whether the encapsulated worm is infected or ruptured, thus indirectly reflecting the activity of the hydatid.

Pulmonary CE has a complex and varied imaging presentation and is difficult to differentiate from peripheral pulmonary cancer and pulmonary abscess. Peripheral pulmonary cancer is most commonly seen in the middle-aged and elderly, with isolated masses in the pulmonary with irregular borders, deep lobulation, spicule sign, and pleural indentation, and often restricted diffusion of the parenchyma on DWI sequences. Pulmonary CE type 2 imaging is even more varied and is very difficult to distinguish from peripheral pulmonary cancer when there is surrounding inflammatory formation. Gas-fluid planes are seen within the pulmonary abscess capsule, and CT-enhanced scans show thick-walled enhancement of the abscess wall, which may be accompanied by a peripheral zone of low-density edema, with

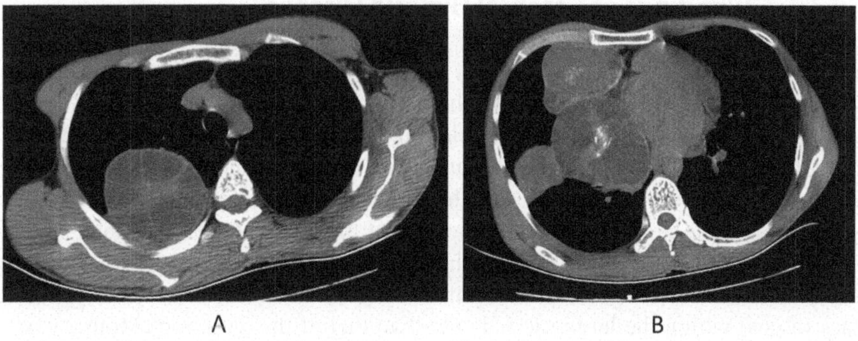


**Fig. 10.3** A 30-year-old male patient presents with intermittent chest pain for half a month. Chest CT showed a round multilocular cystic mass in the upper lobe of the right lung, with a well-defined border and a size of about 67 × 79 mm (**a**), multiple cystic mass shadows in the middle and lower lobes of the right lung, and patchy calcifications in the center of the lesion (**b**)

severe systemic symptoms and a significant rise in white blood cell count and negative serology for echinococcosis. The diagnosis of this disease still depends on pathology.

## 10.3 Cerebral Echinococcosis

Cerebral echinococcosis accounts for 1–2% of systemic encystment (Miriguri and Jia 2010). Cerebral encystment is often concurrent with hepatic and pulmonary encystment, or it can be primary, mostly solitary, located in the supratentorial middle cerebral artery feeding area and involving one lobe of the brain. It can also be multiple and distributed on both sides. The rich blood supply of the brain tissue facilitates the growth of encapsulated cysts. The germ layer of the encapsulated cyst continuously secretes aqueous cystic fluid. It grows into a spherical shape due to the expansion of tension, which can grow into a huge cyst that is relatively stable only after its vitality decreases and can live for more than a decade. The most common sites are frontal and parietal lobes, and temporal and occipital lobes are also more common. The cerebellum, ventricles, epidural and skull base are less common. Cerebral vesicular echinococcosis is extremely rare when it is primary in the brain and is often secondary. Almost 100% of vesicular echinococcosis originates in the liver, and secondary infections in other sites, such as the brain and lungs, tend to arise from bloodstream transfer. Echinococcus vesicularis is honeycomb-shaped, with no cuticle or fibrous layer on the surface of its cysts, and grows and multiplies unrestrictedly outward in the parasitized organ, forming numerous tiny subvacuoles. The vesicles contain gelatinous material and protocephalic larvae, which grow infiltratively in the cerebral matter and can erode the blood vessels in the brain and severely damage the neural tissue to cause granulomatous changes and edema in the surrounding brain tissue. Their metabolites are excreted directly into the surrounding brain tissue, stimulating the production of obvious reactive edema.

### 10.3.1 CT Findings of Cerebral Hydatid Disease

Round or oval multicystic or single cyst swelling in the brain, the intracystic density of single cyst is uniform and slightly higher than cerebrospinal fluid, CT value is −1~+12HU, the wall is thin with circular high density, the boundary is clear, there is no enhancement on enhancement scan and no edema around the cyst, the midline structure is shifted to the opposite side. In polycystic sacs, there are multiple rounded small sacs in the large sac. The high density of the wall between the small sacs is slightly bounded, and some of the neighboring borders are partially fused. The density within the sac is higher than that of the single sac, but the density of the small sacs (ascus) within the large sac is lower than that of the large sac (Mother cyst). Edema is seen around the wall of the large sac, and the midline structures are shifted and hydrocephalus can be seen.

## 10.3.2 MR Manifestations of Cerebral Echinococcosis

### 10.3.2.1 Cystic Echinococcosis

Cystic echinococcosis manifests as round long T1, long T2 signal lesions of variable size in the brain, which can be single or multiple. When the cysts are not ruptured or necrotic, the internal signal is uniform. The signal of internal cyst fluid can be suppressed by a liquid attenuation inversion response (FLAIR) sequence scan, and the foci are low in the DWI sequence, without an edema band around the foci and no enhancement in the enhancement scan. When a ruptured infected encapsulated cyst occurs, the cyst loses its round shape, the internal signal is inhomogeneous, and abnormal circumferential enhancement can be seen on an enhanced scan.

### 10.3.2.2 Echinococcosis Multilocularis

Echinococcosis multilocularis is characterized by multiple T1 and short T2 clusters of abnormal signals in the brain, usually surrounded by edema bands similar to brain tumors. The T2WI signals of the lesion are "coal like" black low signals (Fig. 10.4), numerous small vesicles with a dense and slightly higher signal are seen inside, and the diameter of the small vesicles is about 1–10 mm. The lesion is hypoperfused on the DWI sequence, and irregular abnormal ring reinforcement is seen on the Gd-DTPA enhancement scan. The lesions in cerebral perfusion imaging showed low perfusion changes.

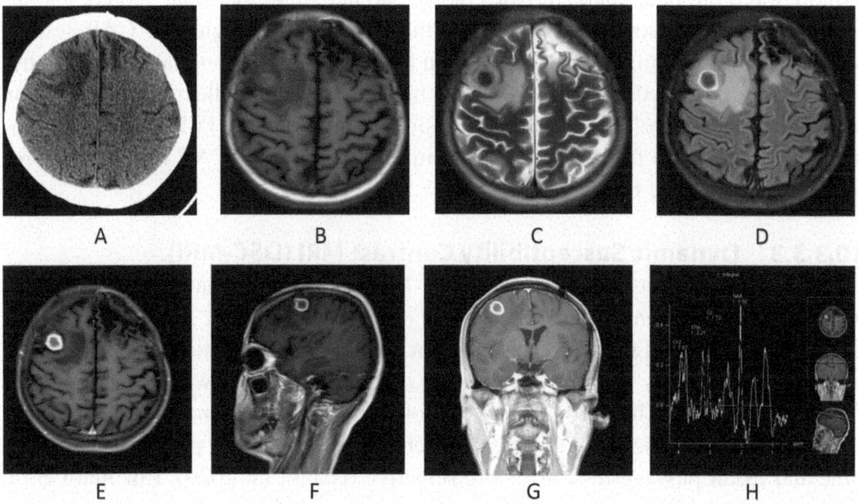


**Fig. 10.4** Imaging findings of alveolar echinococcosis. (**a**) shows a CT scan of the skull without contrast showing a slightly high-density nodule in the right frontal lobe with a large hypodense opacity in the white matter area. MRI showed a round-like nodular shadow in the right frontal lobe, with an equal T1 signal and a short T2 signal (**b**–**c**); T2flair is hypointense, surrounded by patchy edema signal shadows (**d**); Enhanced scan annular enhancement (**e**–**g**) with patchy edema signal shadow around it; (**h**) MRS showed that the NAA peak next to the right frontal lobe lesion was not reduced, and the towering Lip/Lac peak was seen at 1.3 ppm

### 10.3.3 Application of Multimodal MRI Techniques in Cerebral Alveolar Echinococcosis

#### 10.3.3.1 Diffusion Weighted Imaging (DWI)

Ma Juan et al. (2015) used DWI to study CAE lesions and explored the value of the quantitative index (ADC value) in different regions of CAE, and found that the ADC value in the parenchymal area of the lesion is lower than that in the contralateral normal brain tissue. This is because many small vesicles are proliferating in CAE lesions with tightly arranged cells and increased density, thus reducing the diffusion rate of water molecules. In contrast, the ADC value in the edema area increases because CAE lesions push against brain tissue to form vasogenic edema, facilitating the diffusion of water molecules. Chunhui Jiang et al. (2015) used the DWI technique to identify CAE and high-grade glioma and found that the difference in ADC values between the parenchymal and edematous areas of the lesions in both groups is statistically significant ($P < 0.05$). Therefore, the application of the DWI technique in CAE can reflect the diffusion movement rate of water molecules in the parenchymal, marginal and edematous regions of the lesions by ADC values, which indirectly reflects the information such as cell density and microstructure.

#### 10.3.3.2 Intravoxel Incoherent Motion (IVIM)

Intravoxel incoherent motion (IVIM) is a technique to assess tissue perfusion in vivo without using contrast agents. Yang Jie (Yang 2019) used the IVIM technique to explore the biological characteristics of the parenchymal, marginal and edematous regions of CAE lesions and found that the parenchymal region of CAE lesions showed low perfusion, the marginal region had higher perfusion than the parenchymal region, and the edematous region had higher perfusion, reflecting the perfusion level and vascular proliferation of CAE lesions. Therefore, the IVIM technique provides a certain basis for determining the infiltrative zone of CAE lesions and provides some guidance for clinical surgery.

#### 10.3.3.3 Dynamic Susceptibility Contrast MRI (DSC-MRI)

Dynamic susceptibility contrast MRI (DSC-MRI) is a method that uses T2-weighted MR acquisition to enhance the magnetic susceptibility effect of contrast agents (Zhang et al. 2018). DSC-MRI, also known as perfusion-weighted imaging (PWI), has now been widely used in the study of cerebral ischemia, where the degree of reduction in cerebral blood flow and the prolongation of peak time or mean passage time can reflect the severity of cerebral ischemia, and the two parameters of peak time and mean passage time are more sensitive (Hao et al. 2020). Liu Jundi et al. (2019) applied PWI in the differential diagnosis of CAE and brain metastases because CAE is a hypoperfused tumor, and the parenchymal region is hypoperfused due to the relative lack of internal blood supply. In contrast, brain metastases are hyperperfused tumors because the parenchymal region of brain metastases is hyperperfused because brain metastases maintain the same tissue characteristics as the primary tumor, so the quantitative parameters of the PWI technique can be used to distinguish CAE from brain metastases.

#### 10.3.3.4 Dynamic Contrast-Enhanced MRI (DCE-MRI)

Dynamic contrast-enhanced MRI (DCE-MRI) is an MRI perfusion technique using an exogenous contrast agent without any ionizing radiation and has good spatial resolution. Zhang Dan et al. (2019) used the DCE-MRI technique to study CAE, analyzed the correlation between it and the histogram, and found the existence of edge bands in CAE lesions and the destruction of the blood-brain barrier. Therefore, the advantage of DCE applied to CAE is that it can provide a basis for preoperative determination of the extent of resection at the imaging level by clarifying the presence of the marginal zone of the lesion and assessing the extent of blood-brain barrier disruption.

#### 10.3.3.5 Magnetic Resonance Spectroscopy (MRS)

Magnetic resonance spectroscopy (MRS) is a technique that can monitor disease progression and treatment by measuring the concentration of metabolites in tissues. Zhang Dan et al. (2019) and Mao Jun Yue (2016) applied the MRS technique for the differential diagnosis of CAE and brain metastases and brain tuberculoma and found that there are differences in the spectral information of CAE and brain metastases and brain tuberculoma, and the differences are statistically significant ($P < 0.05$). Therefore, the MRS technique can reflect the characteristics of CAE lesions from the perspective of physiological metabolism and provide a basis for the differential diagnosis of CAE and other diseases.

## 10.4 Osseoechinococcosis

Osseoechinococcosis refers to a series of clinical symptoms and signs caused by the larva of granulococcus or multilocular echinococcosis parasiticus in bones, accounting for 0.5–4% of all echinococcosis, of which spine and pelvic echinococcosis are the most common (accounting for more than 60%), followed by shoulder blades, ribs, long bones of limbs and other parts. Osteocystic echinococcosis is the most common clinical disease, while osteovesicular echinococcosis is extremely rare. Hydatids grow slowly inside the bone and can take years or longer to develop symptoms. Due to its insidious onset and special growth pattern of hydatid bone, early clinical diagnosis and treatment are difficult, and patients are often in the middle and late stages of treatment, usually resulting in extremely serious dysfunction and even death.

Steoechinococcosis is mainly a primary infection caused by blood circulation, and secondary osteoechinococcosis is rare. The distribution of echinococcus into the systemic circulation in bone depends on local blood supply and growth activity. The lesions usually develop into hydatid cysts from the rich blood supply and vigorous growth of cancellous bone, diaphysis of the long shaft or bone marrow cavity. Therefore, the incidence of osteohydatid disease is highest in the pelvis and spine, followed by the diaphysis of the limbs, shoulder blades, ribs and other parts. The echinococcosis in cancellous bone grows along the trabecular space with less resistance. The mechanism of bone destruction is different among echinococcosis.

Echinococcus granulosus mainly causes bone absorption, compression and destruction through invasive and expansive growth, which gradually expands and extends the young sacs in cancellous bone.

In contrast, multilocular echinococcus mainly causes bone destruction through erosive and osteolytic growth, similar to bone tuberculosis and tumor, which develop into cystic and vesicular echinococcosis, respectively. The pattern of expansive growth and aggressive growth of cystohydatid and vesicular hydatid determines the characteristics of different osteolytic bone destruction caused by cystic and multilocular conditions (Pedrosa et al. 2000). Cystic and alveolar bone echinococcosis usually no additional outer fiber coated, inside is no typical germinal layer, but increased gradually can break through the bone cyst wall and surrounding tissues, formed around the soft tissues such as liver, lung echinococcosis fiber coated, namely the film outside.

### 10.4.1 CT Examination

Bone hydatid disease appears on CT film as round or oval low-density bone destruction of varying sizes, sharp and clear lesion edges, and calcification within the cyst or cyst wall. The low-density bone destruction area can be single cystic lesions or multicystic lesions of different sizes. In the presence of multiple ascus, CT images showed that multiple cystic dilatative low-density bone defects of different sizes are clustered in a "grape cluster," and the bone cortex is swollen, thin, fractured or defective. After the lesion invades the peripheral or paraspinal soft tissue, it forms a multilocular cystic mass within the soft tissue, which may have a sclerotic ring. Alveolar hydatid disease is rare. CT images mainly show localized lytic bone destruction without typical cystic or multilocular changes, clear edges, less sclerosis, less regularity, and sand granular bone fragments (Fig. 10.5e–f).

### 10.4.2 MRI Examination

MRI plays an important role in the diagnosis of osteohydatid disease, especially in the diagnosis of spinal hydatid. Because hydatid growth is polycystic, T1WI is a saccular, multilocular low signal, the signal of the capsule wall and muscle may be slightly low, and the ascus is significantly low. T2WI showed a medium-high signal, similar to cerebrospinal fluid, and the clusters are "grape clusters." Both the cyst wall and the cyst contents showed a high signal, and the signal of the cyst contents is significantly higher than that of the cyst wall. When the cyst is filled with ascus of different sizes or arranged around the cyst wall, the entire lesion is rosette or wheel-shaped (Fig. 10.5a–d). When hydatid cyst co-infection, T1WI and T2WI significantly enhanced the signal, especially T2WI enhancement. The MRI findings of osteoclastic hydatid disease are osteolytic bone destruction, very similar to tuberculosis and tumors. T1WI showed a mixed low signal and T2WI showed a mixed high signal with marginal ambiguity and no typical cystic or polycystic changes

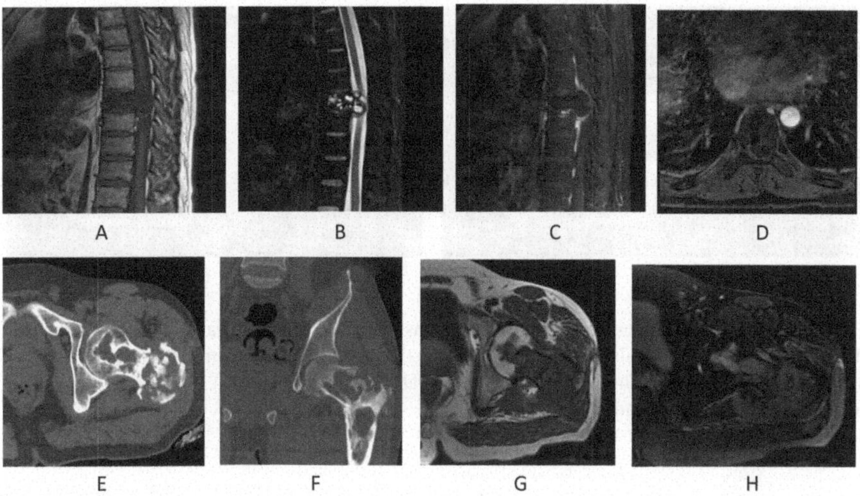


**Fig. 10.5** MRI of vertebral echinococcosis showed vertebral bone destruction, bulging into the spinal canal, spinal cord compression, cystic hypointensity on T1WI (**a**), medium to high intensity on T2WI (**b**), similar to cerebrospinal fluid, hyperintensity on both the cyst wall and cyst contents, and the signal of the cyst contents was significantly higher than that of the cyst wall, and the signal of T2 lipid pressure was hypointense (**c**), and no enhancement was shown on contrast scan (**d**). Imaging of bone echinococcosis in (**e**–**h**), CT showed osteolytic bone destruction at the upper end of the left femur (**e**–**f**); MRI showed lesions showing mixed hypointensity at T1WI (**g**) and T2 lipid pressure showed mixed hyperintensity with blurred margins without typical cystic or polycystic changes (**h**)

(Fig. 10.5g–h). Secondly, when hydatid invades peripheral or parapertebral soft tissues, uniform and continuous low-signal outer capsules can be seen outside the cyst wall, which is similar to the MRI findings of soft tissue hydatid disease (Sun et al. 2022). In addition, compared with X-ray and CT, MRI has better soft tissue imaging ability. For hydatids at various sites, especially those within the spinal canal and paravertebral hydatids, MRI can not only make a clear diagnosis, but also clearly show the number and location of hydatids cysts, providing a basis for further classification and treatment.

## 10.5 Oral and Maxillofacial Cervical Echinococcosis

Oral and maxillofacial-neck hydatid disease is a relatively rare site (about 1–2%), including parotid gland and submandibular gland, tongue, maxillary sinus, infratemporal fossa, parapharyngeal space and pterygopalatine fossa (Yu et al. 2022). Hotz and Gottstein first reported echinococcosis in the parapharyngeal space in 1999 (Hotz and Gottstein 1999). CT and MRI can better display the complex anatomical structure of the mass and its surrounding area (Fig. 10.6a–b), and the special imaging manifestations of echydatid disease (such as ascus sign, streamer sign, cyst wall calcification, etc.).

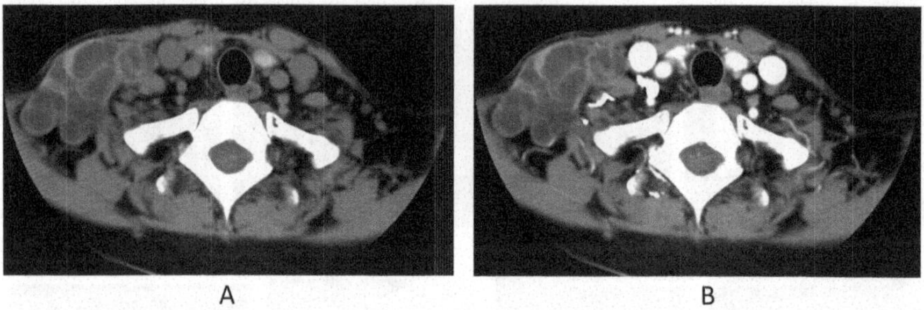


**Fig. 10.6** Imaging findings of cystic echinococcosis in the neck. CT scan showed mixed cystic density clumps in the lateral area of the right cervical root, with a maximum cross-sectional size of about 67 × 36 mm, and multiple thick-walled separations were visible in it (**a**), with mild enhancement of the enhanced septation (**b**)

## 10.6 Adrenal Echinococcosis

The onset of adrenal echinococcosis is insidious, with no obvious clinical symptoms and no abnormal adrenal function. Primary adrenal alveolar echinococcosis is very rare (Jin et al. 2021). On imaging, adrenal echinococcosis is large, can compress the hepatic, grows infiltratively, and is often misdiagnosed as hepatic echinococcosis with adrenal invasion when it is poorly demarcated from the adjacent hepatic parenchyma. Cystic echinococcosis of adrenal gland is characterized by thick walled cystic mass with uniform density or signal. The cystic wall shows circular enhancement on enhanced scan. The imaging features of alveolar echinococcosis are mixed density or signal nodules and masses, with multiple calcifications or small vesicles in them, which may be accompanied by liquefaction areas. Enhanced scan shows no enhancement, and most of them are spread from liver echinococcosis (Fig. 10.7).

## 10.7 Spinal Echinococcosis

Echinococcosis is most often caused by ingesting food contaminated with the eggs of the worms, which enter the portal circulation and enter the hepatic or spread to other areas through the bloodstream. Spinal echinococcosis is rare, with the thoracic spine being the most susceptible (Bai et al. 2021). There are five types of spinal echinococcosis: type 1, intramedullar; type 2, intradural, extramedullar; type 3, extradural; type 4, vertebral; and type 5, paravertebral (Fig. 10.8).

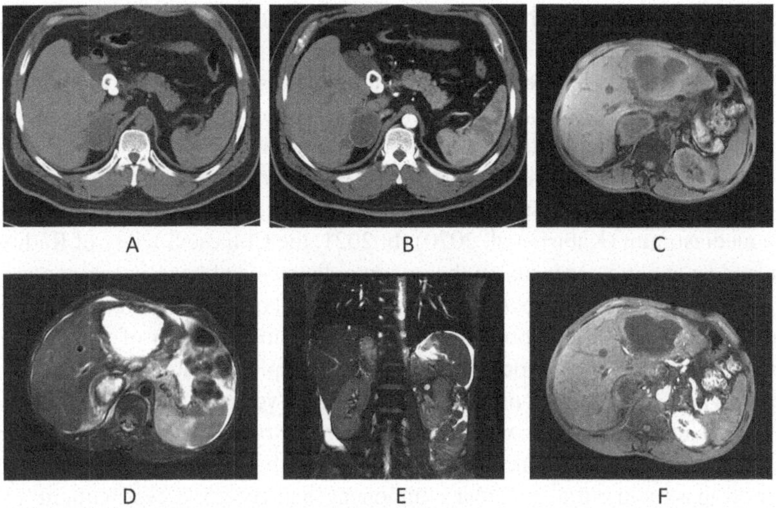


**Fig. 10.7** Imaging findings of adrenal cystic echinococcosis. (**a**–**b**) show that the CT scan shows an irregular cystic predominant mass shadow in the right adrenal gland area, with a superficial lobulated edge and a division within it, and the septum is slightly strengthened after enhancement, with a size of about 50 × 30 mm, adjacent to the right renal vein compression, deformed and displaced. (**c**–**f**) MRI showed a mixed long T2 signal mass shadow in the right adrenal gland (**d**), no enhancement in the lesion on contrast scanning, and irregular mild enhancement at the edge of the lesion (**c**, **e**–**f**)

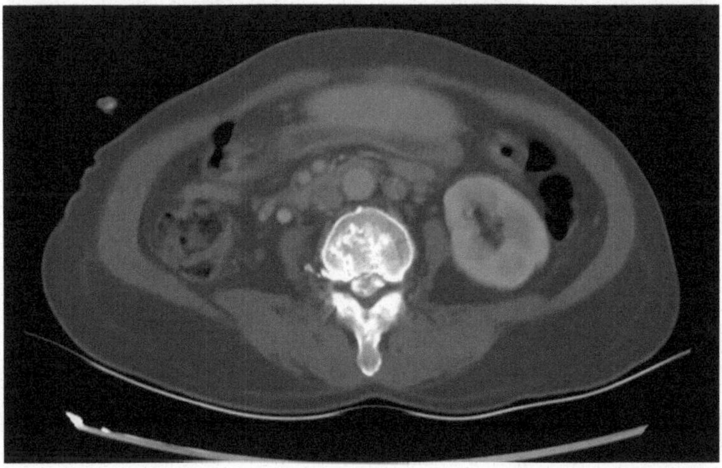

**Fig. 10.8** Spinal echinococcosis CT showed mixed high nodular opacities and spinal stenosis in the lumbar spinal canal

## 10.8 Mediastinal Echinococcosis Disease

echinococcosis is a zoonotic disease with a global distribution, and the prevalence is 0.5–5.0% in the population in the agricultural and pastoral areas of China, one of the most prevalent areas (Liu et al. 2021). It can occur in all body organs, with the liver predominating, followed by the lungs. Intrathoracic extrapulmonary echinococcosis is very rare, with an incidence of 5% to 7%, and occurs in less than 0.1% of the mediastinum (Kabiri et al. 2020). In 2021, the Chinese Society of Radiology organized an expert consensus on the imaging diagnosis of hepatic echinococcosis, which classified echinococcosis into five types based on the imaging characteristics of the cystic and solid components and the distribution pattern of vesicles in the biological evolution of echinococcosis: type 1, multiple vesicles without solid components; type 2, with solid components and multiple vesicles; type 3, surrounded by multiple vesicles; and type 3, with multiple vesicles; type 3, a solid mass with large and/or irregular cysts combined around a center with multiple small vesicles; type 4, a solid mass without a vesicular component; and type 5, a cyst without a solid component. There is no consensus on the staging of vesicular encapsulated disease. The consensus also recommends CT as a routine test for liver encrustation disease and a necessary complementary test for liver encrustation disease (Fig. 10.9).

Mediastinum inclusion disease is rare, and most patients will have clinical symptoms when manifesting an occupying effect. Still, it is not specific, which depends on the size and location of the lesion and the degree of involvement of adjacent structures. Ileri (Ileri et al. 2005) studied 77 patients with mediastinal inclusion

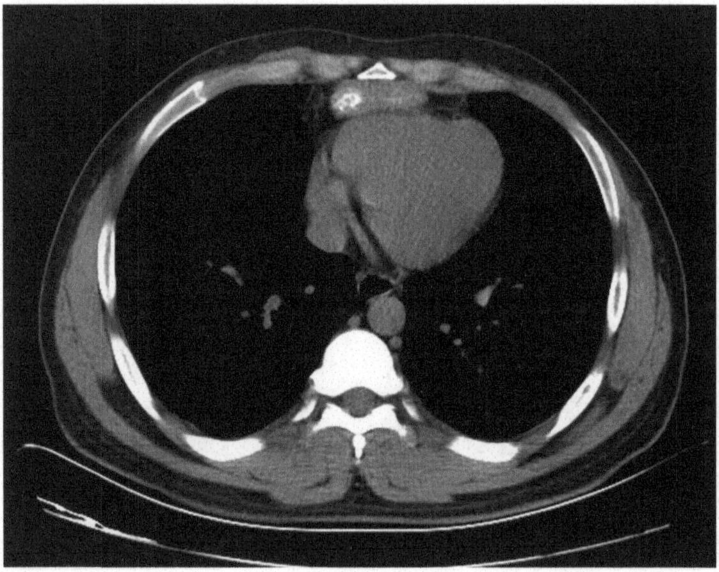

**Fig. 10.9** Mediastinal echinococcosis disease CT showed mixed density mass shadow in the mediastinum with clear borders and calcifications

disease and showed that more than 55% of the lesions are located in the paravertebral sulcus, 36% in the anterior mediastinum, and less than 8% in the heart. It has also been reported in the literature that 36% of the mediastinal inclusion lesions are located in the anterior mediastinum, 45% in the posterior mediastinum, and 18% in the middle mediastinum. These patients are mostly seen for posterior or parasternal pain, cough, swallowing, or dyspnea, and some patients presented with acute pericarditis, acute coronary syndrome, and atrioventricular block as the first symptoms.

The standardized clinical staging of echinococcosis is developed and agreed upon by the WHO expert group on echinococcosis based on Gharbi ultrasound staging, based on its biological evolution (e.g., intracapsular capsule collapse, solidity, or calcification), with which the imaging staging remains largely consistent. In recent years, studies have reported that most echinococcosis in mediastinal encapsulation is predominantly cystic, while other subtypes are less common. CT scan shows intraluminal cystic hypointense shadow, and some of the lesions may show linear calcification in the wall with a significant occupying effect and no enhancement after enhancement.

CMR examination has high soft tissue resolution and can fully reveal the composition and structure of the encapsulated lesions, which can provide more supplementary information for clinical diagnosis and treatment. In this case, the lesion had a thin wall and uniform intracavitary signal, with a low signal on T1WI, a high signal on cine sequences, and a slightly high signal on black blood lipid suppression sequences. The reason for these imaging manifestations is that the composition of the intracavitary component of the lesion is a colorless, transparent fluid with predominantly plasma material mixed with jelly-like material that can be absorbed and concentrated due to degeneration, which is consistent with what is seen intraoperatively. Some studies have reported that the composition of the intraluminal component of the active cysticercosis foci may include the deposition of protoplasmic cephalopodia or cysticercosis sand, and this phenomenon may present a mixed signal in the T1mapping sequence, which is different from the homogeneous signal of the late cysticercosis foci in the biological evolution process. The speculated reason may be the luminal cephalic segments of advanced cystic encysted lesions are dead and mechanized, and the intracavitary signal is more homogeneous than the active encysted lesions, which can be used clinically to determine the active stage of the lesion and guide the use of antiparasitic drugs. However, this needs to be confirmed by further data collection.

In summary, the CT and MRI imaging manifestations of echinococcosis are characteristic and must be combined with clinical history and laboratory tests to make a comprehensive diagnosis, which will help the clinic select the appropriate surgical plan and improve the quality of patient survival.

## References

Bai X, Li M, Xing Q et al (2021) One case of intraspinal epidural echinococcosis. China Med Imaging Technol 37(01):99. https://doi.org/10.13929/j.issn.1003-3289.2021.01.023

Bhutani N, Kajal P (2018) Hepatic echinococcosis: a review [J]. Ann Med Surg (Lond) 36:99–105. https://doi.org/10.1016/j.amsu.2018.10.032

Cao Y (2018) Progress in imaging diagnosis of echinococcosis. J Pract Radiol 34(9):1461–1464. https://doi.org/10.3969/j.issn.1002-1671.2018.09.038

Filippou (2013) History of cystic echinococcosis. CRC Press, Broken Hill

Gharbi HA, Hassine W, Brauner MW et al (1981) Ultrasound examination of the hydatic liver. Radiology 139(2):459–463. https://doi.org/10.1148/radiology.139.2.7220891

Li H (2016) Imaging of parasitic disease. Science Press, Beijing

Hao L, Chen H, Wang H (2020) Evaluation of cerebral blood flow perfusion and collateral circulation establishment in symptomatic stroke patients using MRI-PWI [J]. Magn Reson Imaging 11(10):904–907. https://doi.org/10.12015/issn.1674-8034.2020.10.014

Hotz MA, Gottstein B (1999) Cystic echinococcosis of the parapharyngeal space: case report with a 20-year follow-up. J Oral Maxillofac Surg 57(1):80–83. https://doi.org/10.1016/s0278-2391(99)90641-9

Ileri M, Hisar I, Atak R et al (2005) A pericardial hydatid cyst masquerading as acute inferolateral myocardial infar a case report. Angiology 56:637–640. https://doi.org/10.1177/000331970505600518

Infectious Disease Imaging Group, Radiology Branch, Chinese Medical Association, Infection Imaging Special Committee, Radiology Branch, Chinese Medical Doctor Association (2021) Expert consensus on imaging diagnosis of liver hydatid disease. Chin J Radiol 55(01):5–11. https://doi.org/10.3760/cma.j.cn112149-20191204-00960

Jiang C, Chen H, Wang J et al (2015) The application of diffusion weighted magnetic resonance imaging in the differential diagnosis of alveolar echinococcosis and high-grade gliomas. J Xinjiang Med Univ 38(11):1335–1338. https://doi.org/10.3969/j.issn.1009-5551.2015.11.002

Jin R, Xu Y, Yi W et al (2021) A case report of misdiagnosis of adrenal alveolar echinococcosis Chinese. J Urol 42(12):937–938. https://doi.org/10.3760/cma.j.cn112330-20190707-00314

Kabiri EH, Hammoumi ME, Bhairis M et al (2020) Mediastinal calcified hydatid cyst. Acta Chir Belg 120(2):8. https://doi.org/10.1080/00015458.2020.1722932

Kern P, Wen H, Sato N et al (2006) WHO classification of alveolar echinococcosis: principles and application. Parasitol Int 55(Suppl):S283–S287. https://doi.org/10.1016/j.parint.2005.11.041

Kodama Y, Fujita N, Shimizu T et al (2003) Alveolar echinococcosis: MR findings in the liver. Radiology 228(1):172–177. https://doi.org/10.1148/radiol.2281020323

Li T, Bao H (2020) Progress in the application of multimodal imaging technology in hepatic cystic echinococcosis. Magn Reson Imaging 11(11):1059–1062. https://doi.org/10.12015/issn.1674-8034.2020.11.024

Liu J (2019) Quantitative analysis of magnetic resonance perfusion imaging in the diagnosis and differential diagnosis of cerebral alveolar echinococcosis and brain metastasis [D]. Xinjiang Medical University, Urumqi

Liu W, Jiang Y, Wang J (2021) Expert consensus on imaging diagnosis of hepatic hydatidosis. J Clin Hepatobiliary Dis 37(04):792–797

Lupia T, Corcione S, Guerrera F et al (2021) Pulmonary echinococcosis or lung hydatidosis: a narrative review. Surg Infect 22(5):485–495. https://doi.org/10.1089/sur.2020.197

Ma J, Wang T, Wang J et al (2015) A study on the correlation between quantitative indicators of 1H MRS and DWI in the marginal zone of cerebral alveolar echinococcosis. J Xinjiang Med Univ 38(11):1331–1334. https://doi.org/10.3969/j.issn.1009-5551.2015.11.001

Ma L, Chen DC, Zou SY et al (2020) Epidemiological characteristics of hepatic echinococcosis, concurrent cerebral echinococcosis, and pulmonary echinococcosis in Ganzi County, Sichuan Province, China [J/OL]. Medicine 99(15) [2022-04-28]. https://doi.org/10.1097/MD.0000000000019753

Mao J (2016) The study on the diagnostic value of MRS for cerebral alveolar echinococcosis and cerebral tuberculoma [D]. Qinghai University, Xining

Mao T, Chungda D, Phuntsok L et al (2019) Pulmonary echinococcosis in China. J Thorac Dis 11(7):3146–3155. https://doi.org/10.21037/jtd.2019.07.31

McManus DP, Gray DJ, Zhang W, Yang Y (2012) Diagnosis, treatment, and management of echinococcosis. BMJ 344:e3866. https://doi.org/10.1136/bmj.e3866
Miriguri S, Jia W (2010) MRI manifestations and diagnosis of cerebral echinococcosis. Chinese J Radiol 44(07):700–703. https://doi.org/10.3760/cma.j.issn.1005-1201.2010.07.006
Pedrosa I, Saíz A, Arrazola J et al (2000) Hydatid disease: radiologic and pathologic features and complications. Radiographics 20(3):795–817
Qian MB, Zhou XN (2018) Walk together to combat echinococcosis. Lancet Infect Dis 18(9):946. https://doi.org/10.1016/S1473-3099(18)30466-3
Sun Y, Yan C, Tian D, Zhang C, Zhang Q (2022) Imaging manifestations and misdiagnosis analysis of six cases of bone hydatid disease. Korean J Parasitol 60(6):413–417. https://doi.org/10.3347/kjp.2022.60.6.413
Wen H, Luan M, Yang W et al (2002) Standardized classification and clinical significance of hepatic hydatidosis. J Xinjiang Med Univ 25(2):129–130. https://doi.org/10.3969/j.issn.1009-5551.2002.02.005
Wen H, Vuitton L, Tuxun T et al (2019) Echinococcosis: advances in the 21st Century [J]. Clin Microbiol Rev 32(2). https://doi.org/10.1128/CMR.00075-18
Yang J (2019) The application of IVIM-DWI imaging in cerebral alveolar hydatid disease and its marginal zone. Xinjiang Medical University, Urumqi
Yu P, Li C, Gong Z (2022) A case of oral and maxillofacial echinococcosis [J/CD]. China Clin Case Achiev Database 04(1):E00235–E00235. https://doi.org/10.3760/cma.j.cmcr.2022.e00235
Zhang D (2019) The application and histogram correlation analysis of DCE-MRI in the marginal zone of cerebral alveolar echinococcosis [D]. Xinjiang Medical University, Urumqi
Zhang J, Xia C, Liu Y et al (2018) Comparative study of MR mTI-ASL and DSC-PWI in evaluating cerebral hemodynamics of patients with Moyamoya disease. Medicine 97(41):e12768–e12768. https://doi.org/10.1097/MD.0000000000012768
Zhang D, Liu J, Wang J et al (2019) 1H MRS quantitative analysis for differential diagnosis of cerebral alveolar echinococcosis and brain metastasis. China Med Imaging Technol 35(5):659–663. https://doi.org/10.13929/j.1003-3289.201809067

# Ultrasound in the Diagnosis of Echinococcosis

# 11

Diming Cai and Xuhui Zhang

**Abstract**

Echinococcosis is a parasitic disease caused by the larva of cestode of the genus *Echinococcus* in humans and is a common zoonosis in pastoral areas. Since the liver is the most commonly affected organ in echinococcosis, echinococcosis typically refers to hepatic echinococcosis (HE). Two main types of HE are prevalent in humans and livestock. One is cystic echinococcosis (CE) caused by infection with *Echinococcus granulosus*. The other is alveolar echinococcosis (AE), which is caused by infection with *Echinococcus multilocularis*. Ultrasound is now widely recognized as the preferred imaging modality for diagnosing HE. It is intuitive, non-radiative, inexpensive, and has a high diagnostic accuracy rate of up to 98%. The application of ultrasound to diagnose HE began in the late 1950s and was first reported in the literature in 1964. CE and AE, the two types of echinococcosis, differ significantly in terms of morphology, epidemiology, pathology, clinical course, prognosis, and clinical treatment. Especially for AE, it has a tumor-like growth pattern and is prone to be misdiagnosed as liver cancer. Therefore, it is crucial to understand the ultrasonographic characteristics of CE and AE, respectively, to make a correct diagnosis.

**Keywords**

Echinococcosis · Diagnosis · Ultrasound · Contrast-enhanced ultrasound · Extrahepatic echinococcosis · Differential diagnosis

D. Cai (✉) · X. Zhang
Department of Medical Ultrasound, West China Hospital, Sichuan University, Chengdu, Sichuan Province, China





J. Li et al. (eds.), *Echinococcus: Control and Elimination of Echinococcosis with a Focus on China and Europe*, Parasitology Research Monographs 19,
https://doi.org/10.1007/978-3-031-54479-8_11

**Terminology**
Echinococcus
Echinococcosis
Cystic echinococcosis, CE
Hepatic cystic echinococcosis, HCE
Alveolar echinococcosis, AE
Hepatic alveolar echinococcosis, HAE
*Echinococcus granulosus, E. granulosus*
*Echinococcus multilocularis, E. multilocularis*
Echinococcosis is a parasitic disease caused by the larva of cestode of the genus *Echinococcus* in humans and is a common zoonosis in pastoral areas. Since the liver is the most commonly affected organ in echinococcosis, echinococcosis typically refers to hepatic echinococcosis (HE). Two main types of HE are prevalent in humans and livestock. One is cystic echinococcosis (CE) caused by *Echinococcus granulosus* (*E. granulosus*) infection. The other is alveolar echinococcosis (AE) caused by *Echinococcus multilocularis* (*E. multilocularis*) infection.

Ultrasound is now widely recognized as the preferred imaging modality for diagnosing HE. It is intuitive, non-radiative, inexpensive, and highly accurate, with a diagnostic accuracy rate of up to 98% for CE. The application of ultrasound to diagnose HE began in the late 1950s and was first reported in the literature in 1964.

CE and AE, the two types of echinococcosis, differ significantly in terms of morphology, epidemiology, pathology, clinical course, prognosis, and clinical treatment. Especially for AE, it has a tumor-like growth pattern and is prone to be misdiagnosed as liver cancer. Therefore, it is crucial to understand the ultrasonographic characteristics of CE and AE, respectively, to make a correct diagnosis.

## 11.1 Ultrasound Diagnosis of Cystic Echinococcosis

### 11.1.1 Overview

Cystic echinococcosis (CE) is caused by the larva of *Echinococcus granulosus* (*E. granulosus*) and can develop from a larva in any organ of the human body. More than 80% of patients with CE involved only one organ and eventually form a cystic lesion. According to statistics, 80% of CE cases occur in the liver, and 20% in the lungs. The proportion of CE occurrence in the liver and lungs can range from 2:1 to 7:1, possibly even higher. For CE, the hydatid cyst in the germinal layer gradually expands and develops, eventually forming the protoscolex daughter cyst and entering the central cavity filled with clear fluid in the cystic lesion. The periphery of the CE cyst is surrounded by components of the non-cellular layer, mainly derived from the host response. The size and number of daughter cysts in CE vary greatly, usually appearing inside the mother cyst but can also appear outside (Fig. 11.1).

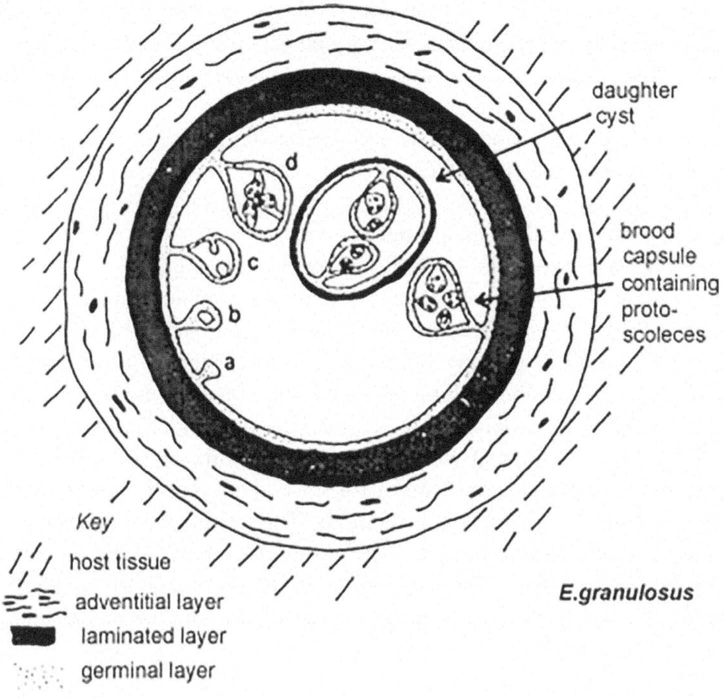


**Fig. 11.1** Schematic diagram of the structure of CE cyst. (Adapted from Brunetti E, Kern P, Vuitton DA. Expert consensus for diagnosing and treating cystic and alveolar echinococcosis in humans. Acta Trop. 2010)

### 11.1.2 Growth and Change Process

Ultrasound examination shows that CE lesion grows at a rate of about 1–50 mm per year or may remain stable in volume for a long time without significant enlargement. During the development of the disease, the cystic lesion may undergo spontaneous rupture, collapse, or even disappear. So far, the development of cysts in the human body is still not fully figured out. In general, CE lesion in the liver develops more slowly than in the lungs. Corresponding clinical symptoms will occur when the cystic lesion compresses the surrounding tissues or ruptures.

### 11.1.3 Diagnosis

The diagnosis of CE mainly includes clinical manifestation, imaging examination, and serological tests. Examination of the protoscolex also requires pathological and histological examination of the cyst fluid.

Ultrasound examination is the basis of CE diagnosis in abdominal locations at individual and population levels. Ultrasound may visualize cysts in other organ locations, including those in the extrahepatic tissues.

There is currently no unified standard for classifying the ultrasonographic features of CE. Most scholars advocate classifying according to the sonographic characteristics of the cyst, while others classify based on the sonographic appearance of the cyst in different stages of natural development. The following principles can be used as a reference for the appropriate and reasonable sonographic classification of hepatic cystic echinococcosis (HCE):

1. Differentiation between single, recurrent, and disseminated lesions. The classification should be based on the primary lesion.
2. Differentiation between single, multiple, and lesion parts.
3. Each sonographic type should have different sonographic characteristics.
4. Able to reflect the natural development and pathological changes at different stages of CE and have guiding significance for choosing surgical plans.

This section will focus on the WHO sonographic classification method for HCE. This method is also the most widely accepted classification standard internationally.

In 1995, the WHO Informal Working Group on Echinococcosis (WHO-IWGE) developed a standardized classification for HCE that could be applied in all settings to replace many previous classifications. The lesions were divided into three groups based on the natural activity within the cyst: active (CE1 and 2), transitional (CE3), and inactive (CE4 and 5). The WHO classification standard differs significantly from that proposed by Gharbi in 1981, which had the most widespread impact. The WHO sonographic classification added a "cystic lesion" (CL) stage and reversed the order of CE2 and 3 (Fig. 11.2). CE3 transitional cysts may be differentiated into CE3a (with detached endocyst) and CE3b (predominantly solid with daughter vesicles). CE1 and CE3a are early stages and CE4 and CE5 are late stages.

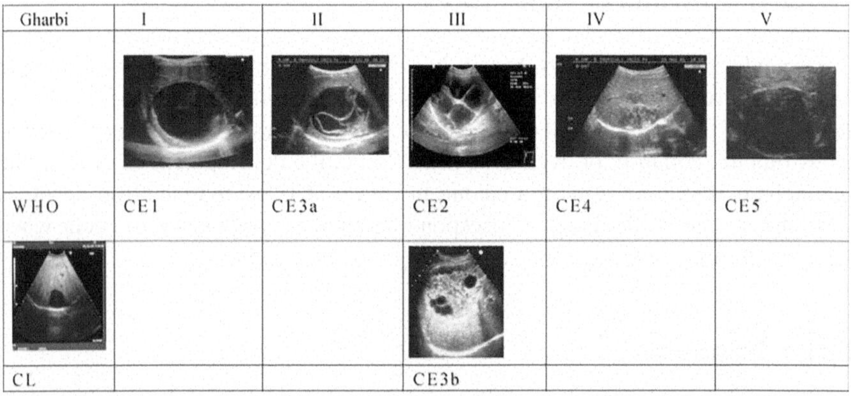


**Fig. 11.2** Comparison of Gharbi's and WHO-IWGE ultrasound classification. CL, as a potentially parasitic cyst, was not in Gharbi's. WHO CE3b had not been explicitly described by Gharbi. CE3b might be classified as Type III, although in the original Gharbi paper, there was no distinction between multivesiculated (honeycomb-like) cysts and cysts with daughter cysts in a solid matrix. (Adapted from Brunetti E, Kern P, Vuitton DA. Expert consensus for the diagnosis and treatment of cystic and alveolar echinococcosis in humans. Acta Trop. 2010)

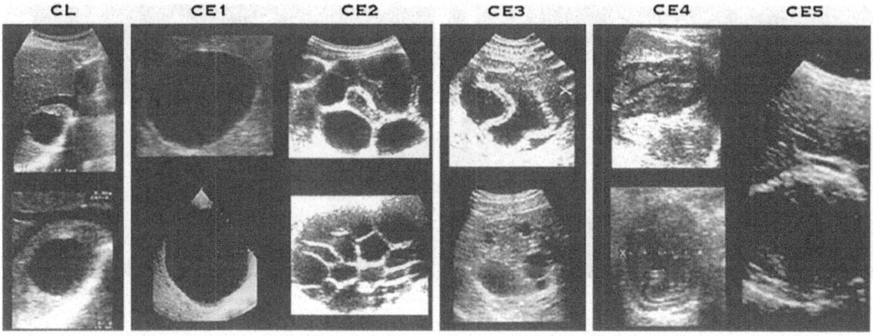


**Fig. 11.3** WHO-IWGE standardized classification. (Adapted from Brunetti E, Kern P, Vuitton DA. Expert consensus for the diagnosis and treatment of cystic and alveolar echinococcosis in humans. Acta Trop. 2010)

In 2010, the WHO-IWGE adopted an ultrasonographic standardized classification for HCE. Based on the sonographic manifestations of HCE, this standard classifies CE lesions into six types: (1) cystic lesion (CL); (2) unilocular cyst (CE1); (3) multivesicular cyst (CE2); (4) ruptured endocyst (CE3); (5) solidified cyst (CE4); (6) calcified cyst (CE5). Among them, Type III can be further divided into CE3a and CE3b subtypes (Fig. 11.3).

The advantage of the WHO classification is that it provides a more relatively accurate prediction of the activity of CE lesions based on ultrasound findings, which can be very helpful for guiding the treatment of CE. The sonographic characteristics of each type will be introduced based on this classification.

### 11.1.4 WHO Classification of HCE

#### 11.1.4.1 Cystic Lesion (CL Type)

A common finding in the liver is a uniformly homogeneous single anechoic area with a round or oval shape, a clear contour and unclear wall, and homogeneous echogenic content inside, with obvious posterior acoustic enhancement. This type is difficult to distinguish from liver cysts. A history of contact with echinococcus-endemic areas is an important basis for diagnosing the CL type, which requires long-term follow-up and observation (Fig. 11.4).

#### 11.1.4.2 Unilocular Cyst (CE1 Type)

This type is the most common, accounting for about 70% of CE cases. It presents as a round or oval-shaped, homogeneous, single anechoic area with a diameter usually larger than 3.0 cm. The cyst has a clear outline, an intact membrane, a clear boundary with liver tissue and a thick wall with posterior acoustic enhancement (Fig. 11.5). The hydatid cyst is filled with clear or slightly turbid fluid. The density difference between the cystic wall and liver tissue is significant, and the cystic wall appears as a clearly circumscribed boundary. The specific imaging feature of this type is a

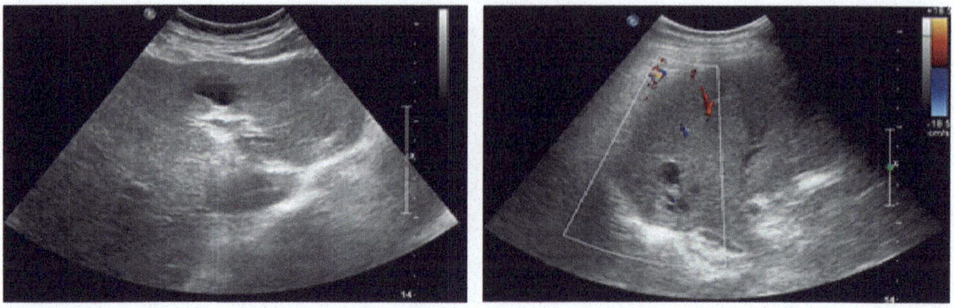

**Fig. 11.4** A 43-year-old male with cystic lesions in the left and right liver of about 1.0 cm. Due to the patient's history of living in a region endemic to echinococcosis, the diagnosis of CL type was made. 1. Ultrasound shows a 1.0 cm cystic lesion in the left liver. 2. Color Doppler flow imaging of the 1.0 cm cystic lesion in the right liver

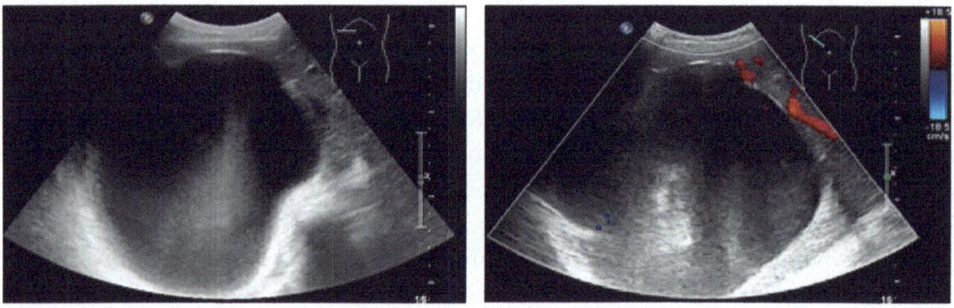

**Fig. 11.5** A 27-year-old female with a 13.4 × 15.4 × 15.6 cm anechoic mass in the right liver, with a clearly circumscribed boundary and a relatively regular shape, without obvious blood flow signals inside. Postoperative pathological results revealed hepatic echinococcosis and *E. granulosus*. 1. Gray-scale image of the giant cystic anechoic mass in the right liver. 2. Color Doppler flow imaging of the mass

potential gap interface between the inner and outer cystic walls, known as the "double wall sign" (Fig. 11.6).

Gray-scale ultrasound can detect the hydatid cyst with obvious posterior acoustic enhancement. When the probe vibrates the cyst, small floating bright spots can be seen in the anechoic area, often called "hydatid sand." Some people also call it "snow flake sign" (Figs. 11.7 and 11.8).

#### 11.1.4.3 Multivesicular Cyst (CE2 Type)

Numerous smaller "daughter cysts" of varying sizes and numbers are found within a unilocular larger "mother cyst." The "daughter cysts" appear as smaller spherical shadows or aureoles with a smooth inner wall and anechoic echogenicity, forming the characteristic "cyst-within-cyst sign" of CE (Fig. 11.9). When there are too many "daughter cysts," they appear as numerous circular aureoles or irregularly and numerously "honeycomb-like" changes separated by multiple thin septa. This presents an ultrasound as a petal-like appearance called a "wheel-like sign" or "honeycomb-like sign" (Fig. 11.10).

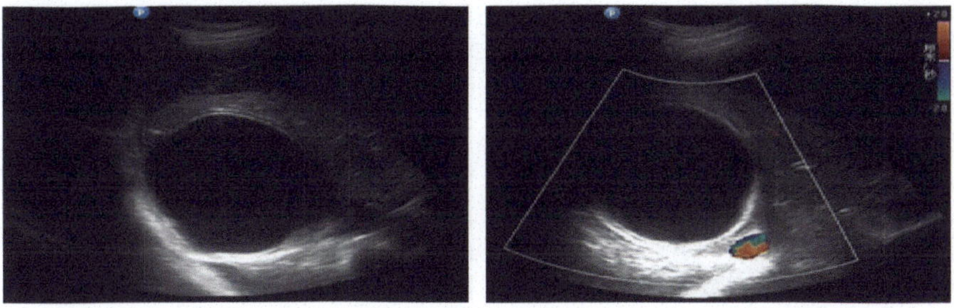

**Fig. 11.6** A 25-year-old female with an 8.7 × 9.0 × 12.6 cm anechoic mass in the right liver. The cystic wall was approximately 4 mm thick, showing the "double wall sign." The boundary was relatively clear and the shape was regular, without blood flow signals. Postoperative pathological results showed hepatic echinococcosis and *E. granulosus*. 1. Gray-scale image of the cystic mass in the right liver. 2. Color Doppler flow imaging of the mass

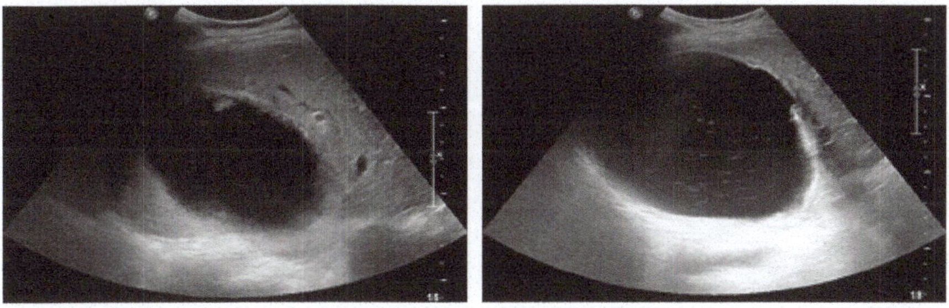

**Fig. 11.7** A 36-year-old female with a 12.0 × 11.0 × 13.0 cm anechoic mass in the right liver. Fine and weak bright spots were seen inside. The wall is not smooth, and nodular echoes with slightly hyperechoic echogenicity were attached to the inner wall, the largest of which was about 1.1 cm. The nodules are relatively regular in shape, without blood flow signals inside the nodules or the mass. Postoperative pathological results showed hepatic echinococcosis and *E. granulosus*. 1. Nodular echoes with slightly hyperechoic echogenicity were attached to the inner wall of the anechoic mass in the right liver. 2. Floating "hydatid sand" within the cyst

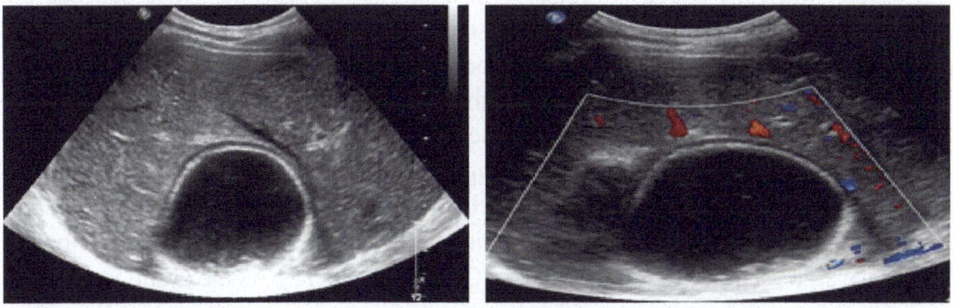

**Fig. 11.8** A 41-year-old male with a 10.0 × 6.5 × 7.2 cm anechoic mass in the right liver. The cystic wall exhibited the "double wall sign" with floating echogenic spots inside, indicating the "snow flake sign." The boundary was relatively clear and the shape was irregular, without obvious blood flow signals inside. Postoperative pathological results showed hepatic echinococcosis and *E. granulosus*. 1. Typical gray-scale image of the anechoic hydatid cyst with the "double wall sign" and "snow flake sign" inside in the right liver. 2. Color Doppler flow imaging of the mass

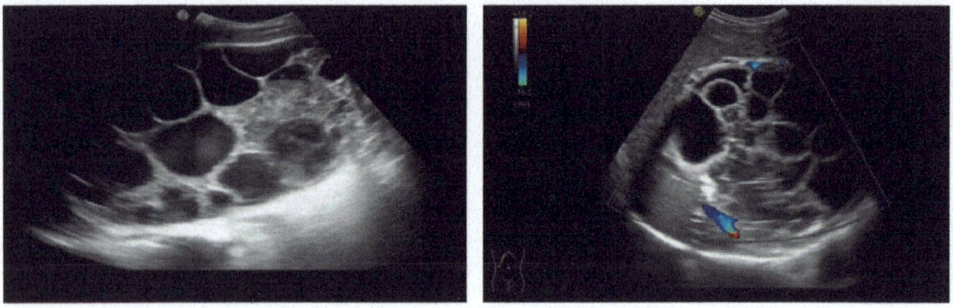

**Fig. 11.9** A 31-year-old male with a 15.0 × 14.0 × 16.0 cm anechoic mass in the liver. The boundary was relatively clear and the shape was relatively regular, without obvious blood flow signals inside. Postoperative pathological results showed hepatic echinococcosis and *E. granulosus*. 1. Gray-scale image of the mass with the typical "cyst-within-cyst sign." 2. Color Doppler flow imaging of the "mother cyst"

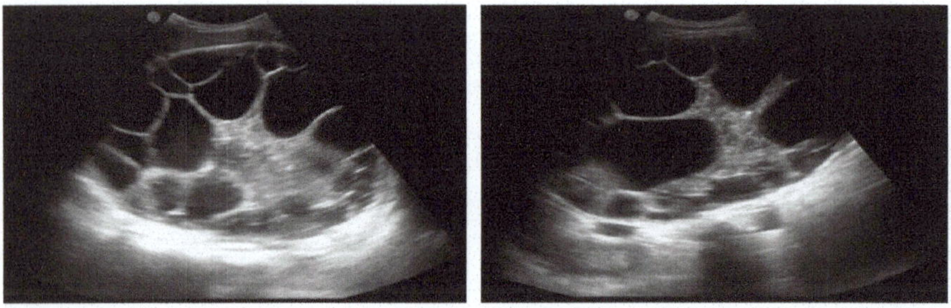

**Fig. 11.10** A 43-year-old male with a 10.0 × 15.0 × 17.0 cm septal mass in the liver. The septa varied in size with clear boundaries. Postoperative pathological results showed hepatic echinococcosis and *E. granulosus*. 1 and 2. Gray-scale images showed the typical "honeycomb-like sign" of the mass

#### 11.1.4.4 Ruptured Endocyst (CE3 Type)

The endocyst wall of the cyst can rupture due to damage from external factors or infection inside. When the endocyst ruptures, the cystic fluid enters the space between the inner and outer cystic walls, separating the inner cystic wall. The gap between the two layers widens, forming an obvious dark band of varying width. Alternatively, the endocyst wall collapses into the cystic fluid with a thickened and rough wall, showing the "nested cyst sign" (Fig. 11.11). If the cystic wall completely separates from the outer cystic wall, a "water-lily sign" is visible (Fig. 11.12), followed by collapse and inward shrinkage of the cystic wall, resulting in curled and folded with twisting and irregularly moving bright or hyperechoic bands within the cystic fluid, known as the "floating membrane sign." This type is referred to as CE3a (Fig. 11.13). As the cystic lesion contains significant solid components, the disease progresses to CE3b (Fig. 11.14).

#### 11.1.4.5 Solidified Cyst (CE4 Type)

As CE lesion gradually degenerates, necrotizes, and calcifies, it still retains a spherical shape, with a significantly thickened cystic wall unevenly and a rough echogenic boundary. The cystic fluid is absorbed and reduced, appearing as a paste-like or cheesy echogenicity with scattered, uneven, and dense hyperechoic spots or areas

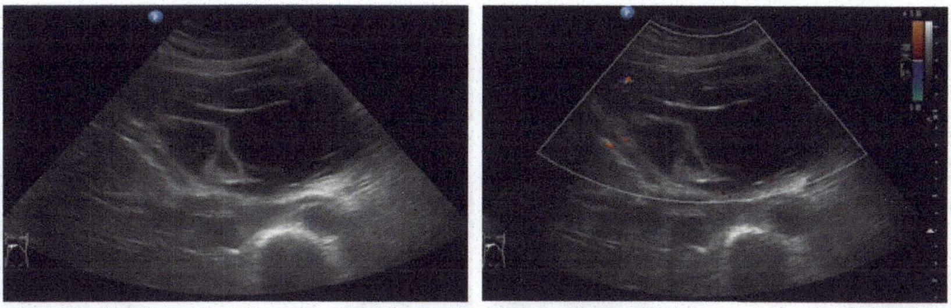

**Fig. 11.11** A 42-year-old female with a 12.5 × 8.5 cm mixed echogenic mass in the left external lobe. The boundary was clear and the shape was regular, without obvious blood flow signals inside. Linear septa and flocculent hypoechoic areas were seen inside. Postoperative pathological results showed hepatic echinococcosis and *E. granulosus*. 1. Gray-scale image of the mass with the typical "nested cyst sign" in the left liver. 2. Color Doppler flow imaging of the mass

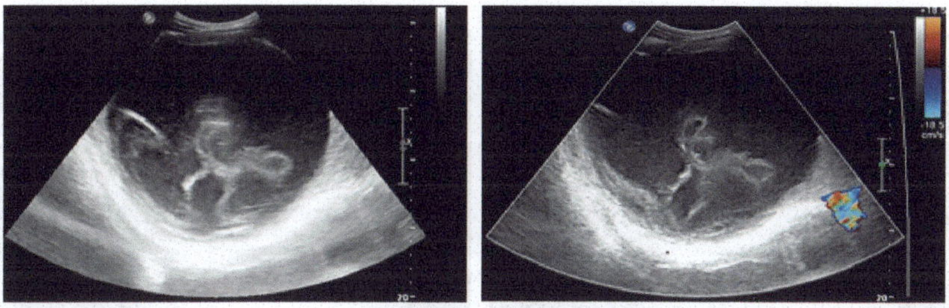

**Fig. 11.12** A 31-year-old male with a 16.0 × 15.0 cm anechoic mass in the right liver. The boundary was clear, and the shape was regular, with curled echogenic bands and no blood flow signals inside. Postoperative pathological results showed hepatic echinococcosis and *E. granulosus*. 1. Gray-scale image of the mass with the typical "water-lily sign" in the right liver. 2. Color Doppler flow imaging of the mass

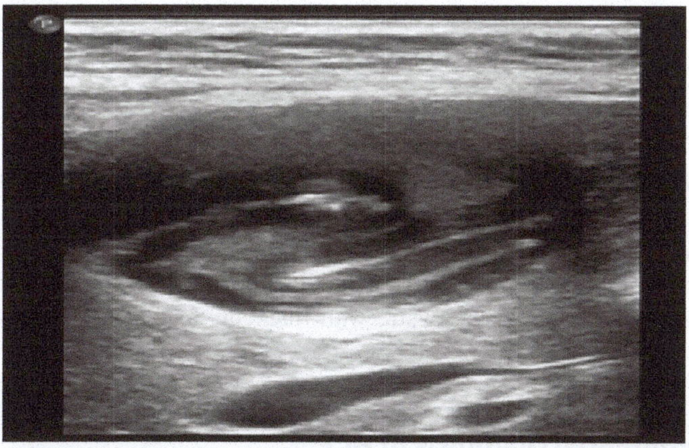

**Fig. 11.13** A 28-year-old female with a 3.3 × 2.8 cm hypoechoic mass in the left external lobe. The boundary was relatively clear, and the shape was relatively regular, with curled hyperechoic bands and no obvious blood flow signals inside. Postoperative pathological results showed hepatic echinococcosis and *E. granulosus*. Gray-scale image of the mass with the typical "floating membrane sign" in the left liver, which is referred to as CE3a

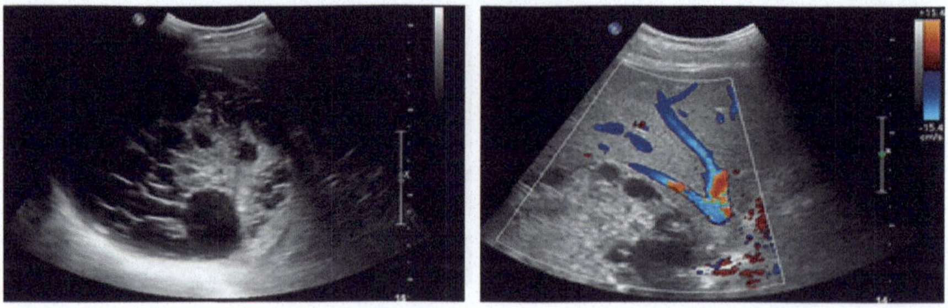

**Fig. 11.14** A 43-year-old male with a 14.5 × 12.0 cm cystic-solid mixed echogenic mass in the liver. The boundary was clear and the shape was relatively regular, without obvious blood flow signals inside. Postoperative pathological results showed hepatic echinococcosis and *E. granulosus*. 1. Gray-scale image of the septal cystic mass with increased solid components, which is referred to as CE3b. 2. Color Doppler flow imaging of the mass

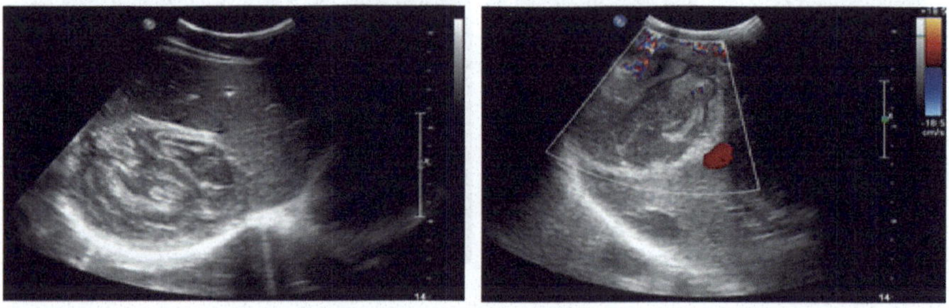

**Fig. 11.15** A 36-year-old male with an 8.5 × 6.5 × 5.5 cm uneven echogenic mass in the right liver. The boundary was relatively clear and the shape was relatively regular. Postoperative pathological results showed hepatic echinococcosis and *E. granulosus*. 1. Gray-scale image of the mass with the typical "gyrus sign" in the right liver. 2. Color Doppler flow imaging of the mass

within the lesion. Ultrasound examination shows a "gyrus sign" with alternating strong and weak density inside (Fig. 11.15).

#### 11.1.4.6 Calcified Cyst (CE5 Type)

The cystic wall of the lesion is thick and rough, showing a hyperechoic spherical aureole or rim-like change, resembling "eggshell" or "tile edge." The contents in the cyst appear as heterogeneous hypoechoic or anechoic echogenicity, or irregular hyperechoic spot, patch, or a combination of both, suggesting the presence of calcifications and that the hydatid is likely dead. Ultrasound shows that the density of the hydatid increases unevenly, the cystic wall thickens flocculently, with broad and lateral shadows (Figs. 11.16 and 11.17).

### 11.1.5 Differential Diagnosis of HCE

The diagnosis of HCE is usually based on the patient's history of living in endemic areas of the disease, exposure to livestock such as cattle, sheep, and dogs, as well as symptoms, local signs, biological tests, and ultrasonographic features. However, differential diagnosis should be considered for the following hepatic diseases.

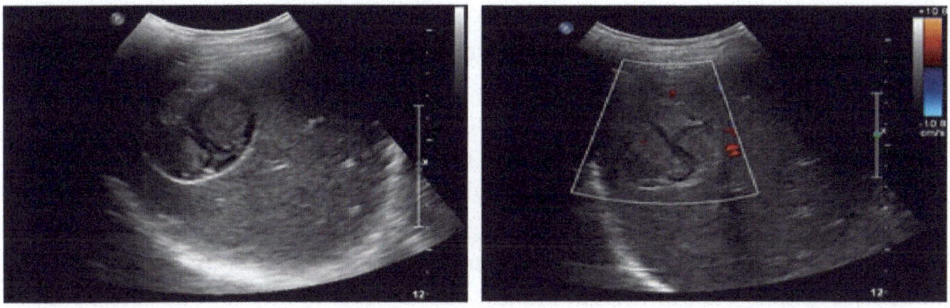

**Fig. 11.16** A 46-year-old male with a 5.0 × 4.5 cm hypoechoic mass in the right posterior lobe. The boundary was clear and the shape was regular, with uneven echo and without obvious blood flow signals inside. Postoperative pathological results showed hepatic echinococcosis and *E. granulosus*. 1. Gray-scale image of the mass in the right liver presented as typical "eggshell-like," with scattered punctiform calcifications inside. 2. Color Doppler flow imaging of the mass

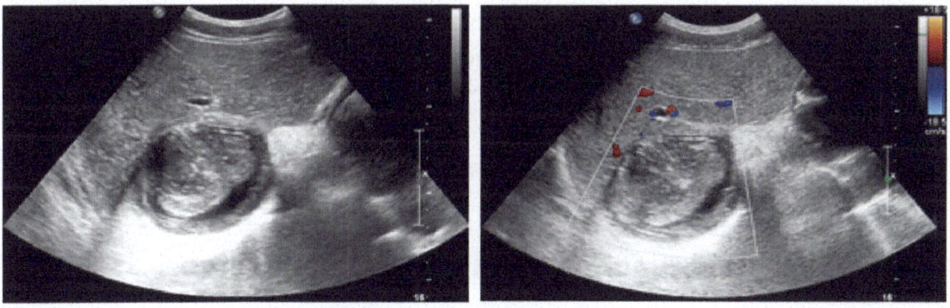

**Fig. 11.17** A 31-year-old male with an 8.2 × 6.7 × 7.5 cm mixed echogenic mass in the right posterior lobe. The boundary was relatively clear and the shape was relatively regular, without obvious blood flow signals inside. Postoperative pathological results showed hepatic echinococcosis and *E. granulosus*. 1. Gray-scale image of the mass in the right liver, which presented as a thickened wall with punctiform and patchy calcifications inside. 2. Color Doppler flow imaging of the mass

1. CL, CE1, and hepatic cyst

Hydatid cysts present with obvious tension, and CE1 cysts are usually larger than 3.0 cm in diameter and have thickened cystic walls. The "double wall sign," a characteristic ultrasonographic feature, may be present in typical CE1 cases. However, an ordinary hepatic cyst usually presents with a thin and smooth cystic wall. It is also important to consider the patient's history of living in an echinococcosis-endemic area, as patients from such areas with thin and smooth cystic walls can be diagnosed with CL and followed up for observation accordingly.

2. CE2 and polycystic liver disease

The structure of the cystic wall and echo inside of the CE2 lesion are inconsistent and significantly different. The morphology of polycystic liver is abnormal, and multiple anechoic masses vary in size within the liver, with normal liver parenchyma in between. Polycystic manifestations in other organs, such as the kidney, spleen, and lungs, are often accompanied. About half of patients with polycystic liver also present with polycystic kidneys, whereas concurrent involvement of the liver and kidneys in primary CE is rare.

3. Giant congenital choledochal cyst

Approximately 70–90% of cases clinically present with intermittent jaundice, abdominal pain and other symptoms. It is important to carefully identify the anatomical relationship between the gallbladder, common bile duct, and portal vein during scanning. The observation of a continuous, anechoic area connected to the biliary tract behind the gallbladder and in front of the portal vein is of great differential diagnostic significance.

4. Large cyst in the right adrenal gland and upper pole of the right kidney

Large hydronephrosis of the right kidney, and CE cyst in the right posterior lobe

Multiple positions and slices should be used to carefully observe the relationship between the cyst and the liver, kidney, and adrenal gland and the structure and echogenicity of the cystic wall. HCE can be easily differentiated from cysts in the right kidney and adrenal gland.

5. CE4 and liver cancer

The internal echo of solidified CE lesion is chaotic, with uneven, dense hyperechoic spots and clusters and a thickened cystic wall. Few color Doppler blood flow signals can be observed inside or around the lesion, and the lesion size often does not match the clinical signs. Combined with epidemiology, clinical manifestations, AFP and other examinations, it is not difficult to distinguish from liver cancer.

## 11.2 Ultrasound Diagnosis of Alveolar Echinococcosis

### 11.2.1 Overview

Alveolar echinococcosis (AE) is a zoonosis caused by the larval stage of the cestode Echinococcus multilocularis, and its terminal hosts are mainly foxes, followed by dogs and cats. The intermediate hosts are mainly voles and wild mice in nature. Man is the intermediate host. Human beings are infected accidentally by eating food contaminated with eggs and develop into alveolar echinococcosis in the liver or other organs. In a majority of cases, AE primarily affects the liver and forms infiltrative alveolar space-occupying lesions, similar to a malignant tumor (Ammann and Eckert 1996; Kern et al. 2017). The parasite is endemic in Germany, France, Austria, and Switzerland, as well as Central Asia and Western China, thus predominantly in the Northern Hemisphere, while cystic echinococcosis occurs mainly in the Southern Hemisphere (Baumann et al. 2019).

### 11.2.2 Clinical Symptoms

After infection with *E. multilocularis*, fox tapeworm disease, is characterized by an asymptomatic progressive course. A period of 5–15 years can pass before a diagnosis of AE is made (Ammann and Eckert 1996; Kern et al. 2017). In over 98% of cases, the liver is the most frequently affected organ (Baumann et al.

2019). From the liver, the larva spreads to other organs by infiltration or metastasis formation. Primary extrahepatic locations of *E. multilocularis* are rare (Kern et al. 2003).

AE is characterized by an initial asymptomatic incubation period of 5–15 years and a subsequent chronic course. The symptoms are primarily cholestatic jaundice (1/3 cases) and/or abdominal pain (1/3 cases). In 1/3 of patients, AE is found incidentally when investigating various symptoms such as fatigue and weight loss, hepatomegaly, and abnormal US or routine laboratory findings (WHO Informal Working Group on Echinococcosis 1996). Mortality is high in nontreated patients, but in Europe, treatment has changed the average life expectancy at diagnosis from 3 years in the 1970s to 20 years in 2005 (Torgerson et al. 2018).

### 11.2.3 Diagnosis

Diagnosis of alveolar echinococcosis is based on clinical findings and epidemiological data, imaging techniques, histopathology and/or nucleic acid detection, and serology.

#### 11.2.3.1 Conventional Ultrasound (US)

US examination is the basis of AE diagnosis in abdominal locations, at the individual and population levels, but needs an experienced examiner (Gottstein et al. 2001). In hepatic alveolar echinococcosis (HAE), the parasitic lesions show highly diverse sonomorphological presentations and are often difficult to diagnose by B-scan sonography (Bartholomot et al. 2002; Romig et al. 1999).

Typical findings (70% of cases) include:

1. The juxtaposition of hyper- and hypoechogenic areas in a pseudo-tumor with irregular limits and scattered calcification (Fig. 11.18)

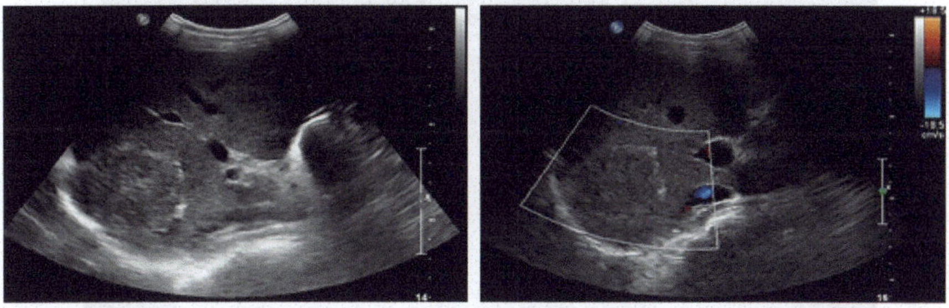

**Fig. 11.18** A 23-year-old female was found to have a hypoechogenic lesion mass of about 5.1 × 4.8 × 5.5 cm in the right liver. The boundary was not clear, the shape was relatively regular, and no obvious blood flow signal was found inside. Postoperative pathological results: hepatic echinococcosis, alveolar echinococcosis. 1. Gray scale ultrasound image showed a slightly stronger echo mass in the right liver with unclear boundary and uneven internal echo. 2. Color Doppler flow diagram of intrahepatic mass

2. Pseudo-cystic appearances due to a large area of central necrosis surrounded by an irregular hyperechogenic ring (Fig. 11.19)

Less typical features (30% of cases) include:

1. Hemangioma-like hyperechogenic nodules as the initial lesion (Fig. 11.20).
2. A small calcified lesion due to either a dead or a small-sized developing parasite (Bresson-Hadni et al. 2003, 2006) (Fig. 11.21)

US with color Doppler provides information on biliary and vascular involvement (Fig. 11.22).

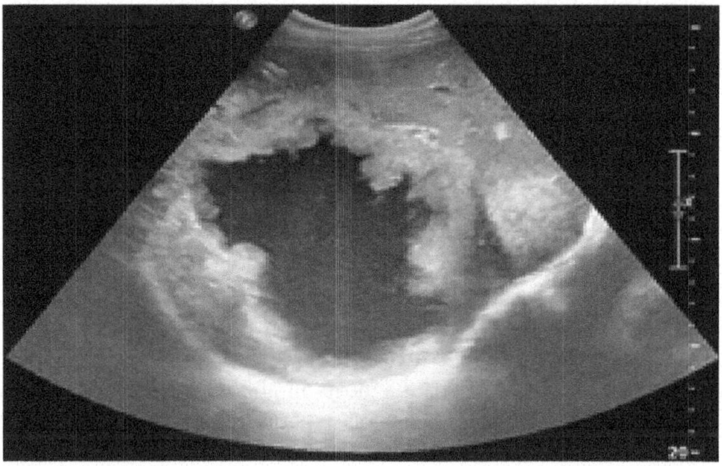

**Fig. 11.19** A 47-year-old male patient with a hyperechogenic mass of about 6.4 × 7.5 × 7.1 cm was found inside the right liver. The boundary was not clear, the shape was irregular, and no obvious blood flow signal was found inside. Pathological results: hepatic echinococcosis (alveolar Echinococcosis). Gray-scale ultrasound showed a large irregular echoless area in the lesion

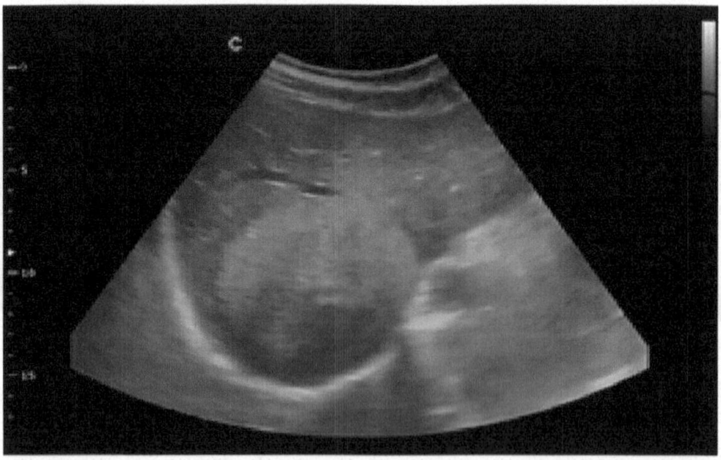

**Fig. 11.20** A 60-year-old male patient showed a hyperechoic mass of about 6 × 6.5 cm with an inhomogeneous inside. The boundary was clear, and the shape was regular. Postoperative pathological results: hepatic echinococcosis, alveolar echinococcosis

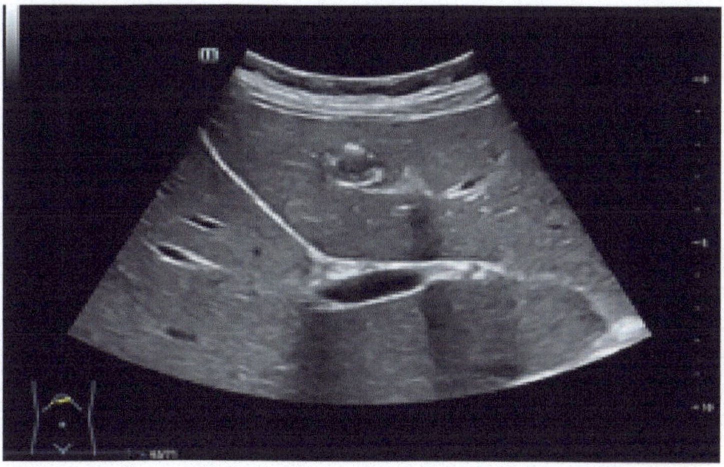

**Fig. 11.21** A 60-year-old male patient showed a mixed echoic mass of about 1.5 × 1.2 cm. The boundary was not clear, the shape was not regular, and the sound shadow was observed behind. Postoperative pathological results: hepatic echinococcosis, alveolar echinococcosis. Gray-scale ultrasonography showed a strong annular echo surrounding the lesion, and punctured calcification was also observed inside the lesion

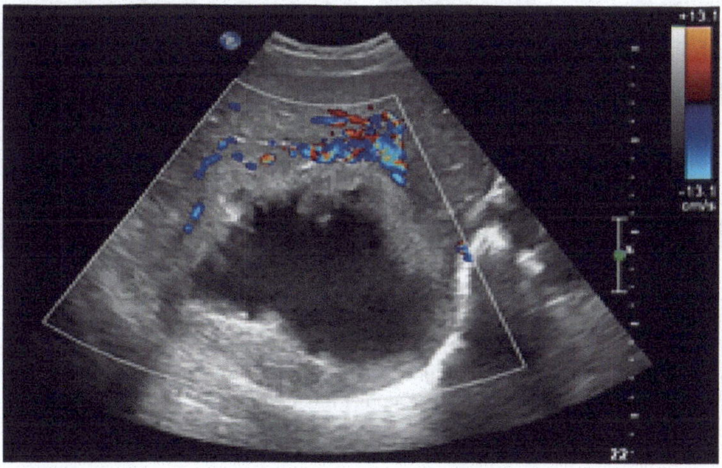

**Fig. 11.22** A hyperechogenic lesion with central necrosis displayed the vascularity of the surrounding lesion by color Doppler mode. Dotted and linear blood flow signals can be seen around the lesion

### 11.2.3.2 Ulm Classification Ultrasound in AE

Kratzer W et al. first reported the research in 2015 about the Ulm classification of the hepatic AE (Kratzer et al. 2015). The Ulm classification, distribution of sonographic patterns among the HAE, was into five types: hailstorm, pseudocystic, ossification, hemangioma-like, and metastasis-like. And the characteristics of the type of ultrasound were described (Schuhbaur et al. 2021) (Fig. 11.23).

| Pattern | Hailstorm pattern | Pseudocystic pattern | Hemangioma-like pattern | Ossification pattern | Metastasis-like pattern |
|---|---|---|---|---|---|
| Example | 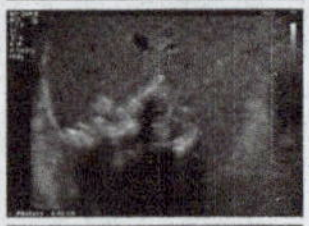 | 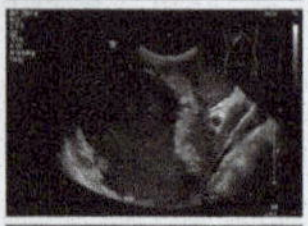 | 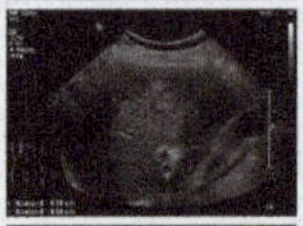 | 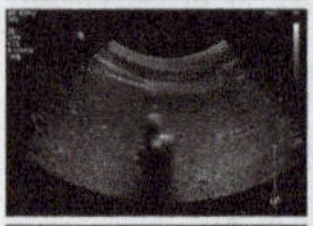 | 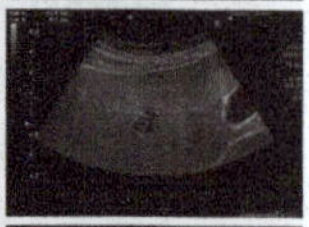 |
| Description | Unclear delimitability-irregular border-inhomogeneous pattern-echo rich texture-with/without sound shadows | Well demarcated, inhomogeneous/ irregular, non-vascularized, echo rich margin-centrally echo poor, inhomogeneous area, with partly echo rich signals-initial or after treatment with benzimidazole | Partially blurred demarcation, inhomogeneous to the surrounding liver tissue-portions of different echogenicity-differential -diagnostic partially thrombosed hemangioma | Circumscribed demarcated lesion with dorsal reverberant shadow-unifocal or multifocal appearance possible-differential diagnosis: Hepatolithiasis, echo-rich metastasis in rectal carcinoma | Low echo lesion-usually without halo-typical sign is a central echo rich structure-difficult differential diagnostic differentiat on from metastases of different primary tumors |

**Fig. 11.23** Illustration of the different B-scan sonographic patterns according to the Echinococcus multilocularis Ulm (EMUC-US) classification. Source: Kratzer W, Weimer H, Schmidberger. Echinokokkose: eine Herausforderung der Lebersonographie. Ultraschall in Med 2021. Graeter T, Ehing F, Oeztuerk S et al. Hepatobiliary complications of alveolar echinococcosis: A long-term follow-up study. World J Gastroenterol 2015; 21: 4925–4932

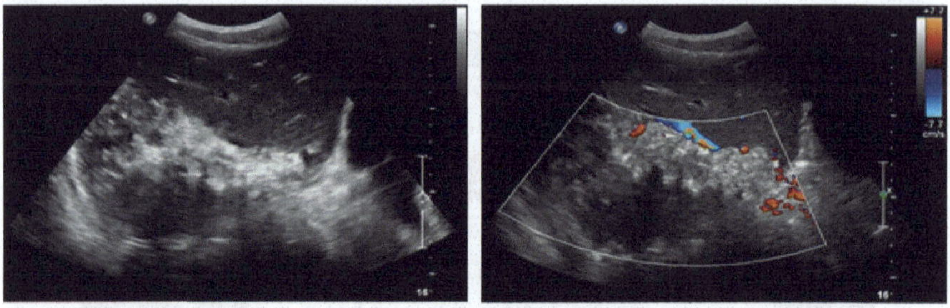

**Fig. 11.24** A 37-year-old male patient was found to have a chaotic echo zone of 12.8 × 8.0 × 9.9 cm in the right liver, with unclear boundary, irregular shape and no obvious blood flow signal. Postoperative pathological results: hepatic echinococcosis, alveolar echinococcosis. 1. gray-scale ultrasonography showed that the lesion presented a chaotic strong echo mass with calcification, and the posterior echo attenuated. 2. Color Doppler blood flow diagram of the lesion, with peripheral point-line blood flow signals

The following section is an ultrasound image of typical AE classified by Ulm classification.

## Hailstorm

The typical hailstorm appearance is characterized by indistinct, irregular boundaries, non-homogeneous patterns and hyperechoic formations, With or Without dorsal acoustic shadow (Fig. 11.24).

### Pseudocystic

Pseudocystic alveolar echinococcosis lesions are primarily characterized by a hyperechoic, irregular, non-homogeneous rim that is non-vascularized at power Doppler and color-coded duplex ultrasonography. It may appear to be >10 mm in thickness. A hypo- or anechoic, often non-homogeneous central zone may contain hyperechoic material (Fig. 11.25).

### Hemangioma-like

These lesions are difficult to distinguish from atypical (e.g., partially thrombosed) hemangiomas and often represent a significant diagnostic challenge. Sonomorphologically, the lesions are a relatively clearly demarcated non-homogeneous tumor that appears hyperechoic compared to the surrounding hepatic parenchyma. Echogenicity ranges from slightly and non-homogeneously hyperechoic to strongly and homogeneous hyperechoic (Fig. 11.26).

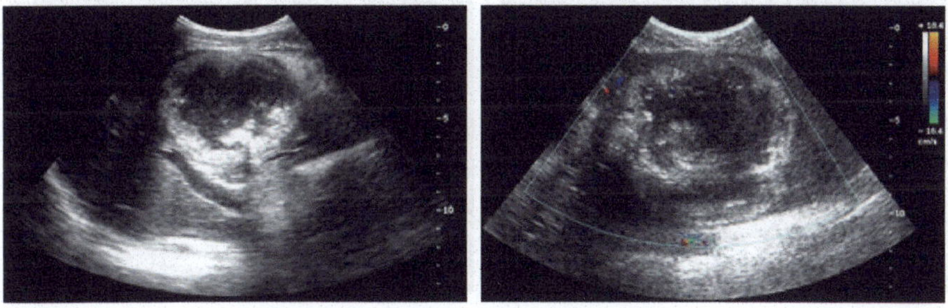

**Fig. 11.25** A 39-year-old male patient was found to have chaotic echo clusters in the liver, the size of which was about 6.8 × 6.3 × 6.6 cm, the wall thickness was about 6.8 × 6.6 cm, most patchy strong echoes could be seen, the boundary was unclear, the shape was irregular, and no obvious blood flow signal was found in the liver. Postoperative pathological results: hepatic echinococcosis, alveolar echinococcosis. 1. Gray-scale ultrasound showed that the lesion was a cystic solid mass with a thick wall and large flake calcification. 2. Color Doppler blood flow mass showing no obvious blood flow signal.

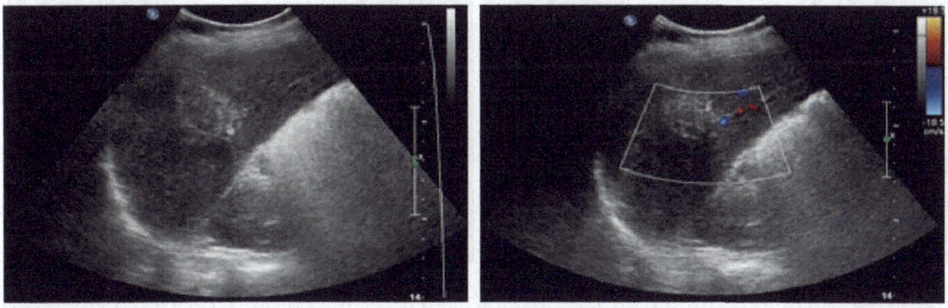

**Fig. 11.26** A 46-year-old female with a slightly strong echoic mass of about 5.1 × 3.8 × 4.5 cm was found in the lower segment of the right posterior lobe of the liver, with uneven internal echo and no obvious blood flow signal. Postoperative pathological results: hepatic echinococcosis, alveolar echinococcosis. 1. Gray-scale ultrasound showed that the focus was a pipe network like a slightly strong echo cluster with irregular shape and unclear boundary. 2. Color Doppler flow diagram of the lesion

## Ossification

The ossification pattern presents with solitary or grouped, mostly sharply delineated lesions with dorsal acoustic shadow. In terms of their differential diagnosis, these alveolar echinococcosis (AE) lesions are often difficult to distinguish from inflammatory or hyperechoic metastases of various carcinomas. Very large ossification-type AE lesions represent a rarity. Both uni- and multifocal involvement is possible (Fig. 11.27).

## Metastasis-Like

Besides the hemangioma-like lesions, the metastasis-like lesions of alveolar echinococcosis represent the greatest diagnostic challenge. Mostly hypoechoic, these lesions exhibit a characteristic compared to typical hepatic metastases (e.g., colorectal cancer)—the absence of the halo phenomenon. Instead, there is a central, hyperechoic, non-homogeneous scar (Fig. 11.28).

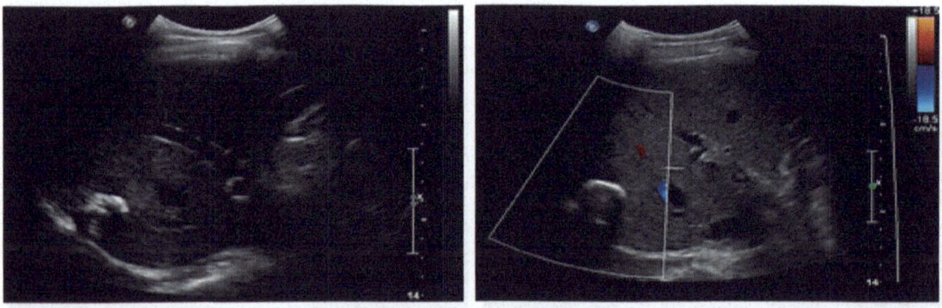

**Fig. 11.27** Male, 22 years old, a strong echo mass about 2.5 × 3.2 cm in size was found in the right posterior lobe, accompanied by a sound shadow in the rear. The boundary was clear and the shape was regular. The wall of the mass was surrounded by patchy, strong echo, and no obvious blood flow signal was found inside. Postoperative pathological results: hepatic echinococcosis, alveolar echinococcosis. 1. Gray-scale ultrasound showed a strong anterior wall echo with a wide loud shadow. 2. Color Doppler flow diagram of the lesion

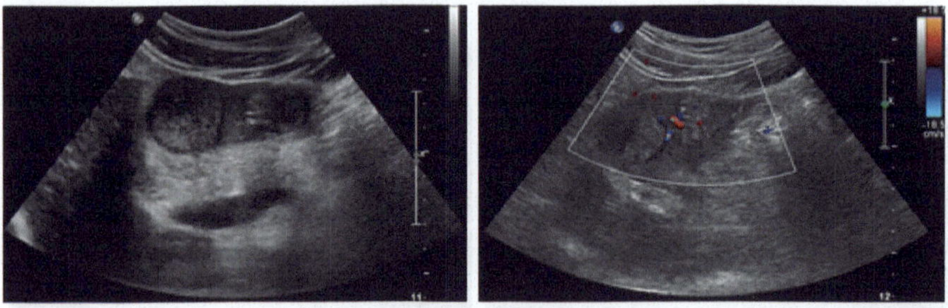

**Fig. 11.28** Male, 48 years old, the left outer lobe of 2.9 × 3.2 × 3.0 cm, 2.4 × 2.4 × 2.2 cm weak echo mass, the boundary is not clear, the shape is relatively regular, the internal echo is not uniform, and the dot linear blood flow signal is found inside—postoperative pathological results: hepatic echinococcosis, alveolar echinococcosis. 1. Gray-scale ultrasound showed a hypoechoic mass with hypoechoic halo and internal punctiform calcification. 2. Color Doppler flow diagram of the lesion showing short linear flow within the lesion

Those with hepatic alveolar hepatic echinococcosis that did not conform to the above five categories were classified as unclassified. According to Kratzer W et al., the incidence of this type in alveolar echinococcosis is less than 5%.

#### 11.2.3.3 WHO Classification of AE

Based on imaging findings, the WHO-IWGE PNM classification system has been established as the international benchmark for standardized evaluation of diagnostic and therapeutic measures (Kern et al. 2017). It denotes the extension of the parasitic mass in the liver (P), the involvement of neighboring organs (N), and metastases (M) (Table 11.1). PNM classification should improve the quality control of current treatment strategies in single centers and uniform evaluation of multicenter studies.

It is worth noting that although AE and CE are two different types of hepatic echinococcosis, patients can be infected with both types of echinococcosis and coexist in the liver (Fig. 11.29).

### 11.2.4 Differential Diagnosis of HAE

According to the epidemiology, there is a history of contact with dogs, cats, cattle, sheep, and foxes and mice from the epidemic area. The diagnosis is not difficult to establish based on the symptoms and signs and epigastric mass, combined with

**Table 11.1** PNM classification of alveolar echinococcosis

| | |
|---|---|
| P | Hepatic localization of the parasite |
| PX | Primary tumour cannot be assessed |
| P0 | No detectable tumour in the liver |
| P1 | Peripheral lesions without proximal vascular and/or biliar involvement |
| P2 | Central lesions with proximal vascular and/or involvement of one lobe[a] |
| P3 | Central lesions with hilar vascular or biliar involvement of both lobes and/or with involvement of two hepatic veins |
| P4 | Any liver lesion with entension along the vessels[b] and the biliary tree |
| N | Extra-hepatic involvement of neighbouring organs [diaphragm, lung, pleura, pericardium, heart, gastric and duodenal wall, adrenal Glands, peritoneum, retroperitoneum, parietal wall (muscles, skin, bone), pancreas, regional lymph nodes, liver ligaments, kidney] |
| NX | Not evaluable |
| N0 | No regional involvement |
| N1 | Regional involvement of contiguous organs or tissues |
| M | The absence of presence of distant metastasis [lung, distant lymphy nodes, spleen, CNS, orbital, bone, skin, muscle, kidney, distant peritoneum and retroperitoneum] |
| MX | Not completely evaluated |
| M0 | No metastasis[c] |
| M1 | Metastasis |

[a]For classification, the plane projecting between the bed of the gall bladder and the inferior vena cava divides the liver into two lobes

[b]Vessels mean inferior vena cava, portal vein and arteries

[c]Chest X-ray and cerebral CT negative

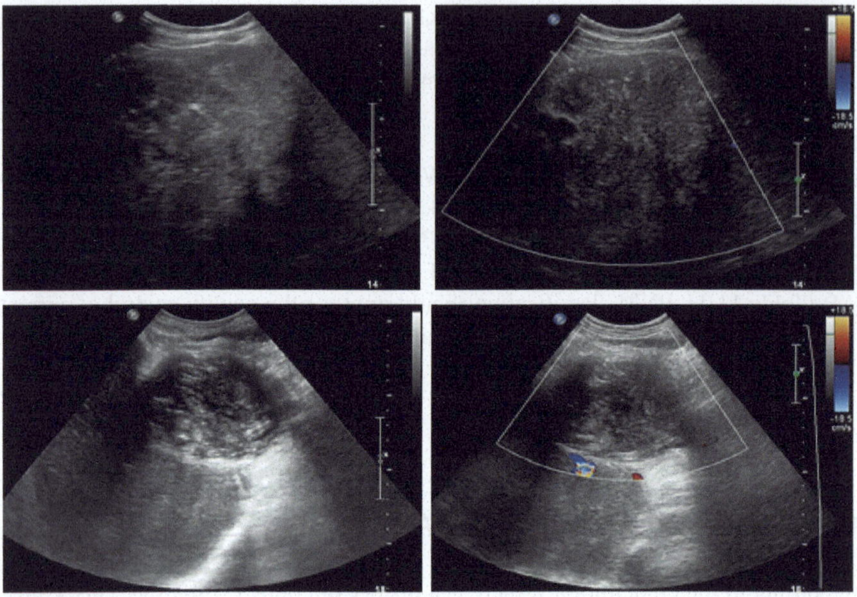

**Fig. 11.29** A 41-year-old female patient had abnormal liver morphology. Examination of the right liver revealed a flake area with a slightly stronger echo, about 11.4 × 11.5 × 11.7 cm, with unclear boundary, irregular shape, uneven internal echo, punctured strong echo, and no obvious blood flow signal. Another hypoechoic mass with a size of 9.5 × 6.8 × 9.5 cm was found in the right liver. The boundary was clear, the shape was irregular, the internal echo was uneven, the punctured strong echo was visible, and no obvious blood flow signal was found. Postoperative pathological results: hepatic echinococcosis, alveolar echinococcosis and granular Echinococcosis. 1. Gray-scale ultrasound showing irregular masses of the right liver with unclear boundaries, irregular shape and punctate calcification. 2. blob color Doppler flow diagram. 3. Gray-scale ultrasound also showed a solid cystic mass with a clear boundary, irregular shape and spot-like calcification foci. 4. Color Doppler flow diagram of the mass

immunological tests and unique sonographic findings. A few cases should be identified with the following liver variations.

1. Hepatic hemangioma. The focus of hepatic hemangioma has a clear boundary, mostly a round internal network shape; there is no coarse sand or linear sound shadow; serum immunology and contrast-enhanced ultrasound can help distinguish it.
2. Hepatocellular carcinoma. There are mostly dark halas around the tumor of hepatocellular carcinoma, and there may be "nodules in nodules," color Doppler can detect color blood flow signals, and there is no characteristic multi-shaped, punctiform, lobulated texture echo clutter and small nodular and small ring sand-like calcification. Combined with epidemiology, AFP test is not difficult to make a differential diagnosis.
3. Hepatic abscess. The ultrasonography of the hepatic abscess showed diffuse enlargement of the liver, uneven hypoecho in the lesion, and liquid dark areas in

some centers, which were completely different from the ultrasonography features of the unique liquefaction cavity of the liver cited, with no envelope echo and insect erosion or lava cave changes.

## 11.3 Contrast-Enhanced Ultrasound (CEUS) Diagnosis in Hepatic Echinococcosis

### 11.3.1 Introduction to CEUS

Ultrasonic diagnostic medical imaging is based on ultrasonic pulse reflection technology as the basic principle. When the difference in the characteristic impedance of the tissue on both sides of the interface reaches 0.1%, there will be ultrasonic reflection. However, due to the complexity of the human tissue interface, ultrasonic instruments have limitations in detection technology. In practical clinical applications, it is often difficult to distinguish finely differentiated tissues by ultrasound. That is, the echo of the lesion is close to the background echo of normal tissue, making it impossible to detect these pathological tissues by ultrasound. Therefore, people try to find a way to make the lesion significantly different from the surrounding normal tissue echo, that is, to enhance the echo of the diseased tissue or reduce the echo of the surrounding background so that there is a clear contrast between the two, so as to bring out the image of the lesion. That is now the contrast-enhanced ultrasonography (CEUS).

There are two types of contrast-enhanced ultrasound used for diagnosis. One is intravascular contrast, that is, after injecting a microbubble contrast agent through peripheral veins or cardiac catheters, the developing power of diseased tissues and the display of some tissue boundaries can be improved within a certain period of time, such as the display of tumors of organs such as liver, spleen and kidney, heart cavity and ventricular wall. The other is non-angiography, that is, the liquid contrast agent is introduced into the human body's pipes and body cavities by oral administration, enema or other means, and the echoless region of the liquid or the scattered echo of tiny particles suspended in the liquid is used as contrast angiography diagnosis, such as gastrointestinal angiography, uterine angiography, urethra angiography, etc. Here, the first type of contrast agent is used for CEUS for hydatid.

### 11.3.2 Ultrasonic Contrast Agent

SonoVue, developed and manufactured by Italian company Bracco, has been approved for clinical use by drug testing authorities in Europe and Asia, including China. The microbubble was composed of carbon hexafluoride as a contrast agent and was coated with a soft phospholipid. The average diameter of the microbubble was 2.5 um, 90% < 8 um. Compared with other contrast agents, SonoVue has the characteristics of high stability, easy preparation, wide application range and high safety. The prepared contrast agent can be stabilized in the bottle for 6 h (less than 6 h) because it is stabilized by a variety of surfactant such as polyoxyethylene, phospholipid, and palmitic acid. SonoVue has a half-life of 6 min and is exhaled by over 80% of the lungs within 11 min. The contrast agent is suitable for enhanced contrast of whole body tissues and organs such as the liver and kidney. At present, it is the only ultrasound contrast agent approved for clinical use in China.

### 11.3.3 Contrast-Enhanced Ultrasound Imaging Technology

The conventional ultrasound imaging system can perform contrast-enhanced ultrasound examination due to its probe frequency, output acoustic power, acoustic transmission rate, post-processing, and other factors. Since the 1990s, many manufacturers engaged in the production of ultrasonic diagnostic instruments began to develop a variety of contrast-enhanced ultrasound technology to meet the needs of high-quality contrast-enhanced ultrasound. This paper will briefly introduce real-time harmonic imaging, which is used most frequently.

Real-time harmonic imaging or low mechanical index imaging is a technology that continuously transmits sound waves and receives harmonic signals under low power or low mechanical index. The second generation of new contrast agents, such as SonoVue, contains stable low soluble inert gas and is coated with soft film. When the mechanical index is about 0.15, the microbubbles have good resonance and produce rich harmonic components, while the rupture of the microbubbles is little. In addition, at a low mechanical index, the harmonic signal of the tissue is very weak, so the signal-to-noise ratio of the microbubble harmonic to the tissue harmonic is high. At this time, the real-time contrast effect can be obtained by using the filter imaging technology, and the quantitative analysis of tissue blood perfusion can be carried out by the corresponding software.

### 11.3.4 Application of Contrast-Enhanced Ultrasound in Hepatic Nodules

Among the tumor detection of substantive organs, the liver tumor is the most mature. Contrast-enhanced ultrasound can improve the detection rate of liver tumor lesions. Still, the sensitivity of early contrast-enhanced ultrasound to tumor lesions was not significantly improved due to the application of fundamental imaging. With the application of various harmonic contrast imaging techniques, the detection rate of lesions has been significantly improved. Bernatik T et al. showed that the detection rate of liver tumors by CEUS increased from 59% to 97% by conventional ultrasound, especially for tumors smaller than 2 cm and those near the diaphragmatic ceiling.

The differential diagnosis of liver tumors by contrast-enhanced ultrasound is mainly by observing tumor enhancement and dynamic enhancement. The liver's blood supply has the characteristics of a hepatic artery and portal vein double blood supply; under normal circumstances, the hepatic artery accounts for 25~30%, and the portal vein accounts for 70~75%. A contrast agent was injected from the peripheral vein. The development of portal vein with contrast agent required more circulation paths than that of hepatic artery, so the beginning time of portal vein development was several seconds later than that of hepatic artery, that is, the development time of most liver tissues was later than that of hepatic artery. Similar to enhanced CT or enhanced MRI, contrast-enhanced ultrasound can define and observe three vascular phases/phases, which basically follow the criteria of enhanced CT. Namely, it is the arterial, portal vein, and hepatic parenchyma phases. The imaging of the hepatic arterial phase began 10~40 s after contrast injection, the portal vein phase was 40~80 s, and the parenchymal phase was 80~240 s. Tissue enhancement caused by hepatic arterial blood supply alone usually begins 10~20 s after peripheral intravenous contrast agent injection and lasts 10~25 s. This is followed by the portal vein phase, which generally lasts until 1.5~2 min after the contrast agent is injected. The delay lasts until the contrast agent is cleared from the hepatic parenchyma, such as 4–6 min or longer after SonoVue injection.

Different hepatic focal lesions have different blood supply characteristics. The differential diagnosis by contrast-enhanced ultrasound is based on the different phases and forms of enhancement. For example, primary liver cancer is mostly tumors with rich hepatic artery blood supply. Compared with liver tissues mainly supplied by the portal vein, contrast-enhanced ultrasound showed that the tumor was significantly enhanced in the early arterial stage. In the portal vein stage, tumor perfusion began to decrease, leading to a weakened echo. In the delayed period of angiography, the perfusion status of tumor vessels subsided quickly. At the same time, the hepatic parenchyma still had perfusion (echo was still enhanced), so the tumor nodules presented a weakly echogenic focus with clear boundaries. The arterial phase of contrast-enhanced ultrasound is of great value in diagnosing hyperperfusion hepatic focal lesions, such as focal nodular hyperplasia, hepatocellular adenocarcinoma, primary liver cancer, and liver metastases. The enhanced features of the portal and delayed phases can improve key information about the nature of the lesion. Compared with the enhanced echo of the liver parenchyma, most malignant lesions in the portal and delayed phases echo less than the peripheral liver parenchyma, while most benign lesions in the portal or delayed phases echo as much or more than the liver parenchyma.

Hepatic hydatid disease, as a space-occupying lesion in the liver, is an inflammatory lesion accompanied by necrosis of liver parenchyma. Contrast-enhanced ultrasound can enhance the difference between the lesion and surrounding parenchyma after angiography, which is helpful for qualitative diagnosis of hepatic hydatid.

### 11.3.5 Indications and Contraindications of Contrast-Enhanced Ultrasound

#### 11.3.5.1 Indications

The European Federation of Ultrasound in Medicine and Biology (EFSUMB) Group revised the guidelines for the use of CEUS in November 2004 and proposed indications.

Contrast-enhanced ultrasound is applicable to all patients with undiagnosed lesions in the liver, especially in the presence of the following clinical conditions. (1) Accidental discovery by conventional ultrasound; (2) lesions or suspected lesions in patients with chronic hepatitis or cirrhosis; (3) lesions or suspected lesions with a history of malignant tumors; (4) when the patient's MRI/CT or cytology/histology findings are inconclusive.

#### 11.3.5.2 Contraindications

1. People who are allergic to any component in ultrasonic contrast media are prohibited from using it.
2. It should not be used in patients with recent symptoms of acute coronary heart disease or clinically confirmed unstable ischemic heart disease. These disorders include myocardial infarction in progress or in the process of onset, typical angina in a quiet state within 7 days, or significantly aggravated heart disease symptoms, recent coronary intervention, or the presence of other unstable factors (such as ECG, laboratory or clinical evidence of aggravation), acute heart failure, and severe arrhythmia.
3. Contraindicated in the following conditions: right-to-left shunt, patients with severe pulmonary hypertension (pulmonary pressure above 90 mmHg), patients with uncontrolled hypertension, and patients with acute respiratory distress. For pregnant and lactating women safety has not been confirmed, so pregnant and lactating women contraindicated use.
4. Caution should be exercised when the use of ultrasound contrast agents in some tissues may damage microvessels and cause serious clinical consequences, such as craneless brains, eyes, and newborns.
5. Avoid using contrast media 24 h before extracorporeal shock wave therapy.

### 11.3.6 Usage

Currently, real-time hepatic CEUS mostly uses real-time scanning imaging techniques with low mechanical index (generally below 0.2). Ultrasound contrast agents with low solubility gas (such as SonoVue) can be used for dynamic imaging and evaluation of the three different vascular phases of the contrast.

#### 11.3.6.1 Use Route and Injection Method

Peripheral venous injection of a contrast agent is routinely used with a new ultrasound contrast agent. Currently, most commercially available contrast agents are bottled powder (SonoVue, etc.), which is prepared into contrast agent solution according to the instructions before injection. The dosage is generally 0.01~0.1 ml/kg and injected through the anterior superficial cubital vein. The injection method can vary according to different examination requirements. In general, the rapid "bolus injection" method is commonly used, that is, a one-time injection within 5 s, which can produce a significant enhanced effect of dynamic phase change of hepatic arterial phase, portal phase and delayed phase, which is conducive to the differential diagnosis of hepatic space-occupying lesions. The continuous injection can also be directly or by means of micro-pump if it is necessary to observe whether there are small lesions in the liver during the delay period, or observe the intravascular contrast agent flowing into the local perfusion.

The use of contrast agent concentration, dose and injection speed affect the performance characteristics of the liver dynamic phase, such as the increase of echo signal enhancement peak and the extension of contrast agent duration caused by high concentration group injection. Therefore, the same criteria should be used to compare a group of cases. The concentration and injection speed of different contrast agents have different requirements and can be followed by instructions.

#### 11.3.6.2 Checking Methods

The general procedure for checking yes is as follows:

1. Routine gray-scale, color and/or energy Doppler ultrasound were performed on the lesions, and attention was paid to any other suspicious lesions in the liver.
2. After the target lesions are identified, the probe is placed in a stable position that continuously displays the target, and the imaging mode is switched to low mechanical index or other specific contrast imaging conditions.
3. If necessary, adjust the mechanical index and probe frequency to obtain adequate tissue suppression and maintain proper penetration depth. The structure of the great vessels and some anatomical markers such as the diaphragm should be maintained at a barely visible level.
4. Ultrasound contrast agent was injected into the peripheral vein by group injection and rinsed with 5 ml of normal saline. After the contrast agent is prepared, it should be shaken again before injection and withdrawn for injection. The diameter of the needle should not be less than 20G to avoid microbubble rupture due to mechanical impact during injection. The meter clock starts when the ultrasound contrast agent is injected.
5. It is recommended to continuously trace the lesions for 60~90 s to obtain images of the arterial phase and portal phase. As for the evaluation of the delay period, the scan may be continuous or intermittent until the absence of the contrast agent from the hepatic microvessels is seen. If the dynamic image is to be analyzed by contrast enhancement time intensity curve, it should be scanned continuously. In addition to observing the target lesions, rapid scanning of the whole liver during the delay period to observe other hypoechoic nodules is an effective method to detect malignant tumors such as liver metastases.
6. Image recording Due to the dynamic characteristics of real-time contrast-enhanced ultrasound, it is suggested to store the detected results in digital form on the hard disk of the device or record the detected results on video media for easy image playback and observation to avoid missing important image information with rapid and temporary changes.

### 11.3.7 Contrast-Enhanced Ultrasound Findings of Hepatic Echinococcosis

The diagnosis of cystic hepatic echinococcosis is relatively easy, which can be accurately diagnosed by gray-scale and color Doppler ultrasound. Therefore, contrast-enhanced ultrasound is rarely used in cystic hepatic echinococcosis. Contrast-enhanced ultrasound is mainly used in the diagnosis of alveolar hepatic echinococcosis due to the large difference in the lesions and the difficulty in differentiating them from hemangioma and liver cancer. There are few relevant literatures at home and abroad. According to the report of Song Tao et al., the typical CEUS mode of hepatic hydatid disease showed that the lesion presented three stages: arterial phase, venous phase and delayed phase with no enhanced "black hole sign." The size of the lesion after CEUS was significantly larger than that of conventional ultrasound, which was closer to the actual size of the lesion (Figs. 11.30 and 11.31).

Contrast-enhanced ultrasound (CEUS) is a new technology that is good for the evaluation of benign and malignant hepatic nodules (Correas et al. 2001; Beissert et al. 2002; Lencioni et al. 2003). CEUS has overcome the limitations of conventional B-mode and color or power Doppler US and can visualize the parenchymal microvasculature. CEUS provides more information for the diagnosis of hepatic nodules, not only distinguishing between benign and malignant nodules but also diagnosing parasitization.

Because of the lesion of hepatic echinococcosis with the highly diverse presentations, especially AE, sometimes it is difficult to diagnose only by grayscale sonography. In recent years, contrast-enhanced ultrasound (CEUS) has emerged as another AE diagnostic technique (Tao et al. 2011; Cai et al. 2017,

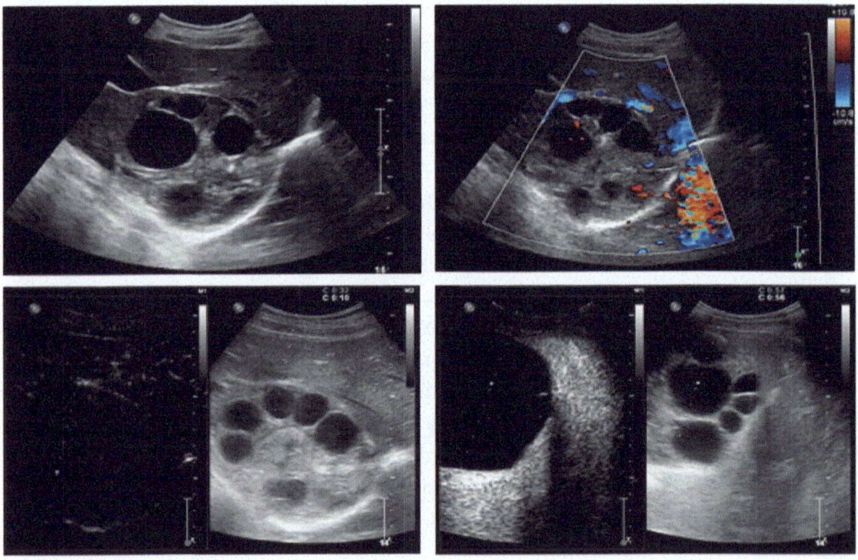

**Fig. 11.30** The CEUS finding in HCE. A 33-year-old female with CE was identified by contrast-enhanced ultrasound (CEUS). A messy echoic mass with a size of 9.7 × 9.7 cm was found in the upper part of the right lobe of the liver with an irregular echoless area. Postoperative pathological results were hepatic echinococcosis and microgranulosis. 1. Gray-scale ultrasound showed separated cystic masses with clear boundaries and regular morphology, consistent with CE2 "honeycomb sign." 2. Color Doppler blood flow diagram showed no obvious blood flow signal in the lesion. 3. Contrast-enhanced ultrasound showed no internal enhancement in the arterial phase. 4. Focus contrast-enhanced ultrasound showed no enhancement at the portal pulse stage

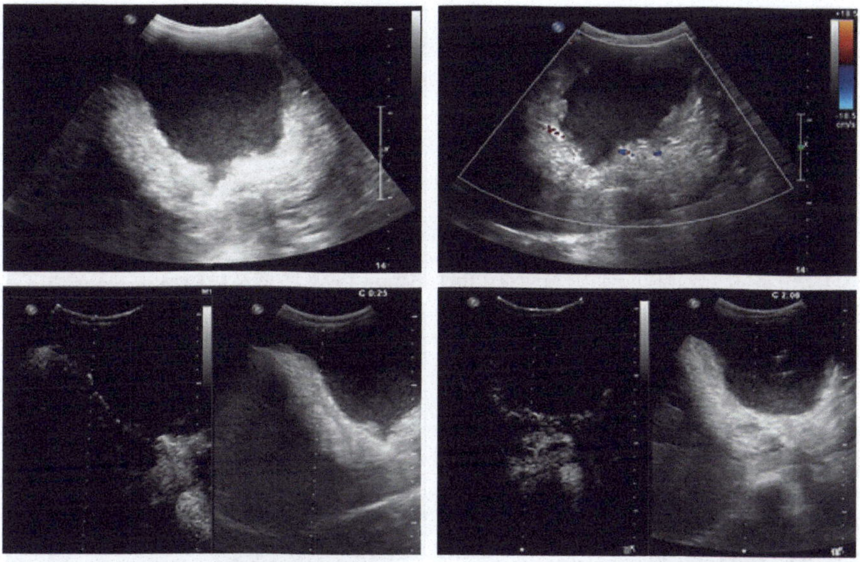

**Fig. 11.31** The CEUS finding in HAE. A male, 23 years old, was found to have a large mixed echo-mass of cystic solid, about 12.0 × 11.9 cm in size, mostly cystic, with uneven wall thickness, about 3.2 cm in thickness, and unclear boundary in the left liver and right anterior lobe of liver. Postoperative pathological results: hepatic echinococcosis, alveolar echinococcosis. 1. Gray-scale ultrasound showed a large cystic mixed echoic mass in the liver with a thick wall and a large anechoic area. 2. Color Doppler blood flow diagram showed only punctured blood flow signals in the posterior wall area of the lesion. 3. Contrast-enhanced ultrasound showed no enhancement in the lesions during the arterial phase. 4 Contrast-enhanced ultrasound showed no enhancement in the delayed lesion

2019; Wa et al. 2020; Zeng et al. 2012). Some researchers believe that contrast-enhanced sonography is a complementary method of choice in hepatic echinococcosis lesions (Dietrich et al. 2013; Horta et al. 2015; Albrecht et al. 2004).

Because CEUS has significantly improved US imaging of vascularization in hepatic lesions, it has been applied to study the vascularization or microvasculature of hepatic AE lesions in recent years. Kaltenbach TE reported that dimensions of AE lesions are displayed more precisely through CEUS than in B-scan (Kaltenbach et al. 2015). US can produce a complex mosaic of high and low echoic areas with an indistinct margin (Correas et al. 2001). It has been difficult to evaluate the real size of the AE lesion. The size of the echinococcosis lesion would be underestimated. CEUS can visualize the AE lesion and surrounding parenchymal microvasculature (Fig. 11.32).

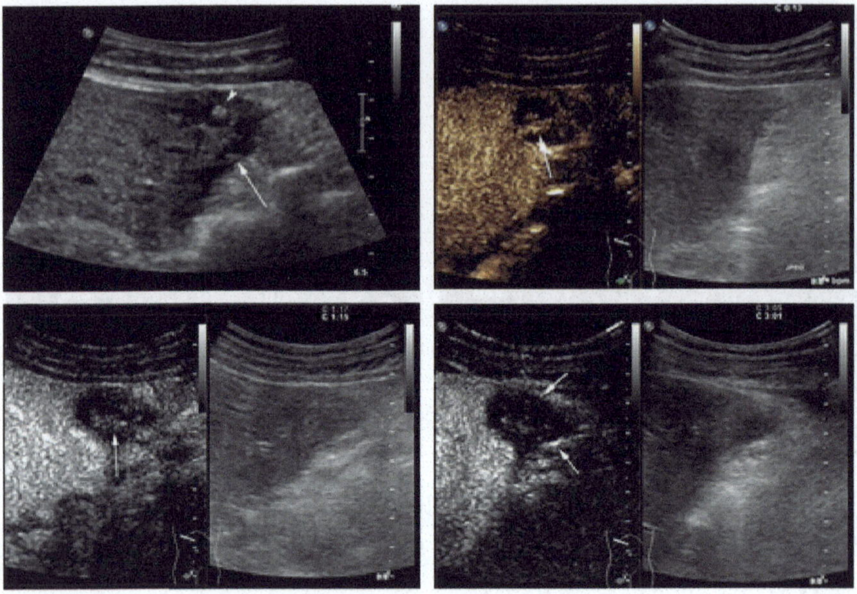

**Fig. 11.32** A 45-year-old male with a mixed echoic mass about 2.9 × 2.2 cm in size was found in the lower part of the right liver. Postoperative pathological results: hepatic echinococcosis, alveolar echinococcosis. 1. Gray-scale ultrasound showed a weakened echo mass with blurred boundaries and uneven echoes, with strong patchy echoes inside. 2. Contrast-enhanced ultrasound showed that the arterial phase was irregularly iso-enhanced with irregular areas of no enhancement. 3. Contrast-enhanced ultrasound showed irregular low enhancement in the portal phase with irregular areas of no enhancement. 4. Contrast-enhanced ultrasound showed irregular low enhancement in the delay period with irregular areas of no enhancement. Source: Cai DM, Wang HY, Wang XL, et al. Ultrasonographic findings of small lesion of hepatic alveolar echinococcosis. Acta Trop. 2017

## 11.4 Extrahepatic Echinococcosis

Extrahepatic echinococcosis refers to echinococcosis in tissues and organs other than the liver. Both AE and CE can occur.

### 11.4.1 Pneumohydatid Disease

The incidence of pulmonary cystic echinococcosis is second only to the liver. About 10~20%. The lung involvement rate is higher in children.

Ultrasonographic findings: hydatid cysts located in the center of the lung are not easy to explore. When hydatid cysts with large volumes on the lung surface area and near the chest wall, liquid dark areas can be detected, the characteristics of hydatid cysts are basically the same as those of hepatic hydatid cysts, and respiratory signs of hydatid cysts can be seen. According to statistics, the cases of multiple lung lesions are about 20%, which are more common in the lower lobe of the right lung, and hydatid cysts of the liver are more common in the right liver. Therefore, when the right lung's lower lobe's hydatid cyst is found, a routine examination of the liver should be carried out. When the hydatid cyst of the right liver is found, a routine scan of the right lung should be performed to detect the coexistence of the two cases in the early stage, and the combined thoracoabdominal incision should be used for one surgical removal.

### 11.4.2 Abdominal Echinococcosis

The incidence of hydatid in abdominal cavity is second only to hydatid in lung, accounting for 12.4% of human hydatid disease. The course of its disease is longer, and the cyst is relatively large when found clinically. The main manifestations were compression symptoms, abdominal touch and elastic mass. Common complications include mechanical ileus and panperitonitis caused by cyst rupture. Hydatid cyst to intestinal canal rupture, the formation of hydatid cyst—intestinal fistula, can be discharged from the stool cyst fluid, ascus and internal cyst fragments.

Ultrasonic findings: The abdominal hydatid mass is generally large, has a circular or quasi-circular anechoic area, and has adjacent organs compression displacement deformation. The course of the disease is very long, common capsular wall calcification is a high echo ring, or the capsular wall is obviously thickened (Fig. 11.33).

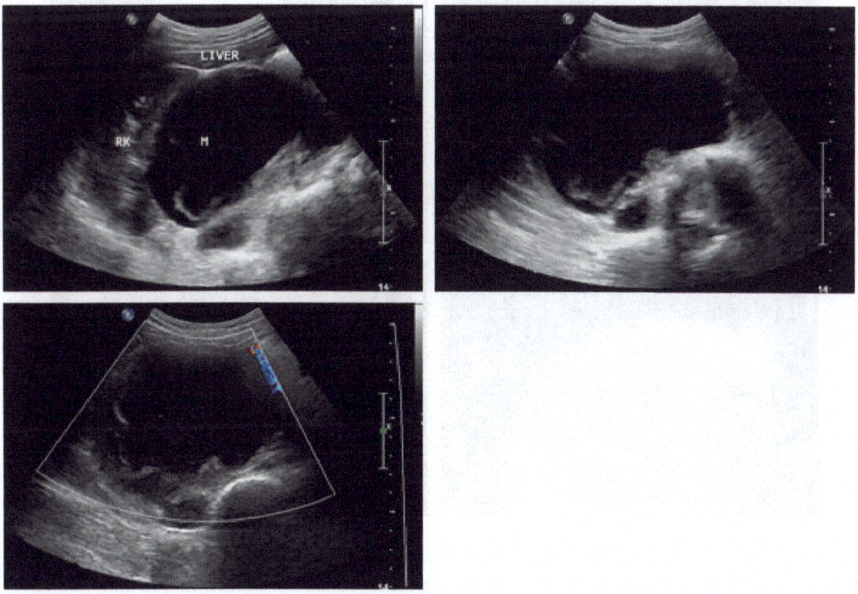


**Fig. 11.33** A 60-year-old male patient has a large, separated thick-walled cystic mass in the right abdomen, about 40 × 12 × 9 cm. The upper part is located behind the right kidney and right liver, and the lower part reaches the inguinal region. The boundary of the mass was not clear, the shape was irregular, and there was no obvious blood flow signal. Postoperative pathological results: hydatid disease, alveolar echinococcosis. 1, 2. Gray-scale ultrasound showed common large delimited cystic masses in the abdomen, with unclear boundaries, irregular shape and large anechoic area. 3. Color Doppler blood flow diagram showed no obvious blood flow signal in the lesion

### 11.4.3 Pelvic Hydatid Disease

Abdominal hydatid mass may involve the pelvic cavity, mostly coexisting with hepatic hydatid mass. The course of the disease is longer, and the clinical findings are larger. The main manifestations are lower abdominal mass and lower abdominal pain.

Ultrasonographic findings: abdominal hydatid mass, located in the pelvic cavity or lower abdominal cavity. The mass is generally large, circular, circular liquid dark area, adjacent organs compression displacement deformation. For pelvic hydatid cysts, transrectal ultrasound can more clearly show the mass (Fig. 11.34).

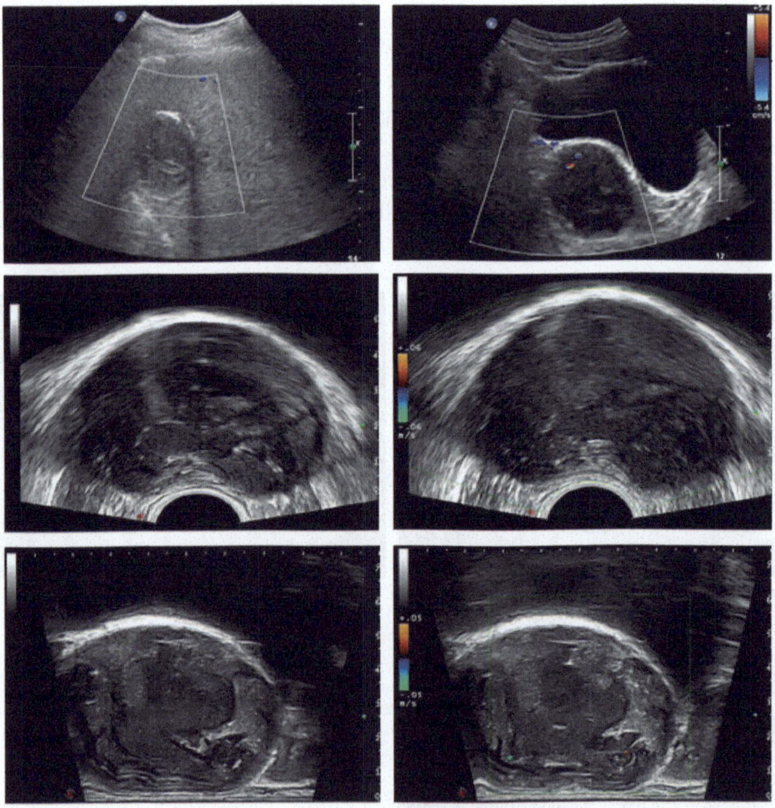

**Fig. 11.34** Male, 39 years old, found a hypoechoic mass of about 6.5 × 5.1 cm in the right liver, with clear boundaries, regular shape, patchy hyperecho, and no obvious blood flow signal. A hypoechoic mass of 5.1 × 5.4 × 4.8 cm was found behind the bladder, with a clear boundary, regular shape, internal heterogeneous, and linear blood flow signal. 1. Gray-scale ultrasound showed an attenuated echoic mass in the right liver with a clear boundary and annular calcification accompanied by sound shadow. 2. Color Doppler ultrasound showing hypoechoic clusters found behind the bladder with clear boundaries, regular shape, uneven internal echo, and linear blood flow signals. 3, 4. Transrectal ultrasound cross-sectional examination, 5, 6. Transrectal ultrasound sagittal-sectional examination

By rectal ultrasonography, we found weak echo clusters with a size of about 63 × 51 × 64 mm in the pelvic cavity above the prostate, behind the bladder and in front of the rectum, with clear boundaries, regular shape, internal heterogeneous, several patchy hyperecho and track hypoecho, and no obvious blood flow signal. When pressed, the probe seemed to see a faint spot echo flow. Postoperative pathological results: hydatid disease, echinococcosis granulosus.

### 11.4.4 Renal Hydatid Disease

The incidence of this disease is very low, accounting for about 0.4~4% of hydatid disease. Still, in the urinary system, hydatid cyst with renal hydatid cyst is the most common, the vast majority of unilateral single, and the upper pole of the kidney is more common than the lower pole.

The ultrasonographic findings showed that the kidney was large and grew up with an abnormal shape. The renal parenchyma showed an echoic mass, round or oval shape, clear boundary, regular edge of the liquid dark area, and thinner cyst wall than the liver hydatid. The collecting system is displaced under compression, the calyces are separated in an arc, and sometimes hydronephrosis is seen. Large cysts that may compress and displace surrounding tissue (Fig. 11.35).

Renal hydatid cysts must be distinguished from hydronephrosis, isolated renal cysts, renal carcinoma with central necrotic bleeding, and liquefaction. Contrast-enhanced ultrasound can be used to diagnose difficult cases.

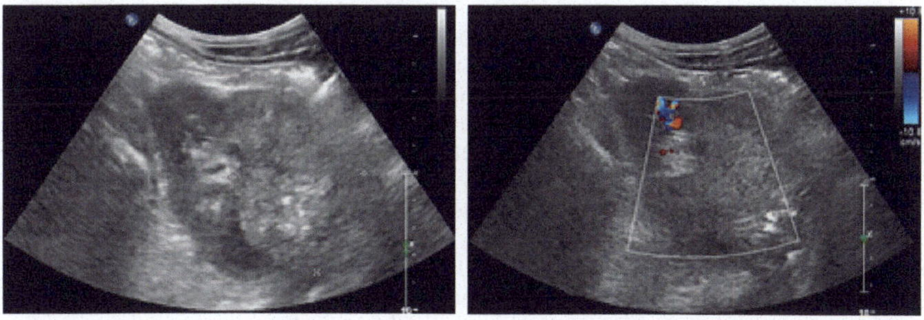

**Fig. 11.35** A 45-year-old male patient had abnormal left kidney morphology. Heterogeneous masses were found in the upper and outer parts of the left kidney, with parenchyma involvement, irregular shape, unclear boundary, uneven internal echo intensity, more scattered hyperechoic dot calcification, and no obvious blood flow signal was found in the mass. Postoperative pathological results: hydatid disease, alveolar echinococcosis. 1. Gray-scale ultrasound showed the abnormal shape of the left kidney. The left kidney showed a hyperechoic mass, unclear boundary, irregular shape, and spot-like calcification. 2. Color Doppler blood flow showed no obvious blood flow signal in the lesion

## References

Albrecht T, Blomley M, Bolondi L, Claudon M, Correas JM, Cosgrove D, Greiner L, Jager K, Jong ND, Leen E, Lencioni R, Lindsell D, Martegani A, Solbiati L, Thorelius L, Tranquart F, Weskott HP, Whittingham T (2004) Guidelines for the use of contrast agents in ultrasound. Ultraschall Med 25(4):249–256

Ammann RW, Eckert J (1996) Cestodes. Echinococcus. Gastroenterol Clin N Am 25:655–689

Bartholomot G, Vuitton DA, Harraga S, Shi DZ, Giraudoux P, Barnish G, Wang YH, MacPherson CN, Craig PS (2002) Combined ultrasound and serologic screening for hepatic alveolar echinococcosis in Central China. Am J Trop Med Hyg 66(1):23–29

Baumann S, Shi R, Liu W, Bao H, Schmidberger J, Kratzer W, Li W, Interdisciplinary Echinococcosis Working Group Ulm (2019) Worldwide literature on epidemiology of human alveolar echinococcosis: a systematic review of research published in the twenty-first century. Infection 47:703–727

Beissert M, Delorme S, Mutze S, Lees WR, Bahner ML, Filimonow S, Gillams AR, Zuna I, Bauer A, Jenett M (2002) Comparison of B-mode and conventional colour/power Doppler ultrasound, contrast-enhanced Doppler ultrasound and spiral CT in the diagnosis of focal lesions of the liver: results of a multicentre study. Ultraschall Med 23(4):245–250

Bresson-Hadni S, Koch S, Miguet JP, Gillet M, Mantion GA, Heyd B, Vuitton DA (2003) Indications and results of liver transplantation for Echinococcus alveolar infection: an overview. Langenbecks Arch Surg 388:231–238

Bresson-Hadni S, Delabrousse E, Blagosklonov O, Bartholomot B, Koch S, Miguet JP, Mantion G, Vuitton DA (2006) Imaging aspects and non-surgical interventional treatment in human alveolar echinococcosis. Parasitol Int 55(Suppl):S267–S272

Cai DM, Wang HY, Wang XL, Jiang Y, Luo Y, Li YZ (2017) Ultrasonographic findings of small lesion of hepatic alveolar echinococcosis. Acta Trop 174:165–170

Cai D, Li Y, Jiang Y, Wang H, Wang X, Song B (2019) The role of contrast-enhanced ultrasound in the diagnosis of hepatic alveolar echinococcosis. Medicine (Baltimore) 98:e14325

Correas JM, Bridal L, Lesavre A, Mejean A, Claudon M, Helenon O (2001) Ultrasound contrast agents: properties, principles of action, tolerance, and artifacts. Eur Radiol 11(8):1316–1328

Dietrich CF, Sharma M, Gibson RN, Schreiber-Dietrich D, Jenssen C (2013) Fortuitously discovered liver lesions. World J Gastroenterol 19:3173–3188

Gottstein B, Saucy F, Deplazes P, Reichen J, Demierre G, Busato A, Zuercher C, Pugin P (2001) Is high prevalence of Echinococcus multilocularis in wild and domestic animals associated with disease incidence in humans? Emerg Infect Dis 7(3):408–412

Horta G, López M, Dotte A, Cordero J, Chesta C, Castro A, Palavecino P, Poniachik J (2015) Benign focal liver lesions detected by computed tomography: review of 1,184 examinations. Rev Med Chile 143:197–202

Kaltenbach TE, Graeter T, Mason RA et al (2015) Determination of vitality of liver lesions by alveolar echinococcosis: comparison of parametric contrast enhanced ultrasound (SonoVue(R)) with quantified 18F-FDG-PET-CT. Nuklearmedizin 54(1):43–49

Kern P, Bardonnet K, Renner E, Auer H, Pawlowski Z, Ammann RW, Vuitton DA, Kern P (2003) European Echinococcosis Registry. European echinococcosis registry: human alveolar echinococcosis, Europe, 1982-2000. Emerg Infect Dis 9(3):343–349

Kern P, Menezes da Silva A, Akhan O, Müllhaupt B, Vizcaychipi KA, Budke C, Vuitton DA (2017) The Echinococcoses: diagnosis, clinical management and burden of disease. Adv Parasitol 96:259–369

Kratzer W, Gruener B, Kaltenbach TE, Ansari-Bitzenberger S, Kern P, Fuchs M, Mason RA, Barth TF, Haenle MM, Hillenbrand A, Oeztuerk S, Graeter T (2015) Proposal of an ultrasonographic classification for hepatic alveolar echinococcosis: Echinococcosis multilocularis Ulm classification-ultrasound. World J Gastroenterol 21(43):12392–12402

Lencioni RA, Allgaier HP, Cioni D, Olschewski M, Deibert P, Crocetti L, FringsH LJ, Zuber I, Blum HE, Bartolozzi C (2003) Small hepatocellular carcinoma in cirrhosis: randomized com-

parison of radio-frequency thermal ablation versus percutaneous ethanol injection. Radiology 228(1):235–240
Romig T, Kratzer W, Kimmig P, Frosch M, Gaus W, Flegel WA, Gottstein B, Lucius R, Beckh K, Kern P (1999) An epidemiologic survey of human alveolar echinococcosis in southwestern Germany. Römerstein Study Group. Am J Trop Med Hyg 61(4):566–573
Schuhbaur J, Schweizer M, Philipp J, Schmidberger J, Schlingeloff P, Kratzer W (2021) Long-term follow-up of liver alveolar echinococcosis using echinococcosis multilocularis ultrasound classification. World J Gastroenterol 27(40):6939–6950
Tao S, Qin Z, Hao W, Yongquan L, Lanhui Y, Lei Y (2011) Usefulness of gray-scale contrast-enhanced ultrasonography (SonoVue®) in diagnosing hepatic alveolar echinococcosis. Ultrasound Med Biol 37:1024–1028
Torgerson PR, Schweiger A, Deplazes P, Pohar M, Reichen J, Ammann RW, Tarr PE, Halkic N, Müllhaupt B (2018) Corrigendum to "Alveolar echinococcosis: from a deadly disease to a well-controlled infection. Relative survival and economic analysis in Switzerland over the last 35 years". J Hepatol 69(5):1208
Wa ZC, Du T, Li XF, Xu HQ, Suo-Ang QC, Chen LD, Hu HT, Wang W, Lu MD (2020) Differential diagnosis between hepatic alveolar echinococcosis and intrahepatic cholangiocarcinoma with conventional ultrasound and contrast-enhanced ultrasound. BMC Med Imaging 20:101
WHO Informal Working Group on Echinococcosis (1996) Guidelines for treatment of cystic and alveolar echinococcosis in humans. Bull World Health Organ 74(3):231–242
Zeng H, Wang J, Xie W, Liu W, Wen H (2012) Assessment of early hepatic echinococcus multilocularis infection in rats with real-time contrast-enhanced ultrasonography. Ultrasound Med Biol 38:1982–1988

# Current Treatments for Echinococcosis What We Have

12

Haining Fan

**Abstract**

Echinococcosis, caused by the larval stage of the cestode *Echinococcus* spp., is a chronic and lethal disease in humans. The current treatment strategy for Echinococcosis is radical resection of the parasitic mass, which is always accompanied by chemotherapy to avoid disease recurrence, especially in the case of alveolar echinococcosis (AE) caused by the metacestode of *Echinococcus multilocularis*. In the past 30 years, extensive research has been conducted on developing novel chemotherapy options against Echinococcosis. In this chapter, we outline a series of chemical compounds against *E. granulosus* and *E. multilocularis*. According to the different physicochemical properties of the compounds, previous studies examined the in vivo and in vitro effects on the mode of action, administration regimen, treatment outcome, and related clinical symptoms. These drugs include imidazole families and their derivatives, targeted chemosynthetic molecules (such as antivirals, antibiotics, antiparasitic, antifungal and antitumor drugs), natural compounds derived from plant and fungal extracts, small inhibitor molecules of key metabolic enzymes or signaling pathways, and combination therapy of drugs. It should be noted that the complexity of Echinococcosis and the heterogeneity of clinical cases make it necessary to deeply understand the host-parasite relationship and parasite phenotype at different developmental stages. This would be greatly helpful for achieving the best therapeutic efficacy of the novel drugs and providing better guidance for clinical treatment.

H. Fun (✉)
Affiliated hospital of Qinghai University, Xining, China
e-mail: fanhaining@medmail.com.cn





J. Li et al. (eds.), *Echinococcus: Control and Elimination of Echinococcosis with a Focus on China and Europe*, Parasitology Research Monographs 19,
https://doi.org/10.1007/978-3-031-54479-8_12

**Keywords**

Echinococcosis · Chemotherapy · Chemosynthetic molecules · Natural compounds · Signaling inhibitors

## 12.1 Introduction

Echinococcosis, also called hydatid disease, is one of the most lethal parasitic zoonoses and remains a public health problem in Middle Asia, West China, South America, Middle Europe and some other epidemic areas (Wen et al. 2019). *Echinococcus granulosus* is the most common species of echinococcosis that causes cystic echinococcosis (CE). *Echinococcus multilocularis* is another common species that causes alveolar echinococcosis (AE), and AE produces tumor-like lesions. In addition, the cancer-like malignant growth of AE may lead to death in most cases within 10–15 years after diagnosis without treatment. CE is caused by the larvae of echinococcus granulosus, the adults of which live in the small intestine of carnivores (canids and felids), and the metacestode is parasitic in the liver, lungs, brain and other tissues of humans or a variety of herbivorous livestock. CE can parasitize almost all parts of the human body; the most common site is the liver, especially in the right lobe, and the lung is the second most common location (Paternoster et al. 2020). Only the larval stage of these species can parasitize the human body; when humans accidentally consume food and water contaminated with parasitic eggs, the eggs that enter the small intestine can hatch into larvae, which can penetrate the small veins or lymphatic vessels of the intestinal wall, and then invade the tissue with the blood circulation, causing an acute inflammatory response (Guo et al. 2021; Pan et al. 2018; Pittini et al. 2019; Wang and Gottstein 2016). Scholars have found that more effective drugs, surgical therapies and combined therapies for the treatment of echinococcosis and optimistic progress has been made in the laboratory and clinical applications (Nicolao et al. 2014; Naseri et al. 2016; Liu et al. 2018; Liu et al. 2015). However, many possible treatments are still under clinical investigation, and this chapter aims to describe current treatments for echinococcosis.

## 12.2 Chemotherapy

### 12.2.1 Old Drugs and Innovation from Old Drugs

Over past decades, albendazole and mebendazole have been the most used drugs in the treatment of both AE and CE, which depletes the glycogen in the germinal layer cells of the echinococcal tissue by inhibiting the uptake of glucose in this tissue. Smego RA et al. found that albendazole has better clinical efficacy than mebendazole, and the recurrence rate of albendazole treatment is lower than mebendazole treatment (Smego and Sebanego 2005). Albendazole was preferred as the anti-infective treatment of choice for echinococcosis at an average dosage of 15 mg/kg/day due to its increased bioavailability and easier administration to patients. The response effect of young patients and patients with small lesion diameter to albendazole treatment is better than that of elderly patients and patients with large lesion diameter, and the treatment effect of CE1 and CE3a cyst type of CE on albendazole is better than that of CE2 and CE3b cyst type of CE (Stojkovic et al. 2009). However, albendazole and mebendazole are insoluble in water, and most organic solvents are absorbed by the gastrointestinal tract (Chai et al. 2021). Some recent studies focused on the absorption and metabolic processes and aimed to increase their absorption, solubility, and biological activity (Hemphill et al. 2014). Specifically, recent studies mainly focus on the new albendazole formulations that show sufficient solubility and biological activity, as some new drug carriers such as self-microemulsion, microspheres, nanoparticles and liposomes showed its advantages when treating echinococcosis (Nair et al. 2020; Lee and Thompson 2017; Panwar et al. 2010; Ghosh Dastidar et al. 2018). However, even though some new carriers are posted in animal and clinical trials, some issues need to be addressed. For instance, in vitro, some carriers may lead to target organ toxicity (Qin et al. 2017). Besides, albendazole and mebendazole are ineffective when treating some cases, and these drugs are parasitostatic rather than parasitocidal. Therefore, further effective drugs for echinococcosis are needed to be explored.

### 12.2.2 Antitumor Therapy

Most antitumor drugs play an important role in interfering with the energy-generating and cell proliferation mechanisms of cancer cells. As echinococcosis cannot obtain energy from the host directly, in vitro and in vivo studies found that some antitumor drugs may improve the current antiparasitic therapy. For instance, metformin is an anti-hyperglycemic and potential anti-cancer agent that may exert its anti-proliferative effects via the induction of energetic stress. Besides, some antitumor drugs such as cyclosporine (Shiee et al. 2021), osthole (Yuan et al. 2016), Taxanes (Huang et al. 2018) atovaquone (Enkai et al. 2020), 5-fluorouracil (Pensel et al. 2014) and bortezomib (Nicolao et al. 2017) have been reported to be effective in the treatment of echinococcosis.

### 12.2.3 Immunotherapies

In past years, much experimental and clinical evidence has found that echinococcosis's growth, reproduction and survival may depend on an effective immune escape mechanism (La et al. 2015). A previous study showed that CE patients with active cysts have proportionally more circulating natural kill cells (Hernández et al. 1999). Moreover, studies have also found that programmed death-1 (PD-1)/programmed death ligand-1 (PD-L1) (B7-H1)/PD-L2 (B7-DC) and CTLA-4/B7–1 are key factors that play roles in immune system regulation, potential immune response, and auto-immunity during parasitic infections (Lepenies and Jacobs 2008; Wilde et al. 2012; Jie et al. 2013; Colley et al. 2005).

Bellanger AP and colleagues have found that in vitro interactions between *Echinococcus multilocularis* vesicular fluid (VF) and different immune checkpoints (PD-1/PD-L1, CTLA-4, LAG-3 and TIM-3) could be a way for *Echinococcus multilocularis* to modulate the host immune response during alveolar echinococcosis (Bellanger et al. 2020). Pang N et al. found that Th9/IL-9 may be involved in the immune response induced by *Echinococcus granulosus* infection (Pang et al. 2014). When immunotherapy is applied in animal tests by regulating adaptive and innate immune cells against echinococcosis, studies suggest that immune checkpoint blockade is efficacious in echinococcosis treatment (Nicolao et al. 2017; Jebbawi et al. 2021). Something that should be noticed is that some immunotherapies such as PD-1/PD-L1 inhibitors may lead to hyperprogression in some cancer patients, because the therapeutic effect of PD-1/PD-L1 inhibitors may affect the aggregation and immunosuppression of T regulatory cells (Yan et al. 2022). When immunotherapies applied, will immunotherapies lead to hyperprogression of echinococcosis or other serious adverse events? Due to limited evidence, further studies are needed to verify those results and hypotheses.

### 12.2.4 Anti-malarial Drugs

Over the past few years, a huge number of drug parasites have been used in clinical practice. Among them, anti-malarial drugs have been found to be sensitive to the blood stage of *P. falciparum*. Furtherly, clinical studies found that several anti-malarial showed activity against echinococcosis (Stadelmann et al. 2016; Küster et al. 2014). For example, mefloquine plays an important role in inhibiting hemozoin formation, an essential step in heme detoxification upon hemoglobin degradation (Egan et al. 1994). Rufener R et al. reported that bi-weekly mefloquine treatment in mice infected orally with E. multilocularis eggs at 100 mg/kg is at least as active as 200 mg/kg ABZ applied 5 days per week (Rufener et al. 2018). Besides, anti-malarial drugs such as nicotinamide phosphoribosyltransferase (identified as a mefloquine binding protein in human cells) and Pyronaridine (results in DNA fragmentation and caspase activation in CE protoscoleces) are also effective during the antiparasitic process (Rufener et al. 2018; Küster et al. 2015).

## 12.3 Invasive Treatment

The treatment for echinococcosis should be discussed according to pathogens, activity of lesions, location of lesions, etc. Figure 12.1 shows the treatment decision-making for echinococcosis.

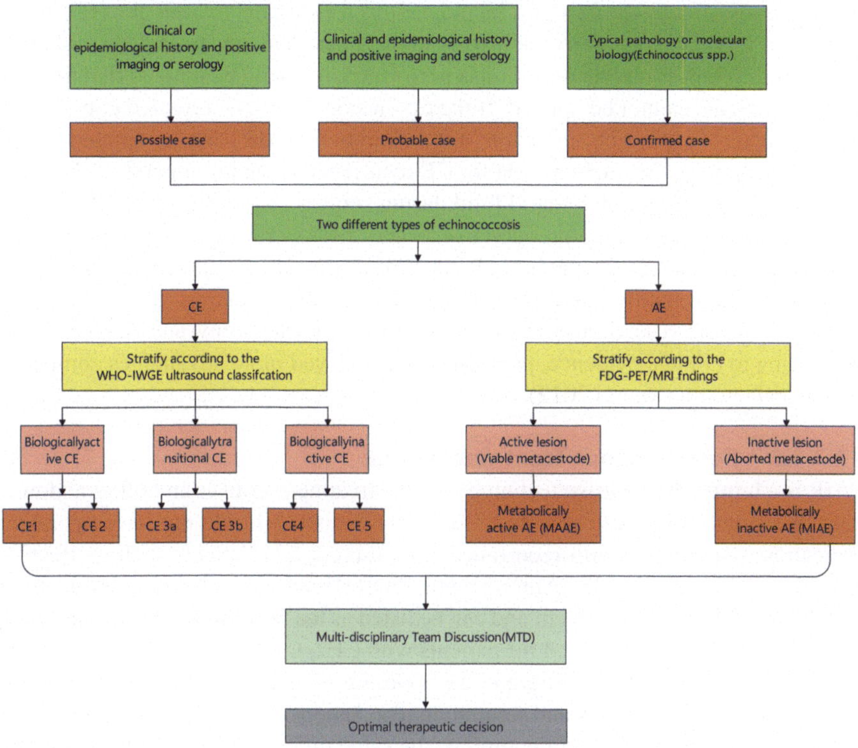


**Fig. 12.1** Treatment decision-making for echinococcosis

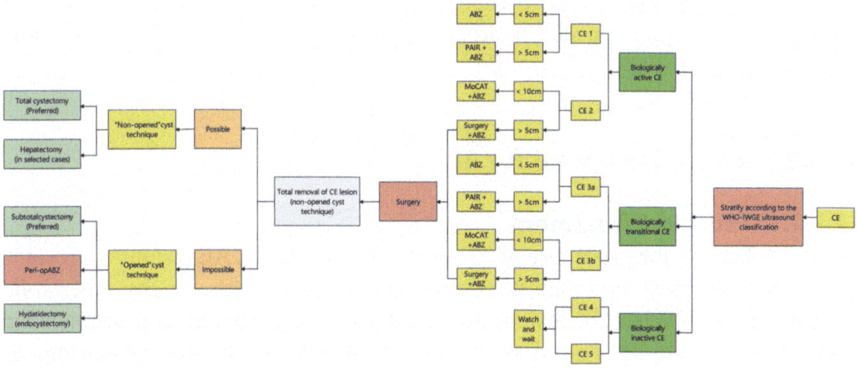


**Fig. 12.2** Procedure of CE treatment

### 12.3.1 Invasive Treatment for CE

#### 12.3.1.1 Surgical Treatment of CE

Figure 12.2 shows the procedure of CE treatment. Complete resection of cysts can be the curative treatment for CE. The indications for surgical treatment of CE are as follows: Cysts compress surrounding organs and percutaneous treatment cannot be performed. When complete resection is performed, it should be noticed that the cyst with its contents cannot be opened. If the cyst lesion cannot be resected completely, PAIR puncture, aspiration, injection, and respiration (PAIR) can be applied (more details see below). The main reason for CE recurrence is the intraoperative dissemination of protoscolex-rich hydatid fluid during surgery or the PAIR. Besides, using anti-echinococcosis drugs preoperatively and postoperatively is recommended (Aktan and Yalin 1996; Dervenis et al. 2005). For hepatic CE, the surgical treatment of hepatic CE is mainly based on WHO-IWGE US classification (McManus et al. 2012). Complete resection is the procedure of pericystectomy and hepatectomy. According to clinical evidence, pericystectomy showed more advantages compared with hepatectomy (Ramia 2012).

#### 12.3.1.2 Percutaneous Treatments of CE

PAIR is a minimally invasive technique for the treatment of liver and other abdominal CE lesions. It is applicable as follows: (1) patients who do not have operative indication. (2) Patients with recurrence after surgery. (3) Patients with ineffective treatment of benzimidazoles. Among them, PAIR+benzimidazole has the best effect on CE1 and CE3a lesions >5 cm and can be listed as the preferred treatment method. PAIR treatment cannot be used for patients with CE2, CE3b, CE4, CE5 and pulmonary echinococcosis (Kabaalioğlu et al. 2006; Giorgio et al. 2008). PAIR has the dual functions of diagnosis and treatment. The lesions with unclear diagnosis can also be clearly diagnosed by puncture. At the same time, compared with surgery, its procedure is simple, with less damage and low cost. The tube drainage should be added based on PAIR, and for multiple cysts, drugs should be given (Kahriman et al. 2017; Zerem and Jusufovic 2006). At present, the commonly used pesticides mainly include hypertonic salt, hydrogen peroxide, ethanol, albendazole solution, etc. (Giorgio et al. 2009).

### 12.3.2 Invasive Treatment for AE

#### 12.3.2.1 Surgical Treatment

Figure 12.3 shows the procedure of AE treatment. Generally, AE lesions are located in the right liver lobe and may invade the lungs, brain, bones, or any other solid organ of the host. When the lesions are not spread, major surgery is recommended in most cases. Complete resection of liver lesions is the most effective and important treatment and the possible curative method. Besides, when complete resection is applied, measures should be taken as follows: (1) anti-echinococcosis drugs such as albendazole must also be taken simultaneously (Torgerson et al. 2008; Ishizu et al. 1997). (2) The resected liver tissue should be at least 2 cm wider than the edge of the lesion. Unfortunately, complete resection cannot be performed for approximately 70% of patients at the time of AE diagnosis because of the progress and the spread of AE, which indicate a rather low resection rate of less than 40% (Ammann

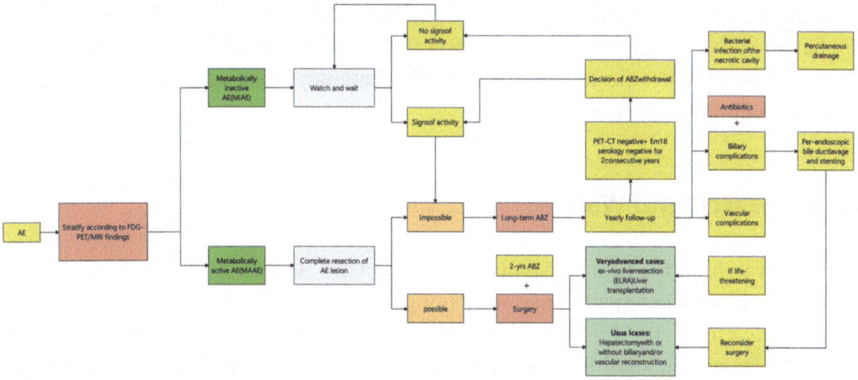


**Fig. 12.3** Procedure of AE treatment

and Eckert 1996; Bresson-Hadni et al. 2000; Sato et al. 1997). If AE has a locally invaded portal vein, common hepatic duct, inferior vena cava, etc., clinical evidence suggested that patients cannot benefit from palliative operations, which may increase the incidence of postoperative adverse events (Kern et al. 2017; Brunetti et al. 2010). Moreover, the study suggested that palliative surgery can only benefit a small number of AE patients with specific surgical indications (Buttenschoen et al. 2009).

For hepatic AE, liver transplantation is another possible curative treatment, which has been performed for those who do not meet the criteria for resection, especially for those with secondary biliary cirrhosis or Budd–Chiari syndrome or recurrent life-threatening cholangitis (Kamiyama 2020). Besides, if the patients have residual or metastatic AE lesions after liver transplantation, the utilization of immunosuppressive drugs can be a reason for the progression of AE, which is fatal in some cases (Bresson-Hadni et al. 2011). With the development of surgical techniques, ex vivo resection and auto-transplantation have brought about alternative options. Some evidence-based studies have concluded that ex vivo resection and auto-transplantation are complex operative procedures for those with unresectable AE lesions in the liver, achieving a satisfactory outcome for both AE and CE (Zawistowski et al. 2020; Baimas-George et al. 2021). However, something should be noticed is that due to the complex procedure of surgery, few cases are applicable to this operation, and fewer institutions can perform ex vivo resection and auto-transplantation. The long-term outcomes need to be further verified.

#### 12.3.2.2 Ablation

Recently, ablation can be used as an alternative therapy for patients with complex hepatic AE and those who are not suitable for surgery. The principle of ablation is to use the principle that high temperature can destroy the alveolar hydatid tissue. Currently, there are not enough reports about radiofrequency and microwave ablation for treating hepatic AE. It is a minimally invasive treatment method guided by ultrasound. As a supplement to conventional resection, intraoperative radiofrequency ablation has also been proposed (Papaconstantinou et al. 2006; Zacharoulis et al. 2006). Currently, there are no reports about radiofrequency ablation for treating hepatic alveolar hydatid disease. Therefore, it is urgent to design an experimental model to explore the feasibility of radiofrequency ablation in treating hepatic alveolar echinococcosis.

## 12.4 Watch and Wait

For CE, watch and wait are usually applied to CE4 and CE5. These two types of lesions have been inactivated and may not grow further. Therefore, when there are no complications, no further treatment is needed. Only regular observation is required. Treatment shall be carried out in case of complications according to specific conditions. For AE, lesions that are massively calcified and/or negative by FDG-PET may benefit from a "watch and wait" approach. However, necessary measures such as anti-echinococcosis or surgery should be taken when the lesion shows signs of activity.

## References

Aktan AO, Yalin R (1996) Preoperative albendazole treatment for liver hydatid disease decreases the viability of the cyst. Eur J Gastroenterol Hepatol 8(9):877–879

Ammann RW, Eckert J (1996) Cestodes. Echinococcus. Gastroenterol Clin N Am 25:655–689

Baimas-George M, Thompson KJ, Watson MD, Iannitti DA, Martinie JB, Baker EH, Levi D, Vrochides D (2021) The technical aspects of ex vivo hepatectomy with autotransplantation: a systematic review and meta-analysis. Langenbeck's Arch Surg 406(7):2177–2200

Bellanger AP, Courquet S, Pallandre JR, Godet Y, Millon L (2020) Echinococcus multilocularis vesicular fluid induces the expression of immune checkpoint proteins in vitro. Parasite Immunol 42(6):e12711

Bresson-Hadni S, Vuitton DA, Bartholomot B, Heyd B, Godart D, Meyer JP et al (2000) A twenty-year history of alveolar echinococcosis: analysis of a series of 117 patients from eastern France. Eur J Gastroenterol Hepatol 12:327–336

Bresson-Hadni S, Blagosklonov O, Knapp J, Grenouillet F, Sako Y, Delabrousse E et al (2011) Should possible recurrence of disease contraindicate liver transplantation in patients with end-stage alveolar echinococcosis? A 20-year follow-up study. Liver Transpl 17:855–865

Brunetti E, Kern P, Vuitton DA (2010) Expert consensus for the diagnosis and treatment of cystic and alveolar echinococcosis in humans. Acta Trop 114:1–16

Buttenschoen K, Gruener B, Carli Buttenschoen D, Reuter S, Henne-Bruns D, Kern P (2009) Palliative operation for the treatment of alveolar echinococcosis. Langenbeck's Arch Surg 394(1):199–204

Chai JY, Jung BK, Hong SJ (2021) Albendazole and Mebendazole as antiparasitic and anti-cancer agents: an update. Korean J Parasitol 59(3):189–225

Colley DG, Sasser LE, Reed AM (2005) PD-L2+ dendritic cells and PD-1+ CD4+ T cells in schistosomiasis correlate with morbidity. Parasite Immunol 27:45–53

Dervenis C, Delis S, Avgerinos C, Madariaga J, Milicevic M (2005) Changing concepts in the management of liver hydatid disease. J Gastrointestinal Surg 9(6):869–877

Egan TJ, Ross DC, Adams PA (1994) Quinoline anti-malarial drugs inhibit spontaneous formation of beta-haematin (malaria pigment). FEBS Lett 352(1):54–57

Enkai S, Inaoka DK, Kouguchi H, Irie T, Yagi K, Kita K (2020) Mitochondrial complex III in larval stage of Echinococcus multilocularis as a potential chemotherapeutic target and in vivo efficacy of atovaquone against primary hydatid cysts. Parasitol Int 75:102004

Ghosh Dastidar D, Saha S, Chowdhury M (2018) Porous microspheres: synthesis, characterisation and applications in pharmaceutical & medical fields. Int J Pharm 548(1):34–48

Giorgio A, de Stefano G, Esposito V, Liorre G, Di Sarno A, Giorgio V et al (2008) Long-term results of percutaneous treatment of hydatid liver cysts: a single center 17 years experience. Infect 36(3):256–261

Giorgio A, Di Sarno A, De Stefano G et al (2009) Percutaneous treatment of hydatid liver cyst. Recent Pat Antiinfect Drug Discov 4(1):29–36

Guo Z, Duo H, Zhang X, Ma Y, Shen X, Fu Y (2021) A pilot study on deworming wild foxes for Echinococcus spp. in Qinghai, China. Iran J Parasitol 16(4):672–677

Hemphill A, Stadelmann B, Rufener R et al (2014) Treatment of echinococcosis: albendazole and mebendazole – what else? Parasite 21:70

Hernández A, O'Connor JE, Mir A (1999) Phenotypic analysis of peripheral lymphocyte subpopulations in hydatid patients. Parasitol Res 85(11):948–950

Huang X, Wiehr S, Wild AM, Voßberg P, Hoffmann W, Grüner B, Köhler C, Soboslay PT (2018) The effects of taxanes, vorinostat and doxorubicin on growth and proliferation of *Echinococcus multilocularis* metacestodes assessed with magnetic resonance imaging and simultaneous positron emission tomography. Oncotarget 9(10):9073–9087

Ishizu H, Uchino J, Sato N, Aoki S, Suzuki K, Kuribayashi H (1997) Effect of albendazole on recurrent and residual alveolar echinococcosis of the liver after surgery. Hepatology 25:528–531

Jebbawi F, Bellanger AP, Lunström-Stadelmann B, Rufener R, Dosch M, Goepfert C, Gottstein B, Millon L, Grandgirard D, Leib SL, Beldi G, Wang J (2021) Innate and adaptive immune responses following PD-L1 blockade in treating chronic murine alveolar echinococcosis. Parasite Immunol 43(8):e12834

Jie HB, Gildener-Leapman N, Li J, Srivastava RM, Gibson SP et al (2013) Intratumoral regulatory T cells upregulate immunosuppressive molecules in head and neck cancer patients. Br J Cancer 109:2629–2635

Kabaalioğlu A, Çeken K, Alimoglu E, Apaydin A (2006) Percutaneous imaging-guided treatment of hydatid liver cysts: do long-term results make it a first choice? Eur J Radiol 59(1):65–73

Kahriman G, Ozcan N, Dogan S, Karaborklu O (2017) Percutaneous treatment of liver hydatid cysts in 190 patients: a retrospective study. Acta Radiol 58(6):676–684

Kamiyama T (2020) Recent advances in surgical strategies for alveolar echinococcosis of the liver. Surg Today 50(11):1360–1367

Kern P, Menezes da Silva A, Akhan O, Mullhaupt B, Vizcaychipi KA, Budke C, Vuitton DA (2017) The echinococcoses: diagnosis, clinical management and burden of disease. Adv Parasitol 96:259–369

Küster T, Kriegel N, Stadelmann B, Wang X, Dong Y, Vennerstrom JL, Keiser J, Hemphill A (2014) Amino ozonides exhibit in vitro activity against Echinococcus multilocularis metacestodes. Int J Antimicrob Agents 43:40–46

Küster T, Stadelmann B, Rufener R, Risch C, Müller J, Hemphill A (2015) Oral treatments of Echinococcus multilocularis-infected mice with the anti-malarial drug mefloquine that potentially interacts with parasite ferritin and cystatin. Int J Antimicrob Agents 46(5):546–551

La X, Zhang F, Li Y et al (2015) Upregulation of PD-1 on $CD4^+CD25^+$ T cells is associated with immunosuppression in liver of mice infected with Echinococcus multilocularis. Int Immunopharmacol 26(2):357–366

Lee Y, Thompson DH (2017) Stimuli-responsive liposomes for drug delivery. Wiley Interdiscip Rev Nanomed Nanobiotechnol 9(5). https://doi.org/10.1002/wnan.1450

Lepenies B, Jacobs T (2008) The role of negative costimulators during parasitic infections. Endocr Metab Immune Disord Drug Targets 8:279–288

Liu C, Zhang H, Yin J, Hu W (2015) In vivo and in vitro efficacies of mebendazole, mefloquine and nitazoxanide against cyst echinococcosis. Parasitol Res 114(6):2213–2222

Liu C, Yin J, Xue J, Tao Y, Hu W, Zhang H (2018) In vitro effects of amino alcohols on echinococcus granulosus. Acta Trop 182:285–290
McManus DP, Gray DJ, Zhang W, Yang Y (2012) Diagnosis, treatment, and management of echinococcosis. BMJ 344:e3866
Nair AR, Lakshman YD, Anand VSK, Sree KSN, Bhat K, Dengale SJ (2020) Overview of extensively employed polymeric carriers in solid dispersion technology. AAPS PharmSciTech 21(8):309
Naseri M, Akbarzadeh A, Spotin A, Akbari NA, Mahami-Oskouei M, Ahmadpour E (2016) Scolicidal and apoptotic activities of albendazole sulfoxide and albendazole sulfoxide-loaded PLGA-PEG as a novel nanopolymeric particle against Echinococcus granulosus protoscoleces. Parasitol Res 115(12):4595–4603
Nicolao MC, Elissondo MC, Denegri GM, Goya AB, Cumino AC (2014) In vitro and in vivo effects of tamoxifen against larval stage Echinococcus granulosus. Antimicrob Agents Chemother 58(9):5146–5154
Nicolao MC, Loos JA, Rodriguez Rodrigues C, Beas V, Cumino AC (2017) Bortezomib initiates endoplasmic reticulum stress, elicits autophagy and death in Echinococcus granulosus larval stage. PLoS One 12(8):e0181528
Pan W, Xu HW, Hao WT et al (2018) The excretory-secretory products of Echinococcus granulosus protoscoleces stimulated IL-10 production in B cells via TLR-2 signaling. BMC Immunol 19(1):29
Pang N, Zhang F, Ma X, Zhang Z, Zhao H et al (2014) Th9/IL-9 profile in human echinococcosis: their involvement in immune response during infection by echinococcus granulosus. Mediat Inflamm 2014:781649
Panwar P, Pandey B, Lakhera PC, Singh KP (2010) Preparation, characterization, and in vitro release study of albendazole-encapsulated nanosize liposomes. Int J Nanomedicine 5:101–108
Papaconstantinou I, Kontos M, Prassas E et al (2006) Radio frequency ablation (RFA)-assisted pericystectomy for hepatic echinococcosis: an alternative technique. Surg Laparosc Endosc Percutan Tech 16(5):338–341
Paternoster G, Boo G, Wang C et al (2020) Epidemic cystic and alveolar echinococcosis in Kyrgyzstan: an analysis of national surveillance data. Lancet Glob Health 8(4):e603–e611
Pensel PE, Albani C, Gamboa GU, Benoit JP, Elissondo MC (2014) In vitro effect of 5-fluorouracil and paclitaxel on Echinococcus granulosus larvae and cells. Acta Trop 140:1–9
Pittini A, Martı'nez-Acosta YE, Casaravilla C, Seoane PI, Ruckerl D, Quijano C et al (2019) Particles from the Echinococcus granulosus laminated layer inhibit CD40 upregulation in dendritic cells by interfering with Akt activation. Infect Immun 87:e00641–e00619
Qin W, Guan D, Ma R, Yang R, Xing G, Shi H, Tang G, Li J, Lv H, Jiang Y (2017) Effects of trigonelline inhibition of the Nrf2 transcription factor in vitro on Echinococcus granulosus. Acta Biochim Biophys Sin Shanghai 49(8):696–705
Ramia JM (2012) Ambispective comparative study of two surgical strategies for liver hydatidosis. World J Gastroenterol 18(6):546
Rufener R, Ritler D, Zielinski J, Dick L, da Silva ET, da Silva AA, Joekel DE, Czock D, Goepfert C, Moraes AM, de Souza MVN, Müller J, Mevissen M, Hemphill A, Lundström-Stadelmann B (2018) Activity of mefloquine and mefloquine derivatives against Echinococcus multilocularis. Int J Parasitol Drugs Drug Resist 8(2):331–340
Sato N, Namieno T, Furuya K, Takahashi H, Yamashita K, Uchino J et al (1997) Contribution of mass screening system to resectability of hepatic lesions involving Echinococcus multilocularis. J Gastroenterol 32:351–354
Shiee MR, Kia EB, Zahabiun F et al (2021) In vitro effects of tropisetron and granisetron against Echinococcus granulosus (s.s.) protoscoleces by involvement of calcineurin and calmodulin. Parasit Vectors 14(1):197
Smego RA, Sebanego P (2005) Treatment options for hepatic cystic echinococcosis. Int J Infect Dis 9(2):69–76

Stadelmann B, Rufener R, Aeschbacher D, Spiliotis M, Gottstein B, Hemphill A (2016) Screening of the open source malaria box reveals an early lead compound for the treatment of alveolar echinococcosis. PLoS Negl Trop Dis 10:e0004535

Stojkovic M, Zwahlen M, Teggi A et al (2009) Treatment response of cystic echinococcosis to benzimidazoles: a systematic review. PLoS Negl Trop Dis 3(9):524

Torgerson PR, Schweiger A, Deplazes P, Pohar M, Reichen J, Ammann RW et al (2008) Alveolar echinococcosis: from a deadly disease to a well-controlled infection. Relative survival and economic analysis in Switzerland over the last 35 years. J Hepatol 49:72–77

Wang J, Gottstein B (2016) Immunoregulation in larval Echinococcus multilocularis infection. Parasite Immunol 38:182–192

Wen H, Vuitton L, Tuxun T, Li J, Vuitton DA, Zhang W, McManus DP (2019) Echinococcosis: advances in the 21st century. Clin Microbiol Rev 32(2):e00075–e00018

Wilde B, Hua F, Dolff S, Jun C, Cai X et al (2012) Aberrant expression of the negative costimulator PD-1 on T cells in granulomatosis with polyangiitis. Rheumatology (Oxford) 51:1188–1197

Yan J, Deng M, Kong S et al (2022) Transarterial chemoembolization in combination with programmed death-1/programmed cell death-ligand 1 immunotherapy for hepatocellular carcinoma: a mini review. iliver 1(4):225–234

Yuan M, Luo Y, Xin Q, Gao H, Zhang G, Jing T (2016) Efficacy of osthole for Echinococcus granulosus in vitro and Echinococcus multilocularis in vivo. Vet Parasitol 226:38–43

Zacharoulis D, Poultsidis A, Roundas C et al (2006) Liver hydatid disease: radiofrequency-assisted pericystectomy. Ann R Coll Surg Engl 88(5):499–500

Zawistowski M, Nowaczyk J, Jakubczyk M, Domagała P (2020) Outcomes of ex vivo liver resection and autotransplantation: a systematic review and meta-analysis. Surgery 168(4):631–642

Zerem E, Jusufovic R (2006) Percutaneous treatment of univesicular versus multivesicular hepatic hydatid cysts. Surg Endosc 20(10):1543–1547

# Novel Chemotherapeutical Approaches Against Echinococcosis: A Swiss Perspective

13

Britta Lundström-Stadelmann, Matías Preza, Marc Kaethner, and Andrew Hemphill

**Abstract**

Metacestodes of *Echinococcus multilocularis* (small fox tapeworm) and *E. granulosus* sensu lato (dog tapeworm) cause the severe and often incurable diseases alveolar echinococcosis (AE) and cystic echinococcosis (CE). Whereas *E. multilocularis* is endemic in Switzerland, *E. granulosus* is not found in animals in Switzerland anymore. Human and animal AE cases are increasing in Switzerland, and so are non-autochthonous human CE cases. Metacestodes consist of an outer laminated layer and an inner germinal layer of living parasite cells. Amongst them, there is a high proportion of undifferentiated stem cells that are refractory to the currently used drugs, benzimidazoles, against AE and CE. Thus, novel drug treatment options are urgently needed. Based on in vitro grown metacestode vesicles of both species, a screening cascade was defined that allows for drug efficacy assessments of repurposed compounds like natural compounds, anti-cancer drugs or anti-infectives. The only identified drugs that so far proofed to be active in subsequent tests in the AE mouse model are benzimidazoles and the anti-malarial mefloquine. In complementary target-based approaches the follow-

B. Lundström-Stadelmann (✉)
Institute of Parasitology, Vetsuisse Faculty, University of Bern, Bern, Switzerland

Multidisciplinary Center for Infectious Diseases, University of Bern, Bern, Switzerland
e-mail: Britta.lundstroem@unibe.ch

M. Preza · A. Hemphill
Institute of Parasitology, Vetsuisse Faculty, University of Bern, Bern, Switzerland

M. Kaethner
Institute of Parasitology, Vetsuisse Faculty, University of Bern, Bern, Switzerland

Graduate School for Cellular and Biomedical Sciences, University of Bern, Bern, Switzerland



J. Li et al. (eds.), *Echinococcus: Control and Elimination of Echinococcosis with a Focus on China and Europe*, Parasitology Research Monographs 19,
https://doi.org/10.1007/978-3-031-54479-8_13

ing potential targets have been identified and studied in vitro: the energy metabolism, in particular the electron transfer chain of the helminth-specific malate dismutation, and the threonine metabolism.

**Keywords**

*Echinococcus multilocularis* · *Echinococcus granulosus* · Drug treatment · Benzimidazole · Anti-malarial · Mefloquine · Energy metabolism · Malate dismutation · Threonine

## 13.1 Life Cycle and Hosts

The zoonotic tapeworms *Echinococcus multilocularis* (small fox tapeworm) and *E. granulosus* sensu lato (dog tapeworm) cause the diseases alveolar echinococcosis (AE) and cystic echinococcosis (CE) in humans and animals. Other medically relevant species are *E. oligarthra*, *E. vogeli,* and *E. shiquicus*. Still, they are not important in Switzerland and therefore not covered in this chapter with a Swiss view. Both *E. multilocularis* and *E. granulosus* exist in predator-prey life cycles, with final hosts being exclusively carnivores (mostly wild and domestic canids) and intermediate hosts covering a wide range of species (Romig et al. 2017). The final host for *E. multilocularis* is mainly the red fox (*Vulpes vulpes*), but in Central Europe, including Switzerland, also the raccoon dog (*Nyctereutes procyonoides*), the golden jackal (*Canis aureus*), the wolve (*Canis lupus*), the domestic dog and the domestic cat can be infected with *E. multilocularis* and thus serve in parasite transmission (Frey et al. 2022). Intermediate hosts in Switzerland are rodents such as mainly voles (e.g., *Microtus arvalis*) or, to a lesser extent, the European beaver (*Castor fiber*), but also other small mammals may serve as intermediate hosts for *E. multilocularis*, and accidentally also a large number of dead-end hosts including humans, captive monkeys, domestic pigs, dogs, and others (Federer et al. 2015). For *E. granulosus s.l.*, which represents a species complex of different strains and species that each show a slightly different intermediate host species range, intermediate hosts are larger mammals such as sheep, goats, cattle, camels, pigs, horses, and others. However, parasites are usually not found in these animals in Switzerland anymore. Final hosts of *E. granulosus s.l.* (in the following termed *E. granulosus*) are dogs that feed on contaminated offal upon home slaughter or improper slaughterhouse hygiene, as well as on livestock carcasses left in the surroundings. However, such transmission dynamics are eradicated in Switzerland, and hardly any autochthonous infections have been reported (Deplazes et al. 2017). Nevertheless, *E. granulosus* does represent an important parasite in Switzerland, as a considerable and rising number of human CE cases have been documented (Casulli et al. 2022).

For both parasite species, eggs represent the infectious stage. *Echinococcus* eggs, 30–36 μm in diameter, contaminate the environment and are taken up orally with food or water by intermediate or accidental hosts. Each *Echinococcus* egg contains a larva, the oncosphere, protected by a multilayered embryophore that renders these eggs highly resistant (Veit et al. 1995). Once the oncosphere has hatched and activated in the intermediate host's stomach and intestinal lumen, it penetrates the intestinal wall, reaches the bloodstream and reaches its prime location, the liver for *E. multilocularis*, or the liver or lung for *E. granulosus*. Here the oncosphere develops into the second larval and disease-causing stage, the metacestode (see Chap. 2). In a natural intermediate host, *E. multilocularis* brood capsules will form through asexual proliferation inside the metacestodes within 2–4 months post-infection, and a large number of new generations of tapeworms, the protoscoleces, develop. Also within *E. granulosus* metacestodes, brood capsules are formed via asexual proliferation, but at a variable rate depending on the intermediate host species within 10–48 months (Thompson 2017). Thus, protoscolex development is much faster in *E. multilocularis*, which is regarded as an evolutionary response to the restricted space for metacestode growth in the respective hosts. Notably, protoscoleces are formed in rodent AE hosts but usually not in human AE patients. However, protoscoleces do form in human CE patients. When an infected rodent with AE, or infected organs with CE, are ingested by a suitable final host, protoscoleces that reach the small intestine will evaginate, attach to the intestinal mucosa, and form adult tapeworms within 20–35 days (Thompson 2017). Adult *Echinococcus* tapeworms live in the intestines of the above-mentioned final hosts, and in contrast to other tapeworm species, *Echinococcus* adults are only a few millimeters in size. They consist of an anterior scolex, a growth zone and proglottids, which produce infectious eggs mainly through self-insemination. *E. multilocularis* produces eggs 28–35 days after infection, *E. granulosus* 34–58 days after infection (Thompson 2017). These are released in the feces of final hosts and are infective for intermediate or accidental hosts, allowing the life cycle to start over again.

## 13.2 Metacestodes

Metacestodes represent the disease-causing stage of *E. multilocularis* and *E. granulosus*. Metacestodes are formed from oncospheres. Within 14 days, oncospheres undergo reorganization and finally form a central cavity as well as a germinal and laminated layer. *Echinococcus* cysts of both species follow the same basic structure. Metacestodes are fluid-filled vesicles that are surrounded by three layers: the laminated layer (LL), the tegument, and the germinal layer (GL) (see Fig. 13.1). The acellular carbohydrate-rich LL is an *Echinococcus*-unique extracellular matrix that forms the outer surface and protects the parasite against host reactions. The LL is periodic acid-Schiff positive, a feature used for diagnostic purposes. Importantly, the LL of *E. granulosus* is much thicker and massive compared to the LL of *E. multilocularis* (Díaz et al. 2011). The syncytial tegument is anchored to the interior surface of the LL via microvilli-like protrusions termed microtriches, and it is followed by the GL. The GL comprises connective tissue, muscle cells, nerve cells, glycogen storage cells, subtegumentary cytons, and undifferentiated stem cells (Koziol et al. 2014). These stem cells are the only proliferative cells in *Echinococcus*, giving rise to metacestodes. The vesicle fluid (VF) located in the interior of these metacestodes plays a role in the nutrition and exchange of metabolites between the parasite and its surroundings. For *E. granulosus*, VF is often called hydatid fluid.

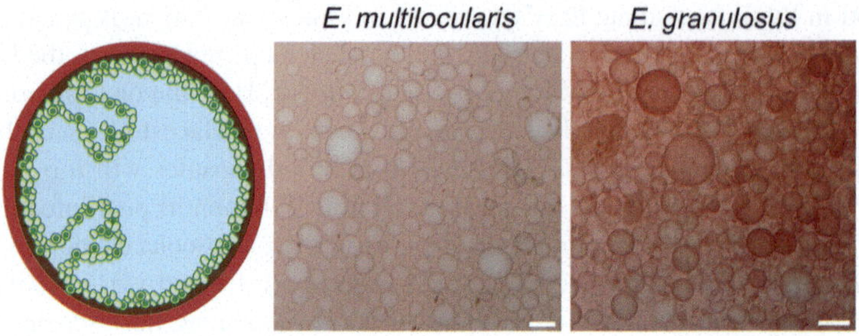


**Fig. 13.1** *Echinococcus* metacestodes. Left: Schematic drawing of an *E. multilocularis* metacestode with the outer laminated layer in red, the inner syncytial tegument in brown and the inner germinal layer containing cells such as muscle cells, nerve cells, glycogen storage cells, subtegumental cytons and undifferentiated stem cells. The metacestodes are filled with vesicle fluid (light blue), which maintains the turgor and, thus, the physical integrity of the metacestodes and plays important roles in the uptake and secretion of metabolites. Figure adapted from Lundström-Stadelmann et al. (2019a). Middle: *E. multilocularis* metacestodes cultured in vitro for 11 weeks. Right: *E. granulosus* metacestodes cultured in vitro for 10 months. Scale bars represent 5 mm

*E. granulosus* cysts are unilocular, grow expansively, and might form secondary internal chambers. They are surrounded by a host fibrous layer, the adventitial layer. In contrast, *E. multilocularis* metacestodes appear more complex (Thompson 2017). *E. multilocularis* metacestodes comprise numerous small vesicles embedded in connective tissue proliferating endogenously and exogenously. *E. multilocularis* metacestodes are networks of cellular GL protrusions that cause the infiltrative growth of the parasite and metastasis formation upon detachment of these protrusions and spread via blood or lymphatic vessels (Thompson 2017). The almost unlimited regenerative potential of metacestodes is due to the presence of stem cells, and they need to be killed to achieve a parasiticidal effect upon drug treatment of the parasite (Brehm and Koziol 2014).

## 13.3 Epidemiology and Impact

*E. multilocularis* is endemic in Central and Eastern Europe, Central and Eastern Asia, and North America. *E. granulosus* is cosmopolitan and highly endemic in pastoral communities. At least 18,235 new human cases of AE occur each year, with the highest prevalence reported in China (Torgerson et al. 2010). Human CE is estimated to account for at least 188,000 cases per year (possibly up to 1.77 million cases). Human AE has an annual impact of more than 688,000 disability-adjusted life years (DALYs), and the impact of human CE is more than 184,000 DALYs (Torgerson et al. 2015). Thus, both neglected diseases AE and CE impose a considerable burden on human and veterinary health and are responsible for high economic losses in livestock industries, particularly for CE, where losses of $2 billion annually were (under)estimated (Budke et al. 2006). In Europe, *E. multilocularis* and *E. granulosus* are the first and fourth most important food-borne parasites

(FAO/WHO 2014; Bouwknegt et al. 2018). In addition, there is a reported emergence of AE in Europe and Canada (Gottstein et al. 2015; Trotz-Williams et al. 2017), of CE in the Middle East (Galeh et al. 2018), and both AE and CE in the East Asia-Pacific region and Central Asia (Torgerson 2013; Hotez 2020).

The annual incidence of AE in Switzerland increased from 0.1 to 0.3 per 100,000 inhabitants. A Swiss echinococcosis network was established in October 2020 (Bresson-Hadni 2023). Preliminary data showed that most AE patients were supported in centers in Zurich and Bern, and at least 98 and 140 new AE cases were identified in Switzerland in 2020 and 2021, respectively. This data suggests that the occurrence of AE in Switzerland was previously underestimated and may be considerably higher, reaching up to 1.3 per 100,000 inhabitants. In Switzerland, peri-urban *E. multilocularis* prevalence levels in foxes are up to 67% and in voles up to 65% (Hofer et al. 2000; Romig et al. 2017). As pigs can be aberrant hosts for *E. multilocularis* metacestodes, slaughter pigs were used in a Swiss-wide surveillance study to analyze factors associated with infection and environmental contamination with *E. multilocularis* eggs. Widespread environmental contamination all over the country was found (Meyer et al. 2020).

A recent systematic review identified approximately 65,000 human CE cases in Europe over the last 25 years (annual incidence of 0.64 cases per 100,000) (Casulli et al. 2022). In Switzerland, all 166 (1997–2021) cases were imported, reaching 10 CE cases per year. In Switzerland, *E. granulosus* infections in dogs and ruminants are only occasionally found (Deplazes et al. 2017).

Within the EU, human AE and CE are mostly reported as "echinococcosis" (EU Decision 2018/945), thus not distinguishing the two diseases. However, since 2019, individual reporting of the two diseases to the European Surveillance System database has been made possible, and this is now done in some of the EU countries (van der Giessen et al. 2021). In Switzerland, human AE and CE are not notifiable diseases since 1997 (Schmidberger et al. 2021). In animals, *E. multilocularis* and *E. granulosus* are notifiable in several of the EU countries, as well as in Switzerland, and they include active and passive surveillance (van der Giessen et al. 2021). An exemplary national registry for echinococcosis has existed in France since 1982 (Bresson-Hadni et al. 2021). Based on this example, a Swiss echinococcosis network was established to register cases in the future (Bresson-Hadni 2023).

## 13.4 Disease, Diagnosis and Treatment

### 13.4.1 The Diseases AE and CE

Both AE and CE are chronic diseases that usually become symptomatic and are diagnosed only at a late stage, years after the infection (Casulli et al. 2022). Based on the general growth characteristics of the parasite, AE is often compared to a malignant tumor and CE to a benign tumor (Bresson-Hadni et al. 2021).

In humans, the disease AE causes severe organ dysfunction, mostly in the liver (predominantly right lobe) but also in other infected organs. *E. multilocularis* metacestodes form cancer-like lumps that can grow up to 15–20 cm and often contain necrotic areas in the center. At a later stage of infection, metacestodes can form metastases and spread to other body locations. In animals such as dogs or captive monkeys, the clinical picture is very similar to humans (Corsini et al. 2015; Wenker et al. 2019). AE is a chronic disease, which exhibits severe symptoms only after up to 10–15 years. In the progressive stage of human AE, non-specific symptoms such as abdominal pain, cholestatic jaundice, hepatomegaly, fever, anemia, weight loss, and pleural pain appear (Kern et al. 2017). In addition, chronic disease induces high psychological burdens in patients, ranging from depression and anxiety to reduced physical quality of life (Schmidberger et al. 2018; Nikendei et al. 2019). Finally, at an advanced stage, and if not treated properly or if treatment fails, AE will lead to the death of the patient (Lundström-Stadelmann et al. 2019a).

Historically, CE, known as hydatid disease or hydatidosis, follows very different growth characteristics. Human CE most commonly affects the liver, lungs, and other organs. In animals, CE is very similar to human CE (Ohiolei et al. 2022). *E. granulosus* metacestodes grow as fluid-filled, isolated, concentrically growing, but not metastasizing, cysts (Casulli et al. 2022). They usually do not induce symptoms before having reached a certain size. As metacestodes compress neighboring organs, this may result in non-specific symptoms depending on the location of the cyst. CE cysts grow up to 1–50 mm per year, but growth can also completely halt, or cysts can spontaneously rupture, collapse, or disappear (Brunetti et al. 2010).

### 13.4.2 Diagnosis of AE and CE

AE and CE are usually incidental findings from imaging examinations for other reasons. Clinical findings, in combination with imaging techniques like ultrasonography, computed tomography, magnetic resonance imaging, and serology, are the basis for the diagnosis of echinococcosis. Based on imaging, *E. multilocularis* metacestodes are classified by the PNM system, *E. granulosus* cysts are classified as CL or CE1–CE5 into different stages (Brunetti et al. 2010). Serology for human AE and CE is based on purified and recombinant antigens with high diagnostic sensitivity and specificity (Siles-Lucas et al. 2017), and serology for canine AE is adapted from these protocols (Frey et al. 2017). Serology of AE or CE in other animals is also based on tests adapted from human serology.

### 13.4.3 Treatment of AE and CE

Without treatment, human AE mortality is >90% within 10–15 years, and human CE mortality is significantly lower at around 2–4% (Kern et al. 2017). Still, these values must be taken cautiously as they also highly depend on the availability and accessibility of medical care.

The only curative treatment for AE is invasive surgical resection of the whole parasite tissue. However, radical surgery is applied in only up to 50% of all human

AE cases (Kern et al. 2017) and can often not be performed when diagnosed at a late stage of infection with the parasite growing highly invasively (Grüner et al. 2017; Salm et al. 2019). Surgery has to be combined with temporal chemotherapy and long-term monitoring to treat and follow up eventual parasite residuals (Kern et al. 2017). In cases where complete resection is not possible, the AE patient is left with chemotherapeutical treatment based on the benzimidazoles (BMZ) albendazole (ABZ, 10–15 mg/kg/day, in 2 doses) and/or mebendazole (MBZ, 40–50 mg/kg/day, in 3 doses) (Brunetti et al. 2010). Based on the present treatment options, 10-year survival rates of AE patients increased drastically (Ammann and Eckert 1996; Grüner et al. 2017). However, as reported in a German study, disease progression due to treatment failure was described in up to 16% of all AE cases. BMZ are not always well tolerated and cause side effects in 54.5% of the patients, including 6.9% of severe adverse effects such as life-threatening hepatotoxicity (Grüner et al. 2017). Further, BMZ can halt the progression of AE but do not act parasiticidally. Therefore, BMZ have only limited potential to bring about a cure for AE, and massive doses of these drugs usually have to be administered throughout life (Kern et al. 2017). Furthermore, BMZ are potentially teratogenic (at least shown in rodents, human fetal exposure expected to be smaller). All these shortcomings underline the urgency of developing alternative chemotherapeutic options against AE.

Liver transplantation was applied in patients with inoperable lesions or chronic liver failure, as well as ex vivo resection with auto-transplantation. However, as immunosuppression supports the regrowth of the parasite, these options are generally not recommended (Bresson-Hadni et al. 2003; Aji et al. 2018; Beldi et al. 2019). Immunosuppression poses an increasing problem also for the treatment and diagnostics of human AE patients (Lachenmayer et al. 2019; Gottstein et al. 2019; Bresson-Hadni et al. 2021).

Swiss experts have adapted the protocols from human AE to treat animals with AE such as dogs and monkeys, but more studies and developments are needed for species-specific treatment protocols (Corsini et al. 2015; Wenker et al. 2019).

Treatment of CE is based on a stage-based therapeutic strategy and may include (a) percutaneous treatment, (b) surgery, (c) anti-infective drug treatment, or (d) watch and wait. (a) Percutaneous treatment, mostly by PAIR (puncture, aspiration, injection, reaspiration), allows removing the parasite material and confirming the diagnosis while being minimally invasive and easier to perform than full surgery (Brunetti et al. 2010). (b) Surgery is performed if possible, but protoscolex spillage must be avoided, as it can spread the parasite via re-differentiation of these protoscoleces to the metacestodes stage. (c) Inoperable CE patients, or patients with multiple cysts or cysts in multiple organs, are treated by BMZ, ABZ being the drug of choice and MBZ being used as an alternative. The same dosages are recommended for AE, as optimal dosages for CE have never been assessed (Brunetti et al. 2010). BMZ are contraindicated when there is a risk for cyst rupture. Treatment with BMZ is not effective in large cysts and should be avoided in patients with pre-conditions such as chronic hepatic diseases besides AE or CE or bone-marrow depression, as well as in pregnancy. In contrast to human AE, ABZ treatment in human CE cases has resulted in a curative outcome in around 30% of the cases (Brunetti et al. 2010).

It is not clear to date whether this reflects the fact that BMZ exhibit parasiticidal activity against *E. granulosus*. (d) In uncomplicated human CE cases, long-term follow-up of patients is performed, and treatment is put on hold in the watch-and-wait approach. However, this method needs further evaluation. In conclusion, even though a spontaneous cure is possible, and cysts might stop growing, there is no safe cure against CE, and alternative chemotherapeutic options against CE should be developed.

On the veterinary side, CE is responsible for extensive economic losses, especially in the sheep industry, and therefore control programs employing the safe and efficient vaccine Eg95 in combination with regular dog deworming were developed (Lightowlers et al. 1996). The first programmes were established in New Zealand and Tasmania, and they were very successful in eliminating CE (Larrieu et al. 2019). Continental control programmes in South America were moderately effective (Larrieu et al. 2019). Due to veterinary CE cases being hardly present in Switzerland, respective programs have not been implemented.

Though infections of final hosts are not the topic of this chapter, it should be noted here that regular deworming of dogs is crucial in the control of echinococcosis, as it not only reduces the adult worm burden in final hosts but also the resulting environmental egg contamination is reduced. For *E. multilocularis*, specific deworming programs for foxes were successful, albeit not sustainably, allowing worm levels to be kept low (Umhang et al. 2019). Also in Switzerland, respective studies were conducted in small areas around Zürich, and a monthly baiting period resulted in a decrease of *E. multilocularis* prevalence in foxes and rodents, and thus lowered the infection pressure (Hegglin and Deplazes 2008).

Taken together, there are possibilities to control the infection pressure for both parasites, but no curative drugs for AE or CE are available to date. BMZ treatments lead to improved survival rates and disease outcomes, but they are not always active, do not act parasiticidally, and sometimes, parasites resume growth. Furthermore, BMZ can induce considerable side effects that lead to treatment discontinuation, and they are not safe during pregnancy. As they can only halt parasite growth, BMZ have to be taken daily and for extended periods, which in turn increases the risk of side effects and poses a heavy psychological and economic burden to patients. Thus, current drug treatment options against AE and CE are suboptimal and novel drug-treatment options against echinococcosis are needed (Lundström-Stadelmann et al. 2019a; Wang et al. 2022).

### 13.4.4 The Mode of Action of BMZ

BMZ have diverse effects at the cellular and biochemical levels. In nematodes the primary mode of action is mediated through binding to the beta-tubulin subunit and thereby impairment of the microtubule network (Borgers et al. 1975; Dawson et al. 1984), which leads to reduced nutrient uptake and parasite growth (Lacey 1988). Moreover, reduced BMZ-tubulin-binding occurs in BMZ-resistant nematodes (Sangster et al. 1985). Various other cellular mechanisms are affected by BMZ, such as glucose uptake and storage, carbohydrate metabolism, acetylcholine esterase secretion, etc. (Lacey 1988). However, these are generally thought to be secondarily affected targets, leading to a biochemical imbalance in the worm and parasite death (Lacey 1988).

Inferring from the findings in nematodes, beta-tubulin was postulated as the main target in *E. multilocularis*. Beta-tubulin genes from *E. multilocularis* were cloned and three isoforms were identified (Tub1, 2 and 3) (Brehm et al. 2000). *E. multilocularis* stem cells mainly express the Tub2 isoform, which harbors an amino acid substitution at its C-terminus directly at the BMZ-binding site and does not bind to BMZ under the dosages used in standard treatment of AE. Therefore, stem cells are refractory to treatment (Brehm and Koziol 2014). This could partially explain the parasitostatic rather than parasiticidal activity of BMZ, together with the limited uptake and half-life of the drug (Brehm and Koziol 2014). However, the role of beta-tubulin as a primary BMZ drug target in *E. multilocularis* has not been experimentally demonstrated. Electron microscopy studies showed that BMZ led to a rapid degeneration of microtriches; the cytoskeleton comprised actin filaments but not microtubules (Küster et al. 2014b). Thus, BMZ might affect other primary targets than microtubules, but this needs to be further investigated. Largely excluded from studies on the mode of action of BMZ were potential host-mediated effects. In the *E. multilocularis* mouse model, it was shown that ABZ was only efficacious in wild-type mice with a functional immune system. In contrast, in athymic nude mice lacking a mature T-cell immune response, it was not active (Wang et al. 2020). In human AE patients, ABZ treatment increases the host immune response against the parasite, implying that the action of BMZ could also influence, or be dependent on, host immunity (Ricken et al. 2017).

For *E. granulosus* several additional studies investigated the mode of action of BMZ: The tubulins were cloned and polymerized in 2018, displaying virtually identical isoforms, but surprisingly no effects of BMZ on tubulin were reported in this study (Liu et al. 2018). In *E. granulosus* metacestodes, ABZ treatment reduced glucose uptake and glycogen levels, and changes in amino acid composition were described (Xiao et al. 1990, 1992; Yao et al. 1994). Further, treatment of mice with BMZ affected the activity of several enzymes of the *E. granulosus* cyst wall (acid phosphatase, ATPase, malate dehydrogenase, pyruvate kinase, phosphoenolpyruvate carboxykinase, superoxide dismutase, alanine aminotransferase, glucosephosphate isomerase), but it was not further assessed whether some of them were primary drug targets (Feng et al. 1992, 1995; Xiao et al. 1993, 1994, 1995; You et al. 1997).

## 13.5 Novel Treatment of Echinococcosis

### 13.5.1 Identification and Assessment of Novel Compounds Against Echinococcosis

To discover novel drugs with anti-echinococcal properties, as for any other antiparasitic drug, models of the respective disease are needed in vitro and in vivo. For in vitro drug screening and testing, two basic principles are generally applied: (1) phenotypic screening based on whole organisms or (2) mechanism-/target-based screening (Geary et al. 2015; Müller and Hemphill 2016). Thanks to markedly improved in vitro culture techniques for *E. multilocularis* metacestodes and primary (stem) cells (Spiliotis et al. 2004, 2008), novel medium-throughput drug-screening assays were developed. The proliferative capacity of *E. multilocularis* in vitro is higher than that of *E. granulosus*, and therefore, past screening efforts have focused more on *E. multilocularis*. As a result, an in vitro screening pipeline for *E. multilocularis* was established that allows screening hundreds of compounds (Lundström-Stadelmann et al. 2019a), such as the medicines for malaria venture (MMV) drug libraries (Stadelmann et al. 2016; Rufener et al. 2018b). This in vitro screening cascade has recently been adapted and applied for *E. granulosus* (Kaethner et al. 2023).

Three parasite stages are included: *Echinococcus* metacestodes, GL cells, and protoscoleces: (1) The metacestode is the actual disease-causing stage that needs to be assessed for drug-impact; (2) GL cell cultures include the undifferentiated stem cells of the parasite, that ensure a high regenerative potential; (3) protoscoleces are in particular important in the case of *E. granulosus*, as upon cyst rupture they are often responsible for spread of the parasite and there is the need for protoscolicidal substances. The overall screening cascade depicted in Fig. 13.2 shows all the assays currently included in in vitro drug screening for *Echinococcus*. The assays used in the screening cascade can monitor (1) the physical impact of drugs on metacestodes as measured by the PGI-assay (Stadelmann et al. 2010), (2) the effects on metacestode viability assessed by Alamar blue assay (Stadelmann et al. 2016), (3) effects on the viability of isolated parasite GL cells by ATP measurements (Stadelmann et al. 2016), (4) effects on in vitro vesicle formation (Hemer and Brehm 2012; unpublished, Lundström-Stadelmann lab, University of Bern), and (5) effects on drugs on protoscolex motility (Ritler et al. 2017). Other assessments include (6) morphological/ultrastructural effects that are investigated by electron microscopy (Stadelmann et al. 2016; Rufener et al. 2018b), (7) effects on mammalian cell viability by Alamar blue assay (Rufener et al. 2018b), and (8) the structure-activity relationship of derivatives from selected compounds is determined (Rufener et al. 2018b). Finally, (9) the mode of action of selected compounds is investigated in vitro (Rufener et al. 2018b). In case in vitro data are indicative of a therapeutic window, (10) the efficacy of selected compounds is assessed in experimentally infected mice.

Animal models are labor-intensive and expensive; they allow only a very low throughput of compounds to be screened, which involves ethical downsides; therefore, animal models should no longer be the first line of screening (Paveley and Bickle 2013). However, animal models are needed to validate compounds once they have proven activity and a therapeutic window has been demonstrated in vitro, particularly for immunological studies.

Experimentally, immunotherapy was shown to be a promising approach for treating AE. Human immunosuppressed patients and mice under immunosuppression have increased susceptibility to AE (Lachenmayer et al. 2019), and the parasite-induced Th2-immunity shift allows for uncontrolled parasite proliferation. The first

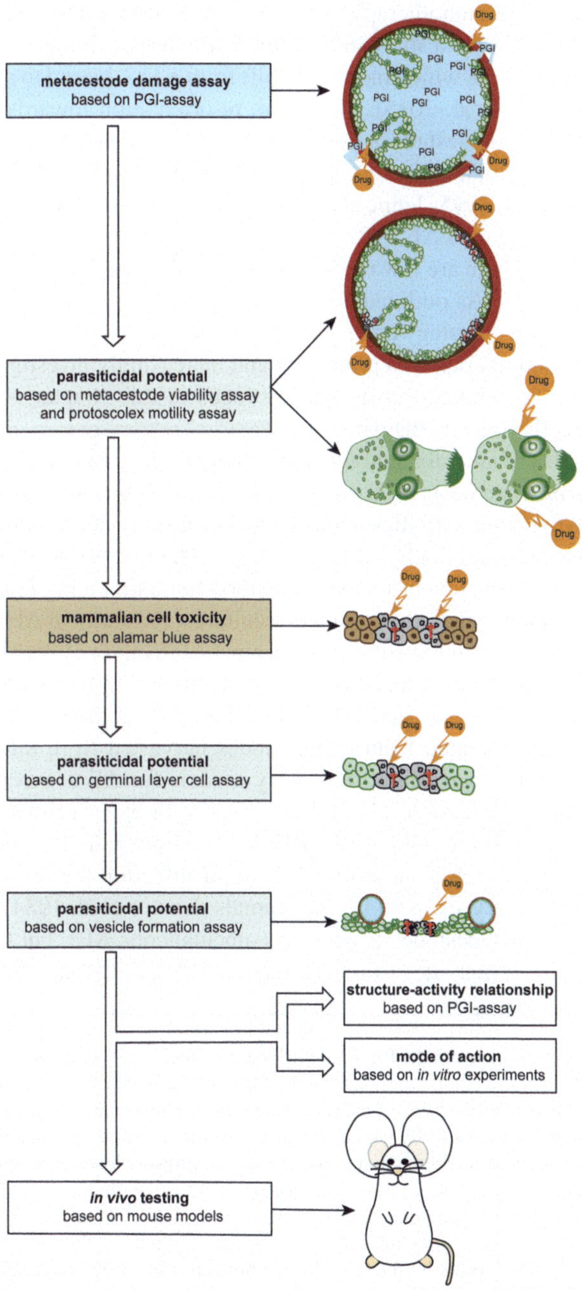
metacestode damage assay
based on PGI-assay
parasiticidal potential
based on metacestode viability assay
and protoscolex motility assay
mammalian cell toxicity
based on alamar blue assay
parasiticidal potential
based on germinal layer cell assay
parasiticidal potential
based on vesicle formation assay
structure-activity relationship
based on PGI-assay
mode of action
based on *in vitro* experiments
*in vivo* testing
based on mouse models
Drug
PGI

evidence that active immunotherapy may represent a sustainable route for the control of AE was provided in a study in experimentally infected mice treated with the recombinant antigen P29, which induced a shift from a Th2-biased to a Th1-oriented immunity (Boubaker et al. 2015). Therefore, re-orientation towards a more Th1-based immunity was suggested for application in AE patients (Wang et al. 2018b), and Th1-inducing cytokines were applied in human patients, albeit with limited efficacy (Schmid et al. 1995; Jenne et al. 1998). It was shown that the activity of the current drug in use against AE, ABZ, was dependent on an intact immune system (Wang et al. 2020). There are few reports of immunotherapeutic approaches against CE (Chen et al. 2020). As outlined above, the Eg95 vaccine has been successfully applied as a preventive strategy against CE in animals, but it is not used in humans. Even though immunotherapy is a promising and increasingly investigated approach for the treatment of echinococcosis, further focus in this chapter is laid on drug therapy that directly impacts the parasite.

Rodent models are ideal for studying the activity of drugs against AE, since mice are natural intermediate hosts of *E. multilocularis*, and therefore represent a perfect model for studying drug activities against AE. For CE, mice are not natural hosts, but mice can be experimentally infected as well. The use of laboratory mice and their intraperitoneal infection is a standard method for maintaining *E. multilocularis* and for the subsequent generation of in vitro cultures, the study of AE immunology, as well as the in vivo testing of substances against AE (Smyth 1990; Hemphill et al. 2014; Wang et al. 2018a, b; Lundström-Stadelmann et al. 2019a). For *E. granulosus*, parasite strains may be maintained in laboratory animals, but more often, in vitro cultures are initiated from metacestodes harvested from infected animals from abattoirs. Four experimental laboratory models for AE are available, which differ in the route of infection: (1) oral infection with eggs (primary AE), which represents the natural route (Ohbayashi 1960), (2) intrahepatic injection of a metacestode suspension, which is an artificial mode of infection that relies on surgical manipulation and is very stressful for the animals (Liance et al. 1984), (3) subcutaneous injection of metacestode suspension (subcutaneous AE), but the parasite is then growing at irrelevant sites (Yamashita et al. 1957; Ali-Khan and Siboo 1980;

**Fig. 13.2** Drug screening cascade for *E. multilocularis* and *E. granulosus*. Compounds are screened for potential anti-echinococcal activity by a metacestode damage assay in which vesicle fluid containing PGI is released by the metacestodes upon structural damage. In the next step, the parasiticidal potential is evaluated by testing the effect of compounds on the viability of metacestodes and whether they lead to a reduction of the motility of protoscoleces. A potential therapeutic window is evaluated by performing cell toxicity assays against mammalian cells. If further therapeutic potential is suggested, compounds are tested against isolated GL cells of these parasites via a cell viability assay. Active compounds are furthermore tested against GL cells at subtoxic concentrations for their ability to inhibit the formation of novel metacestode vesicles. Promising compounds can further be investigated for an improved molecular structure by performing structure-activity relationship studies against metacestodes and protoscoleces. The mode of action of compounds can also be studied via different target-based in vitro experiments. Finally, highly promising compounds are tested in vivo for primary and secondary AE in different mouse models. Figure adapted from Lundström-Stadelmann et al. (2019b, 2020), Kaethner et al. (2023)

Küster et al. 2013), and (4) intraperitoneal infection of a metacestode suspension (secondary AE), which mimics the chronic stage of AE (Hinz 1973). Most often, the intraperitoneal or the peroral infection models are applied. Corresponding mouse models for CE are (1) primary, (2) secondary, or (3) subcutaneous (Colli and Schantz 1974; Dempster et al. 1991; Zhang et al. 2001; Li et al. 2019, 2023). For infection with activated oncospheres, the route of infection determines the region in which cysts will develop, which is the lungs and oral cavity, peritoneal cavity or under the skin for intravenous, intraperitoneal, or subcutaneous infection, respectively (Dempster et al. 1991; Zhang et al. 2001). For infection with protoscoleces, mouse models exist that allow the growth of cysts within the peritoneal cavity (Heath 1970) or that mimic the natural route of infection with cysts in the liver by injecting protoscoleces into the portal vein (Li et al. 2019).

### 13.5.2 New Compound Classes with Activity Against Echinococcosis

AE and CE are neglected diseases with a low direct market-return, and therefore, pharmaceutical companies are not engaging in preclinical drug development for echinococcosis. For this reason, the above-mentioned drug tests are largely focused on repurposing existing drugs or compound series that are developing for other indications. Drug repurposing (also called drug reprofiling, drug repositioning or drug redirecting) is a general concept that results overall in lower costs, less risk of failure, and faster time to the market within the drug development process (Andrews et al. 2014; Panic et al. 2014; Hernandez et al. 2018). Repurposed drugs against echinococcosis fall into the three major categories of natural products (Sect. 13.5.2.1), anti-cancer compounds (Sect. 13.5.2.2) and anti-infective drugs (Sect. 13.5.2.3).

#### 13.5.2.1 Natural Products Against Echinococcosis

A continuously growing list of natural products and plant extracts has been tested for a potential application in treating *E. granulosus* infection (Siles-Lucas et al. 2018; Ali et al. 2020). In contrast, the number of extracts of natural products assessed for activity against *E. multilocularis* is still sparse, with thymol, osthole, origanum, anacardic acid from Brazilian cashew, thyme, and ampelopsin showing in vitro activity against the parasite (Albani et al. 2014, 2015; Yuan et al. 2016; Hizem et al. 2019; Xin et al. 2019). In an interesting reverse approach, Karpstein et al., 2021, showed that a single promising case of spontaneous cure from human AE upon self-medication with Maca was not confirmed by in vitro activity against the parasite. Therefore, it was speculated that plant extracts like Maca might have rather immunomodulatory effects beneficial in the treatment of echinococcosis. Repurposed drugs from plants and natural products from traditional medicine have a huge potential for rediscovery (Tu 2016). Still, future studies should include toxicological profiles and also work toward identifying fractions or compounds of major anti-echinococcal activity.

#### 13.5.2.2 Anti-cancer Compounds Against Echinococcosis

Due to very similar growth characteristics of AE and malignant cancers, the unlimited proliferative capacity of the parasite, metastasis formation, and the induction of blood vessels (Klinkert and Heussler 2006), a variety of studies investigated effects of anti-cancer compounds against *E. multilocularis*. As *E. granulosus* does not follow these growth characteristics, no studies have investigated the effects of anti-cancer compounds against this parasite.

Studies on *E. multilocularis* included investigations on isoflavonoids and genistein (Naguleswaran et al. 2006), 2-methoxyestradiol (Spicher et al. 2008a), methotrexate, navelbine and vincristine (Hübner et al. 2010; Huang et al. 2018), the p38 mitogen-activated kinase inhibitors ML3403 and SB202190 (Gelmedin et al. 2008), the Raf-inhibitor sorafenib (BAY 43–9006) (Gelmedin et al. 2010), the MEK1/2 inhibitor PD184352 (Gelmedin et al. 2010), the ABL-like kinase inhibitor imatinib (Hemer and Brehm 2012) and the polo-like kinase inhibitor BI 2536 (Schubert et al. 2014), the proteasome-inhibitor bortezomib (Stadelmann et al. 2014), kinase inhibitors everolimus and nilotinib (Joekel et al. 2018), docetaxel, paclitaxel and vorinostat (Huang et al. 2018), the tyrosine-kinase inhibitor BIBF 1120 (Förster et al. 2019), JNK ihibitors (Stoll et al. 2021), the EGF-R inhibitor BIBW2992 and the MEK inhibitor U0126 (Cheng et al. 2020) and the PIM-kinase inhibitor Z196138710 (Koike et al. 2022). However, of all these potentially active inhibitors, only 2-methoxyestradiol (Spicher et al. 2008a), bortezomib (Stadelmann et al. 2014), everolimus and nilotinib (Joekel et al. 2018), 2-Deoxy-D-Glucose with ABZ (Xin et al. 2022), BIBW2992 and U0126 (Cheng et al. 2020) were tested in the murine AE model, and only 2-Deoxy-D-Glucose with ABZ, BIBW2992 and U0126were (slightly) more active than ABZ alone, yet not shown to act parasiticidally nor to be specific enough for safe treatment of patients.

#### 13.5.2.3 Anti-infective Compounds Against Echinococcosis

A plethora of anti-infective compounds were tested against *E. multilocularis* and *E. granulosus* in rodent models and in vitro in the past (Siles-Lucas et al. 2018; Lundström-Stadelmann et al. 2019a), therefore in this paragraph only some more promising compounds are discussed: amphotericin B (Reuter et al. 2003a, 2006), nitazoxanide (Stettler et al. 2003; Stadelmann et al. 2010), BMZ (Küster et al. 2014b; Siles-Lucas et al. 2018), and the anti-malarial mefloquine (Lundström-Stadelmann et al. 2020). However, to date only the BMZ were shown to be active also against human AE.

Two drugs that have reached clinical application against AE over the last few years are the anti-fungal agent amphotericin B and the broad-spectrum anti-parasitic nitazoxanide. Amphotericin B showed promising activity in vitro (Reuter et al. 2003a, 2006), mouse studies were not conducted, and it induced nephrotoxicity and was not acting parasiticidally in patients, and therefore mostly discontinued for clinical use against AE (Kern et al. 2008; Tappe et al. 2009). However, there are newer case reports stating some activity of amphotericin B in patients (Reuter et al. 2003b; Burkert et al. 2022; Jelicic et al. 2023). Nitazoxanide showed promising results in vitro and in vivo in *E. multilolcularis*-infected mice, but it was not active against human AE (Kern et al. 2008; Tappe et al. 2009). Nitazoxanide was also active in vitro against *E. granulosus* metacestodes and protoscoleces (Walker et al. 2004), and clinical studies in human patients indicated that combinations of nitazoxanide with praziquantel and ABZ could potentially be useful in patients with disseminated CE (Pérez-Molina et al. 2011).

The mode of action of BMZ and the lack of parasiticidal activity was already discussed above (Sect. 13.4.4). Due to their slow-acting mechanisms, BMZ are often claimed to be inactive in vitro, but effects are visible after longer incubation times. Apart from ABZ and MBZ, other BMZ such as febendazole, oxfendazole, or flubendazole were successfully applied for their anti-echinococcal activity in mice (Siles-Lucas et al. 2018). However, they are not expected to exert different modes of action than ABZ or MBZ and thus do not provide real alternatives to these drugs. More recent studies suggest improving the absorption and oral bioavailability of BMZ by developing new formulations such as BMZ salt formulations (Cirilli et al. 2017), chitosan-based microcyst formulations (Torabi et al. 2018) or nanocrystals (Pensel et al. 2018; Bakhtiar et al. 2019) with activity against murine AE and/or CE. There are many more studies on the way to help improve the bioavailability of BMZ in the future.

Other attempts focus on combination therapies of BMZ, which per se have limited activity against echinococcosis, with other compounds to achieve synergistically enhanced therapeutic performance. The only drugs that led to an improved efficacy of low-dose treatment with ABZ in murine AE and CE are metformin (Loos et al. 2017a, 2022) and atovaquone for murine AE (Enkai et al. 2021), as outlined both in Sect. 13.5.3.

A larger group of anti-infective drugs that showed promising in vitro activities against *E. multilocularis* are the anti-malarials: Artemisinin-derivatives and artesunate (Spicher et al. 2008b), synthetic amino-ozonides (Küster et al. 2014a), mefloquine (Küster et al. 2011, 2015; Stadelmann et al. 2011; Gorgas et al. 2017; Rufener et al. 2018a; Lundström-Stadelmann et al. 2020), compounds from the medicines for malaria venture (MMV) malaria box like MMV665807 (Stadelmann et al. 2016). Of these compounds, only mefloquine also exhibited anti-echinocococcal activity in primary and secondary AE mouse models (Küster et al. 2011, 2015; Rufener et al. 2018a). Excitingly, mefloquine plasma levels reached in AE mice were in a range achievable by a standard malaria-prophylactic dose of mefloquine in humans (Rufener et al. 2018a). Therefore, the drug was mentioned for potential salvage treatment in AE patients, even though it may induce neuropsychiatric side effects in some of the patients. In the few human AE patients where mefloquine was applied as a salvage treatment, such adverse effects were not observed (Burkert et al. 2022), but also only in one case the treatment appeared to be curative, and that was in combination with amphotericin B (Jelicic et al. 2023). The current problem in mefloquine therapy is that the drug acts only parasitostatically and thus does not represent an improvement compared to BMZ therapy. Another highly interesting anti-malarial compound, MMV665807 was ten times more active than mefloquine in vitro and potentially parasiticidal, but it was not active in the AE mouse model (Stadelmann et al. 2016). MMV665807 is a salicylanilide-derivative, with high structural similarity to the already marketed drug niclosamide. Niclosamide is a known anthelmintic drug with activities against a variety of other diseases, including cancer. When tested in vitro, this drug had a magnitude of 100 times more activity than mefloquine. As the drug is poorly absorbed in vivo, ongoing studies focus on novel niclosamide formulations for a successful AE treatment (Lundström-Stadelmann lab, unpublished).

### 13.5.3 Targeting the Energy Metabolism for Future Tretments

Like any pathogen, parasites depend on energy and molecular building blocks from their hosts for proliferation and survival (Haanstra et al. 2017). Therefore, energy metabolism is an interesting target for drug treatment. Inhibition of the central carbon metabolism—including glycolysis, tricarboxylic (TCA) cycle, and electron transport chain (ETC)—results in reduced ATP levels and can induce cell death (Timson 2016). Also *Echinococus* spp. employ these central energy-generating pathways (McManus and Smyth 1978, 1982).

Glucose is an essential nutrient *Echinococcus* takes from its host as a prime energy source like animal species (Dalton et al. 2004; Tielens and van Hellemond 2006). Cestodes lost the ability to perform gluconeogenesis, which makes them completely dependent on the uptake of glucose from the host (Tielens and van Hellemond 2006). They stock large amounts of glucose in the form of glycogen as an internal energy storage. This might be a direct consequence of cestodes living in nutrient-rich environments rather than being a prerequisite for growth (Halton 1997; Dalton et al. 2004). In vitro cultured *Echinococcus* and metacestodes absorb high amounts of glucose (Jeffs and Arme 1988; Ritler et al. 2019), and in murine AE, high glycogen consumption was described (Novak et al. 1993). Further, *Echinococcus* spp. can interact with host insulin through the insulin-specific-receptor kinase EmIR1 expressed in glycogen storage cells. This leads to an increased glucose uptake by metacestodes (Hemer et al. 2014). A second receptor, EmIR2, is mainly expressed in *Echinococcus* stem cells and might be involved in cell proliferation and differentiation (Hemer et al. 2014). By drug repurposing of the antidiabetic drug metformin, it was shown that this compound inhibits the growth of *Echinococcus* by inducing carbohydrate starvation (Loos and Cumino 2015), and it improved the efficacy of low-dose ABZ treatment in mice infected with *E. multilocularis* or *E. granulosus* (Loos et al. 2017a, 2022). Metformin is used against type II diabetes, polycystic ovarian syndrome, and other diseases, and in general, it inhibits mitochondrial complex I, the AMPK signaling pathway and the insulin signaling pathway (Morales and Morris 2015). Against *E. multilocularis* metformin was shown to target the TOR pathway (Loos et al. 2020). Also the hypoglycemic drug glibenclamide was shown to be active against *E. granulosus* in vitro and in mice (Loos et al. 2017b).

Another energy-generating pathway, the ETC, has received attention over the last years, as on other parasites selectively active compounds were applied successfully to inhibit this pathway (Sen and Majumder 2008; Hayward and van Dooren

2019; Fisher et al. 2020). Buparvaquone, atovaquone and endochin-like quinolones (ELQs) were shown to exhibit in vitro activity against *E. multilocularis* and it was shown that they inhibit the complex III (*bc*1 complex) of this parasite (Rufener et al. 2018b; Enkai et al. 2020; Chaudhry et al. 2022). However, they failed to be active in the mouse model. One study investigating the activity of a small library of ELQs showed that these molecules were only able if a second energy-generating pathway in the mitochondria, the malate dismutation (MD), was inhibited simultaneously (Chaudhry et al. 2022).

The MD pathway is regarded as an adaptation to the anaerobic environments of *Echinococcus*. It is found in all helminths, marine invertebrates, as well as euglenids, and similar pathways are even found in *Trypanosoma* spp. and bacteria, but it is not used in mammalian cells (Tielens 1994; Komuniecki and Harris 1995; Besteiro et al. 2002; Kita et al. 2003). MD in helminths is an oxygen-independent mitochondrial pathway applied for energy generation, and it includes cytosolic as well as mitochondrial enzymes with a partial inversion of the citric acid cycle. The pathway consists of two arms, the (a) acetate and (b) succinate branch based on malate that gets imported to the mitochondria. In part (a), malate is oxidized to pyruvate, acetyl-CoA and acetate as the final electron acceptor. Therefore, acetate is a strongly released metabolite by *Echinococcus* metacestodes, in particular under anaerobic conditions (Agosin and Repetto 1963; McManus and Smyth 1978; Novak et al. 1992; Hosch et al. 2008; Ritler et al. 2019). In part (b), malate is reduced to fumarate and succinate as a final electron acceptor, and this product is strongly released as well (Ritler et al. 2019). Whereas several conserved enzymes are employed in the MD, some parts are unique, not found in mammals, and could be good drug targets. These are, for example, the fumarate reductase activity of complex II and the alternative electron carrier rhodoquinone that transfers electrons between complexes I and II and its synthesis. The experimental inhibition of the MD of *E. multilocularis* protoscoleces was shown by quinazoline and derivatives (Matsumoto et al. 2008), but these experimental compounds are no drugs for clinical application.

Overall, the energy balances of energy-generating pathways show that aerobic respiration, including glycolysis, citric acid cycle and ETC, have the highest energy production (total 38 mole of ATP per mole of glucose). However, oxygen-independent fermentation and MD also produce some energy (2 or 5 moles of ATPs per mole glucose) (Tielens and van Hellemond 2006). This allows *Echinococcus*, like all helminths, to generate energy independent of oxygen availability and with fewer reaction steps, but less efficient in ATP generated per glucose molecule.

### 13.5.4 Threonine Metabolism as a Potential *Echinococcus* Drug Target

Recent studies on the metabolic footprint of *E. multilocularis* have shown that the parasite consumes high amounts of threonine in vitro (Ritler et al. 2019). Our studies showed that indeed threonine is used for *Echinococcus* metacestode growth and the formation of new metacestodes from GL cells (Kaethner et al. 2022). Flux analyses showed that the parasite employs one pathway for the metabolization of threonine, the threonine dehydrogenase (TDH) pathway. Interestingly, in mouse embryonic stem cells, threonine is used via the same pathway, and it is essential for rapid cell division (Wang et al. 2009). In Trypanosomes, threonine is used by this pathway for later fatty acid synthesis (Cross et al. 1975; Gilbert et al. 1983; Millerioux et al. 2013, 2018). *E. multilocularis* actively expresses an EmTDH that is functionally active. Strikingly, the same enzyme is expressed as a non-functional pseudogene in humans (Edgar 2002), rendering EmTDH a highly interesting drug target. In vitro screening of TDH inhibitors is ongoing, and the future will show if inhibition of this pathway will also affect the parasite in an in vivo setting (Kaethner et al. 2022).

## 13.6 Conclusions and Outlook

In Switzerland, *E. multilocularis* and *E. granulosus* are medically relevant species. *E. multilocularis* is endemic in Switzerland, whereas cases of *E. granulosus* infection are mostly considered to be imported. Human echinococcosis cases are not notifiable, whereas veterinary cases must be reported. Switzerland's recently established multi-disciplinary echinococcosis network is a huge step in the right direction that will also allow the reporting of human and animal cases in the future in a one health approach.

Medical management of human or animal echinococcosis is easier in Switzerland than in countries with less accessible healthcare infrastructure, but drug-based treatment is still problematic. BMZ are the current drugs in use that, however, do not exhibit parasiticidal activity. In vitro and in vivo models are set in place for larger-scale screening and testing of potential novel compounds with both scolicidal and cysticidal activity. As AE and CE are neglected diseases and orphan drugs would hardly be developed, drug repurposing is the most frequently applied approach to identify novel chemotherapeutically interesting compounds. To improve treatment, a better understanding of the modes of action of BMZ is needed. Novel formulations of BMZ, or combined treatments with other drugs, might result in the development of improved therapies. None of the other repurposed drugs have been superior to BMZ, but ongoing studies on compounds such as niclosamide have revealed promising in vitro results.

Immunotherapies might offer a new avenue for the treatment of echinococcosis, but more studies on the immunomodulatory paths of *Echinococcus* are needed, as well as is a more basic understanding of the parasite and its metabolism needed to identify novel targetable pathways. The energy-generating pathways of *Echinococcus* might offer good future targets.

In the future it will be important that echinococcosis is not seen isolated as a human disease, but also the veterinary aspects should be included. In particular, dogs and captive primates are increasingly diagnosed with AE in Switzerland, and improved treatment regimens are needed for these animals and human patients.

Treatment should be more individualized and consider the patient-specific growth characteristics of the parasite.

## References


Agosin M, Repetto Y (1963) Studies on the metabolism of Echinococcus granulosus. Vii. Reactions of the tricarboxylic acid cycle in E. granulosus scolices. Comparative Biochem Physiol 9:245–261

Aji T, Dong J-H, Shao Y-M, Zhao J-M, Li T, Tuxun T, Shalayiadang P, Ran B, Jiang T-M, Zhang R-Q, He Y-B, Huang J-F, Wen H (2018) Ex vivo liver resection and autotransplantation as alternative to allotransplantation for end-stage hepatic alveolar echinococcosis. J Hepatol 69:1037–1046. https://doi.org/10.1016/j.jhep.2018.07.006

Albani CM, Denegri GM, Elissondo MC (2014) Effect of different terpene-containing essential oils on the proliferation of Echinococcus granulosus larval cells. Interdiscipl Perspect Infect Dis 2014:746931. https://doi.org/10.1155/2014/746931

Albani CM, Pensel PE, Elissondo N, Gambino G, Elissondo MC (2015) In vivo activity of albendazole in combination with thymol against Echinococcus multilocularis. Vet Parasitol 212:193–199. https://doi.org/10.1016/j.vetpar.2015.06.030

Ali R, Khan S, Khan M, Adnan M, Ali I, Khan TA, Haleem S, Rooman M, Norin S, Khan SN (2020) A systematic review of medicinal plants used against Echinococcus granulosus. PLoS One 15:e0240456. https://doi.org/10.1371/journal.pone.0240456

Ali-Khan Z, Siboo R (1980) Pathogenesis and host response in subcutaneous alveolar hydatidosis. I. Histogenesis of alveolar cyst and a qualitative analysis of the inflammatory infiltrates. Zeitschrift Für Parasitenkunde (Berlin, Germany) 62:241–254

Ammann RW, Eckert J (1996) Cestodes. Echinococcus. Gastroenterol Clin North Am 25:655–689

Andrews KT, Fisher G, Skinner-Adams TS (2014) Drug repurposing and human parasitic protozoan diseases. Int J Parasitol Drugs Drug Resist 4:95–111. https://doi.org/10.1016/j.ijpddr.2014.02.002

Bakhtiar NM, Akbarzadeh A, Casulli A, Mahami-Oskouei M, Ahmadpour E, Nami S, Rostami A, Spotin A (2019) Therapeutic efficacy of nanocompounds in the treatment of cystic and alveolar echinococcoses: challenges and future prospects. Parasitol Res 118:2455. https://doi.org/10.1007/s00436-019-06416-5

Beldi G, Vuitton D, Lachenmayer A, Heyd B, Dufour J-F, Richou C, Candinas D, Bresson-Hadni S (2019) Is ex vivo liver resection and autotransplantation a valid alternative treatment for end-stage hepatic alveolar echinococcosis in Europe? J Hepatol 70:1030–1031. https://doi.org/10.1016/j.jhep.2018.12.011

Besteiro S, Biran M, Biteau N, Coustou V, Baltz T, Canioni P, Bringaud F (2002) Succinate secreted by Trypanosoma brucei is produced by a novel and unique Glycosomal enzyme, NADH-dependent fumarate reductase. J Biol Chem 277:38001–38012. https://doi.org/10.1074/jbc.M201759200

Borgers M, De Nollin S, De Brabander M, Thienpont D (1975) Influence of the anthelmintic mebendazole on microtubules and intracellular organelle movement in nematode intestinal cells. Am J Vet Res 36:1153–1166

Boubaker G, Hemphill A, Huber CO, Spiliotis M, Babba H, Gottstein B (2015) Prevention and immunotherapy of secondary murine alveolar echinococcosis employing recombinant EmP29 antigen. PLoS Negl Trop Dis 9:e0003795. https://doi.org/10.1371/journal.pntd.0003795

Bouwknegt M, Devleesschauwer B, Graham H, Robertson LJ, van der Giessen JW, The Euro-Fbp workshop participants (2018) Prioritisation of food-borne parasites in Europe, 2016. Euro Surveillance: Bulletin Europeen Sur Les Maladies Transmissibles = European Communicable Disease Bulletin 23. https://doi.org/10.2807/1560-7917.ES.2018.23.9.17-00161

Brehm K, Koziol U (2014) On the importance of targeting parasite stem cells in anti-echinococcosis drug development. Parasite (Paris, France) 21:72. https://doi.org/10.1051/parasite/2014070

Brehm K, Kronthaler K, Jura H, Frosch M (2000) Cloning and characterization of beta-tubulin genes from Echinococcus multilocularis. Mol Biochem Parasitol 107:297–302

Bresson-Hadni S (2023) Alveolar echinococcosis in Switzerland. Universimed

Bresson-Hadni S, Koch S, Miguet J-P, Gillet M, Mantion G-A, Heyd B, Vuitton D-A, European group of clinicians (2003) Indications and results of liver transplantation for Echinococcus alveolar infection: an overview. Langenbeck's Arch Surg 388:231–238. https://doi.org/10.1007/s00423-003-0394-2

Bresson-Hadni S, Spahr L, Chappuis F (2021) Hepatic alveolar echinococcosis. Semin Liver Dis 41:393. https://doi.org/10.1055/s-0041-1730925

Brunetti E, Kern P, Vuitton DA (2010) Expert consensus for the diagnosis and treatment of cystic and alveolar echinococcosis in humans. Acta Trop 114:1–16. https://doi.org/10.1016/j.actatropica.2009.11.001

Budke CM, Deplazes P, Torgerson PR (2006) Global socioeconomic impact of cystic echinococcosis. Emerg Infect Dis 12:296–303. https://doi.org/10.3201/eid1202.050499

Burkert S, Peters L, Bloehdorn J, Grüner B (2022) Salvage therapy for alveolar echinococcosis-a case series. Pathogens (Basel, Switzerland) 11:333. https://doi.org/10.3390/pathogens11030333

Casulli A, Abela-Ridder B, Petrone D, Fabiani M, Bobić B, Carmena D, Šoba B, Zerem E, Gargaté MJ, Kuzmanovska G, Calomfirescu C, Rainova I, Sotiraki S, Lungu V, Dezsényi B, Herrador Z, Karamon J, Maksimov P, Oksanen A, Millon L, Sviben M, Shkjezi R, Gjoni V, Akshija I, Saarma U, Torgerson P, Šnábel V, Antolová D, Muhovic D, Besim H, Chereau F, Belhassen García M, Chappuis F, Gloor S, Stoeckle M, Müllhaupt B, Manno V, Santoro A, Santolamazza F (2022) Unveiling the incidences and trends of the neglected zoonosis cystic echinococcosis in Europe: a systematic review from the MEmE project. Lancet Infect Dis S1473-3099(22):00638–00637. https://doi.org/10.1016/S1473-3099(22)00638-7

Chaudhry S, Zurbriggen R, Preza M, Kämpfer T, Kaethner M, Memedovski R, Scorrano N, Hemphill A, Doggett JS, Lundström-Stadelmann B (2022) Dual inhibition of the Echinococcus multilocularis energy metabolism. Front Vet Sci 9:981664. https://doi.org/10.3389/fvets.2022.981664

Chen C, Gao Q, Luo Y, Zhang G, Xu X, Li Z, Wang J, He Q, Sheng L, Ma X (2020) The immunotherapy with hMASP-2 DNA nanolipoplexes against echinococcosis in experimentally protoscolex-infected mice. Acta Trop 210:105579. https://doi.org/10.1016/j.actatropica.2020.105579

Cheng Z, Xu Z, Tian H, Liu F, Li X, Luo D, Wang Y (2020) In vitro and in vivo efficacies of the EGFR/MEK/ERK signaling inhibitors in the treatment of alveolar echinococcosis. Antimicrob Agents Chemother 64:e00341–e00320. https://doi.org/10.1128/AAC.00341-20

Cirilli R, Guglielmi P, Formica FR, Casulli A, Carradori S (2017) The sodium salt of the enantiomers of ricobendazole: preparation, solubility and chiroptical properties. J Pharm Biomed Anal 139:1–7. https://doi.org/10.1016/j.jpba.2017.01.057

Colli CW, Schantz PM (1974) Growth and development of Echinococcus granulosus from Embryophores in an abnormal host (Mus musculus). J Parasitol 60:53–58. https://doi.org/10.2307/3278678

Corsini M, Geissbühler U, Howard J, Gottstein B, Spreng D, Frey CF (2015) Clinical presentation, diagnosis, therapy and outcome of alveolar echinococcosis in dogs. Vet Rec 177:569. https://doi.org/10.1136/vr.103470

Cross GA, Klein RA, Linstead DJ (1975) Utilization of amino acids by Trypanosoma brucei in culture: L-threonine as a precursor for acetate. Parasitology 71:311–326

Dalton JP, Skelly P, Halton DW (2004) Role of the tegument and gut in nutrient uptake by parasitic platyhelminths, vol 82, p 211

Dawson PJ, Gutteridge WE, Gull K (1984) A comparison of the interaction of anthelmintic benzimidazoles with tubulin isolated from mammalian tissue and the parasitic nematode Ascaridia galli. Biochem Pharmacol 33:1069–1074

Dempster RP, Berridge MV, Harrison GBL, Heath DD (1991) Echinococcus granulosus: development of an intermediate host mouse model for use in vaccination studies. Int J Parasitol 21:549–554. https://doi.org/10.1016/0020-7519(91)90059-G

Deplazes P, Rinaldi L, Alvarez Rojas CA, Torgerson PR, Harandi MF, Romig T, Antolova D, Schurer JM, Lahmar S, Cringoli G, Magambo J, Thompson RCA, Jenkins EJ (2017) Global distribution of alveolar and cystic echinococcosis. Adv Parasitol 95:315–493. https://doi.org/10.1016/bs.apar.2016.11.001

Díaz A, Casaravilla C, Irigoín F, Lin G, Previato JO, Ferreira F (2011) Understanding the laminated layer of larval Echinococcus I: structure. Trends Parasitol 27:204–213. https://doi.org/10.1016/j.pt.2010.12.012

Edgar AJ (2002) The human L-threonine 3-dehydrogenase gene is an expressed pseudogene. BMC Genet 3:18

Enkai S, Inaoka DK, Kouguchi H, Irie T, Yagi K, Kita K (2020) Mitochondrial complex III in larval stage of Echinococcus multilocularis as a potential chemotherapeutic target and in vivo efficacy of atovaquone against primary hydatid cysts. Parasitol Int 75:102004. https://doi.org/10.1016/j.parint.2019.102004

Enkai S, Kouguchi H, Inaoka DK, Irie T, Yagi K, Kita K (2021) In vivo efficacy of combination therapy with albendazole and atovaquone against primary hydatid cysts in mice. Eur J Clin Microbiol Infect Dis 40:1815–1820. https://doi.org/10.1007/s10096-021-04230-5

FAO/WHO (2014) Multicriteria-based ranking for risk management of food-borne parasites

Federer K, Armua-Fernandez MT, Hoby S, Wenker C, Deplazes P (2015) In vivo viability of Echinococcus multilocularis eggs in a rodent model after different thermo-treatments. Exp Parasitol 154:14–19. https://doi.org/10.1016/j.exppara.2015.03.016

Feng JJ, Xiao SH, Guo HF, Ren L, Jiao PY, Yao MY, Chai JJ (1992) Effects of mebendazole, albendazole, and praziquantel on alkaline phosphatase, acid phosphatase, and adenosine triphosphatase of Echinococcus granulosus cysts harbored in mice. Zhongguo Yao Li Xue Bao = Acta Pharmacol Sin 13:497–501

Feng JJ, Guo HF, Yao MY, Xiao SH (1995) Effects of mebendazole, albendazole, and praziquantel on glutathione S-transferase and superoxide dismutase of Echinococcus granulosus cyst wall harbored in mice. Zhongguo Yao Li Xue Bao = Acta Pharmacol Sin 16:297–300

Fisher N, Meunier B, Biagini GA (2020) The cytochrome bc1 complex as an antipathogenic target. FEBS Lett 594:2935–2952. https://doi.org/10.1002/1873-3468.13868

Förster S, Koziol U, Schäfer T, Duvoisin R, Cailliau K, Vanderstraete M, Dissous C, Brehm K (2019) The role of fibroblast growth factor signalling in Echinococcus multilocularis development and host-parasite interaction. PLoS Negl Trop Dis 13:e0006959. https://doi.org/10.1371/journal.pntd.0006959

Frey CF, Marreros N, Renneker S, Schmidt L, Sager H, Hentrich B, Milesi S, Gottstein B (2017) Dogs as victims of their own worms: Serodiagnosis of canine alveolar echinococcosis. Parasit Vectors 10:422. https://doi.org/10.1186/s13071-017-2369-0

Frey CF, Basso WU, Zürcher-Giovannini S, Marti I, Borel S, Guthruf S, Gliga D, Lundström-Stadelmann B, Origgi FC, Ryser-Degiorgis M-P (2022) The golden jackal (Canis aureus): a new host for Echinococcus multilocularis and Trichinella britovi in Switzerland. Schweizer Archiv Fur Tierheilkunde 164:71–78. https://doi.org/10.17236/sat00338

Galeh TM, Spotin A, Mahami-Oskouei M, Carmena D, Rahimi MT, Barac A, Ghoyounchi R, Berahmat R, Ahmadpour E (2018) The seroprevalence rate and population genetic structure of human cystic echinococcosis in the Middle East: a systematic review and meta-analysis. Int J Surg 51:39–48. https://doi.org/10.1016/j.ijsu.2018.01.025

Geary TG, Sakanari JA, Caffrey CR (2015) Anthelmintic drug discovery: into the future. J Parasitol 101(2):125–133. https://doi.org/10.1645/14-703.1

Gelmedin V, Caballero-Gamiz R, Brehm K (2008) Characterization and inhibition of a p38-like mitogen-activated protein kinase (MAPK) from Echinococcus multilocularis: antiparasitic activities of p38 MAPK inhibitors. Biochem Pharmacol 76(9):1068–1081. https://doi.org/10.1016/j.bcp.2008.08.020

Gelmedin V, Spiliotis M, Brehm K (2010) Molecular characterisation of MEK1/2- and MKK3/6-like mitogen-activated protein kinase kinases (MAPKK) from the fox tapeworm Echinococcus multilocularis. Int J Parasitol 40:555–567. https://doi.org/10.1016/j.ijpara.2009.10.009

Gilbert RJ, Klein RA, Miller PG (1983) The role of threonine in the metabolism of acetyl coenzyme a by Trypanosoma brucei brucei. Comparative Biochem Physiol B, Comparative Biochem 74:277–281

Gorgas D, Marreros N, Rufener R, Hemphill A, Lundström-Stadelmann B (2017) To see or not to see: non-invasive imaging for improved readout of drug treatment trials in the murine model of secondary alveolar echinococcosis. Parasitology 144:937–944. https://doi.org/10.1017/S0031182017000051

Gottstein B, Stojkovic M, Vuitton DA, Millon L, Marcinkute A, Deplazes P (2015) Threat of alveolar echinococcosis to public health – a challenge for Europe. Trends Parasitol 31:407–412. https://doi.org/10.1016/j.pt.2015.06.001

Gottstein B, Lachenmayer A, Beldi G, Wang J, Merkle B, Vu XL, Kurath U, Müller N (2019) Diagnostic and follow-up performance of serological tests for different forms/courses of alveolar echinococcosis. Food Waterb Parasit 16:e00055. https://doi.org/10.1016/j.fawpar.2019.e00055

Grüner B, Kern P, Mayer B, Gräter T, Hillenbrand A, Barth TFE, Muche R, Henne-Bruns D, Kratzer W, Kern P (2017) Comprehensive diagnosis and treatment of alveolar echinococcosis: a single-center, long-term observational study of 312 patients in Germany. GMS Infect Dis 1–12. https://doi.org/10.3205/id000027

Haanstra JR, Gerding A, Dolga AM, Sorgdrager FJH, Buist-Homan M, du Toit F, Faber KN, Holzhütter H-G, Szöör B, Matthews KR, Snoep JL, Westerhoff HV, Bakker BM (2017) Targeting pathogen metabolism without collateral damage to the host. Sci Rep 7:40406. https://doi.org/10.1038/srep40406

Halton D (1997) Nutritional adaptations to parasitism within the Platyhelminthes. Int J Parasitol 27:693–704. https://doi.org/10.1016/S0020-7519(97)00011-8

Hayward JA, van Dooren GG (2019) Same same, but different: uncovering unique features of the mitochondrial respiratory chain of apicomplexans. Mol Biochem Parasitol 232:111204. https://doi.org/10.1016/j.molbiopara.2019.111204

Heath DD (1970) The development of *Echinococcus granulosus* larvae in laboratory animals. Parasitology 60:449–456. https://doi.org/10.1017/S0031182000078252

Hegglin D, Deplazes P (2008) Control strategy for Echinococcus multilocularis. Emerg Infect Dis 14:1626–1628. https://doi.org/10.3201/eid1410.080522

Hemer S, Brehm K (2012) In vitro efficacy of the anticancer drug imatinib on Echinococcus multilocularis larvae. Int J Antimicrob Agents 40:458–462. https://doi.org/10.1016/j.ijantimicag.2012.07.007

Hemer S, Konrad C, Spiliotis M, Koziol U, Schaack D, Förster S, Gelmedin V, Stadelmann B, Dandekar T, Hemphill A, Brehm K (2014) Host insulin stimulates Echinococcus multilocularis insulin signalling pathways and larval development. BMC Biol 12:5. https://doi.org/10.1186/1741-7007-12-5

Hemphill A, Stadelmann B, Rufener R, Spiliotis M, Boubaker G, Müller J, Müller N, Gorgas D, Gottstein B (2014) Treatment of echinococcosis: albendazole and mebendazole - what else? Parasite (Paris, France) 21:70. https://doi.org/10.1051/parasite/2014073

Hernandez HW, Soeung M, Zorn KM, Ashoura N, Mottin M, Andrade CH, Caffrey CR, de Siqueira-Neto JL, Ekins S (2018) High throughput and computational repurposing for neglected diseases. Pharm Res 36:27. https://doi.org/10.1007/s11095-018-2558-3

Hinz E (1973) Die Aufbereitung des Infektionsmaterials für die intraperitoneale Infektion der Maus mit Echinococcus multilocularis. Z Tropenmed Parasitol 4:387–390

Hizem A, Lundström-Stadelmann B, M'rad S, Souiai S, Ben Jannet H, Flamini G, Ascrizzi R, Ghedira K, Babba H, Hemphill A (2019) Activity of Thymus capitatus essential oil components against in vitro cultured Echinococcus multilocularis metacestodes and germinal layer cells. Parasitology 1–12:956. https://doi.org/10.1017/S0031182019000295

Hofer S, Gloor S, Müller U, Mathis A, Hegglin D, Deplazes P (2000) High prevalence of Echinococcus multilocularis in urban red foxes (Vulpes vulpes) and voles (Arvicola terrestris) in the city of Zürich, Switzerland. Parasitology 120(Pt 2):135–142

Hosch W, Junghanss T, Stojkovic M, Brunetti E, Heye T, Kauffmann GW, Hull WE (2008) Metabolic viability assessment of cystic echinococcosis using high-field [1] H MRS of cyst contents. NMR Biomed 21:734–754. https://doi.org/10.1002/nbm.1252

Hotez PJ (2020) The rise or fall of neglected tropical diseases in East Asia Pacific. Acta Trop 202:105182. https://doi.org/10.1016/j.actatropica.2019.105182

Huang X, Wiehr S, Wild A-M, Voßberg P, Hoffmann W, Grüner B, Köhler C, Soboslay PT (2018) The effects of taxanes, vorinostat and doxorubicin on growth and proliferation of Echinococcus multilocularis metacestodes assessed with magnetic resonance imaging and simultaneous positron emission tomography. Oncotarget 9:9073–9087. https://doi.org/10.18632/oncotarget.24142

Hübner C, Wiehr S, Kocherscheidt L, Wehrl H, Pichler BJ, Schmid A, Kern P, Soboslay PT (2010) Effects of in vitro exposure of Echinococcus multilocularis metacestodes to cytostatic drugs on in vivo growth and proliferation of the parasite. Parasitol Res 107:459–463. https://doi.org/10.1007/s00436-010-1892-0

Jeffs SA, Arme C (1988) Glucose transport in protoscoleces of Echinococcus granulosus (cestoda). Comp Biochem Physiol A Physiol 91:203–207. https://doi.org/10.1016/0300-9629(88)91617-9

Jelicic K, Papic N, Viskovic K, Vince A (2023) Case report: amphotericin B and Mefloquine as a salvage treatment of alveolar echinococcosis. Am J Trop Med Hyg tpmd220465:581. https://doi.org/10.4269/ajtmh.22-0465

Jenne L, Kilwinski J, Radloff P, Flick W, Kern P (1998) Clinical efficacy of and immunologic alterations caused by interferon gamma therapy for alveolar echinococcosis. Clin Infect Dis 26:492–494

Joekel DE, Lundström-Stadelmann B, Müllhaupt B, Hemphill A, Deplazes P (2018) Evaluation of kinase-inhibitors nilotinib and everolimus against alveolar echinococcosis in vitro and in a mouse model. Exp Parasitol 188:65–72. https://doi.org/10.1016/j.exppara.2018.04.002

Kaethner M, Grossenbacher P, Lochner M, Schürch S, Regnault C, Ramírez Villalobos D, Lundström-Stadelmann B (2022) The Achilles' heel of the fox tapeworm? -investigation of the threonine metabolism of Echinococcus multilocularis. Copenhagen, Denmark

Kaethner M, Preza M, Kaempfer T, Zumstein P, Tamponi C, Varcasia A, Hemphill A, Brehm K, Lundström-Stadelmann B (2023) Establishment and application of unbiased in vitro drug screening assays for the identification of compounds against Echinococcus granulosus sensu stricto. PLoS Negl Trop Dis 17:e0011343. https://doi.org/10.1371/journal.pntd.0011343

Karpstein T, Chaudhry S, Bresson-Hadni S, Hayoz M, Boubaker G, Hemphill A, Rufener R, Kaethner M, Schindler I, Aebi Y, Cunha AS, Largiadèr CR, Lundström-Stadelmann B (2021) Maca against echinococcosis?-a reverse approach from patient to in vitro testing. Pathogens (Basel, Switzerland) 10:1335. https://doi.org/10.3390/pathogens10101335

Kern PA, Abboud P, Kern W, Stich TA, Bresson-Hadni S, Guerin B, Buttenschoen K, Gruener B, Reuter S, Hemphill A (2008) Critical appraisal of nitazoxanide for the treatment of alveolar echinococcosis. Am J Trop Med Hyg 79:119

Kern P, Menezes da Silva A, Akhan O, Müllhaupt B, Vizcaychipi KA, Budke C, Vuitton DA (2017) The Echinococcoses: diagnosis, clinical management and burden of disease. Adv Parasitol 96:259–369. https://doi.org/10.1016/bs.apar.2016.09.006

Kita K, Nihei C, Tomitsuka E (2003) Parasite mitochondria as drug target: diversity and dynamic changes during the life cycle. Curr Med Chem 10:2535–2548

Klinkert M-Q, Heussler V (2006) The use of anticancer drugs in antiparasitic chemotherapy. Mini Rev Med Chem 6:131–143

Koike A, Becker F, Sennhenn P, Kim J, Zhang J, Hannus S, Brehm K (2022) Targeting Echinococcus multilocularis PIM kinase for improving anti-parasitic chemotherapy. PLoS Negl Trop Dis 16:e0010483. https://doi.org/10.1371/journal.pntd.0010483

Komuniecki R, Harris BG (1995) Carbohydrate and energy metabolism in helminths. In: Biochemistry and molecular biology of parasites. Academic Press Inc, San Diego

Koziol U, Rauschendorfer T, Zanon Rodríguez L, Krohne G, Brehm K (2014) The unique stem cell system of the immortal larva of the human parasite Echinococcus multilocularis. EvoDevo 5:10. https://doi.org/10.1186/2041-9139-5-10

Küster T, Stadelmann B, Hermann C, Scholl S, Keiser J, Hemphill A (2011) In vitro and in vivo efficacies of mefloquine-based treatment against alveolar echinococcosis. Antimicrob Agents Chemother 55:713–721. https://doi.org/10.1128/AAC.01392-10

Küster T, Hermann C, Hemphill A, Gottstein B, Spiliotis M (2013) Subcutaneous infection model facilitates treatment assessment of secondary alveolar echinococcosis in mice. PLoS Negl Trop Dis 7:e2235. https://doi.org/10.1371/journal.pntd.0002235

Küster T, Kriegel N, Stadelmann B, Wang X, Dong Y, Vennerstrom JL, Keiser J, Hemphill A (2014a) Amino ozonides exhibit in vitro activity against Echinococcus multilocularis metacestodes. Int J Antimicrob Agents 43:40–46. https://doi.org/10.1016/j.ijantimicag.2013.09.012

Küster T, Stadelmann B, Aeschbacher D, Hemphill A (2014b) Activities of fenbendazole in comparison with albendazole against Echinococcus multilocularis metacestodes in vitro and in a murine infection model. Int J Antimicrob Agents 43:335–342. https://doi.org/10.1016/j.ijantimicag.2014.01.013

Küster T, Stadelmann B, Rufener R, Risch C, Müller J, Hemphill A (2015) Oral treatments of Echinococcus multilocularis-infected mice with the antimalarial drug mefloquine that potentially interacts with parasite ferritin and cystatin. Int J Antimicrob Agents 46:546–551. https://doi.org/10.1016/j.ijantimicag.2015.07.016

Lacey E (1988) The role of the cytoskeletal protein, tubulin, in the mode of action and mechanism of drug resistance to benzimidazoles. Int J Parasitol 18:885–936

Lachenmayer A, Gebbers D, Gottstein B, Candinas D, Beldi G (2019) Elevated incidence of alveolar echinococcosis in immunocompromised patients. Food Waterb Parasitol 16:e00060. https://doi.org/10.1016/j.fawpar.2019.e00060

Larrieu E, Gavidia CM, Lightowlers MW (2019) Control of cystic echinococcosis: background and prospects. Zoonoses Public Health 66:889–899. https://doi.org/10.1111/zph.12649

Li Z, Zhang C, Li L, Bi X, Li L, Yang S, Zhang N, Wang H, Yang N, Abulizi A, Aini A, Lin R, Vuitton DA, Wen H (2019) The local immune response during Echinococcus granulosus growth in a quantitative hepatic experimental model. Sci Rep 9:19612. https://doi.org/10.1038/s41598-019-56098-3

Li J, Ma G, Yan M, Liu H, Zhang K, Lin R, Lü G (2023) Subcutaneous infection mouse model could be applied into real time monitoring the efficacy of anti-cystic echinococcosis drug in vivo. Acta Trop 238:106802. https://doi.org/10.1016/j.actatropica.2022.106802

Liance M, Vuitton DA, Guerret-Stocker S, Carbillet JP, Grimaud JA, Houin R (1984) Experimental alveolar echinococcosis. Suitability of a murine model of intrahepatic infection by Echinococcus multilocularis for immunological studies. Experientia 40:1436–1439

Lightowlers MW, Lawrence SB, Gauci CG, Young J, Ralston MJ, Maas D, Heath DD (1996) Vaccination against hydatidosis using a defined recombinant antigen. Parasite Immunol 18:457–462. https://doi.org/10.1111/j.1365-3024.1996.tb01029.x

Liu C, Yao J, Yin J, Xue J, Zhang H (2018) Recombinant α- and β-tubulin from Echinococcus granulosus: expression, purification and polymerization. Parasite 25:62. https://doi.org/10.1051/parasite/2018063

Loos JA, Cumino AC (2015) In vitro anti-Echinococcal and metabolic effects of metformin involve activation of AMP-activated protein kinase in larval stages of Echinococcus granulosus. PLoS One 10:e0126009. https://doi.org/10.1371/journal.pone.0126009

Loos JA, Dávila VA, Rodrígues CR, Petrigh R, Zoppi JA, Crocenzi FA, Cumino AC (2017a) Metformin exhibits preventive and therapeutic efficacy against experimental cystic echinococcosis. PLoS Negl Trop Dis 11:e0005370. https://doi.org/10.1371/journal.pntd.0005370

Loos JA, Churio MS, Cumino AC (2017b) Anthelminthic activity of glibenclamide on secondary cystic echinococcosis in mice. PLoS Negl Trop Dis 11:e0006111. https://doi.org/10.1371/journal.pntd.0006111

Loos JA, Dávila VA, Brehm K, Cumino AC (2020) Metformin suppresses development of the Echinococcus multilocularis larval stage by targeting the TOR pathway. Antimicrob Agents Chemother 64:e01808–e01819. https://doi.org/10.1128/AAC.01808-19

Loos JA, Coccimiglio M, Nicolao MC, Rodrigues CR, Cumino AC (2022) Metformin improves the therapeutic efficacy of low-dose albendazole against experimental alveolar echinococcosis. Parasitology 149:138–144. https://doi.org/10.1017/S0031182021001633

Lundström-Stadelmann B, Rufener R, Ritler D, Zurbriggen R, Hemphill A (2019a) The importance of being parasiticidal… an update on drug development for the treatment of alveolar echinococcosis. Food Waterb Parasitol 15:e00040. https://doi.org/10.1016/j.fawpar.2019.e00040

Lundström-Stadelmann B, Rufener R, Ritler D, Zurbriggen R, Hemphill A (2019b) The importance of being parasiticidal… an update on drug development for the treatment of alveolar echinococcosis. Food Waterborne Parasitol 15:e00040. https://doi.org/10.1016/j.fawpar.2019.e00040

Lundström-Stadelmann B, Rufener R, Hemphill A (2020) Drug repurposing applied: activity of the anti-malarial mefloquine against Echinococcus multilocularis. Int J Parasitol Drugs Drug Resistance 13:121–129. https://doi.org/10.1016/j.ijpddr.2020.06.002

Matsumoto J, Sakamoto K, Shinjyo N, Kido Y, Yamamoto N, Yagi K, Miyoshi H, Nonaka N, Katakura K, Kita K, Oku Y (2008) Anaerobic NADH-fumarate reductase system is predominant in the respiratory chain of Echinococcus multilocularis, providing a novel target for the chemotherapy of alveolar echinococcosis. Antimicrob Agents Chemother 52:164–170. https://doi.org/10.1128/AAC.00378-07

McManus DP, Smyth JD (1978) Differences in the chemical composition and carbohydrate metabolism of Echinococcus granulosus (horse and sheep strains) and E. Multilocularis. Parasitology 77:103–109

McManus DP, Smyth JD (1982) Intermediary carbohydrate metabolism in protoscoleces of *Echinococcus granulosus* (horse and sheep strains) and *E. Multilocularis*. Parasitology 84:351–366. https://doi.org/10.1017/S0031182000044899

Meyer A, Olias P, Schüpbach G, Henzi M, Barmettler T, Hentrich B, Gottstein B, Frey CF (2020) Combined cross-sectional and case-control study on Echinococcus multilocularis infection in pigs in Switzerland. Vet Parasitol: X 4:100031. https://doi.org/10.1016/j.vpoa.2020.100031

Millerioux Y, Ebikeme C, Biran M, Morand P, Bouyssou G, Vincent IM, Mazet M, Riviere L, Franconi J-M, Burchmore RJS, Moreau P, Barrett MP, Bringaud F (2013) The threonine degradation pathway of the *Trypanosoma brucei* procyclic form: the main carbon source for lipid biosynthesis is under metabolic control: metabolic adaptation for acetate production. Molecular Microbiol 90:114. https://doi.org/10.1111/mmi.12351

Millerioux Y, Mazet M, Bouyssou G, Allmann S, Kiema T-R, Bertiaux E, Fouillen L, Thapa C, Biran M, Plazolles N, Dittrich-Domergue F, Crouzols A, Wierenga RK, Rotureau B, Moreau P, Bringaud F (2018) De novo biosynthesis of sterols and fatty acids in the Trypanosoma brucei procyclic form: carbon source preferences and metabolic flux redistributions. PLoS Pathog 14:e1007116. https://doi.org/10.1371/journal.ppat.1007116

Morales DR, Morris AD (2015) Metformin in cancer treatment and prevention. Annu Rev Med 66:17–29. https://doi.org/10.1146/annurev-med-062613-093128

Müller J, Hemphill A (2016) Drug target identification in protozoan parasites. Expert Opin Drug Discov. https://doi.org/10.1080/17460441.2016.1195945

Naguleswaran A, Spicher M, Vonlaufen N, Ortega-Mora LM, Torgerson P, Gottstein B, Hemphill A (2006) In vitro metacestodicidal activities of genistein and other isoflavones against Echinococcus multilocularis and Echinococcus granulosus. Antimicrob Agents Chemother 50:3770–3778. https://doi.org/10.1128/AAC.00578-06

Nikendei C, Greinacher A, Berkunova A, Junghanss T, Stojkovic M (2019) Psychological burden and resilience factors in patients with alveolar echinococcosis - a cross-sectional study. PLoS Negl Trop Dis 13:e0007082. https://doi.org/10.1371/journal.pntd.0007082

Novak M, Hameed N, Buist R, Blackburn BJ (1992) Metabolites of alveolar Echinococcus as determined by [31P]- and [1H]-nuclear magnetic resonance spectroscopy. Parasitol Res 78:665–670

Novak M, Modha A, Blackburn BJ (1993) Metabolic alterations in organs of Meriones unguiculatus infected with Echinococcus multilocularis. Comparative Biochem Physiol B, Comparative Biochem 105:517–521

Ohbayashi M (1960) Studies on echinococcosis X. : Histological observations on experimental cases of Multilocular echinococcosis. Jpn J Vet Res 8:134–160

Ohiolei JA, Yan H-B, Odeniran PO, Li L, Shumuye NA, Qurishi SA, Isaac C, Fu B-Q, Jia W-Z (2022) Echinococcus granulosus sensu lato in animal intermediate hosts: what is with the organ location? Vet Parasitol 304:109695. https://doi.org/10.1016/j.vetpar.2022.109695

Panic G, Duthaler U, Speich B, Keiser J (2014) Repurposing drugs for the treatment and control of helminth infections. Int J Parasitol Drugs Drug Resistance 4:185–200. https://doi.org/10.1016/j.ijpddr.2014.07.002

Paveley RA, Bickle QD (2013) Automated Imaging and other developments in whole-organism anthelmintic screening. Parasite Immunol 35(9–10):302–313. https://doi.org/10.1111/pim.12037

Pensel P, Paredes A, Albani CM, Allemandi D, Sanchez Bruni S, Palma SD, Elissondo MC (2018) Albendazole nanocrystals in experimental alveolar echinococcosis: enhanced chemoprophylactic and clinical efficacy in infected mice. Vet Parasitol 251:78–84. https://doi.org/10.1016/j.vetpar.2017.12.022

Pérez-Molina JA, Díaz-Menéndez M, Gallego JI, Norman F, Monge-Maillo B, Ayala AP, López-Vélez R (2011) Evaluation of Nitazoxanide for the treatment of disseminated cystic echinococcosis: report of five cases and literature review. Am J Trop Med Hyg 84:351–356. https://doi.org/10.4269/ajtmh.2011.10-0513

Reuter S, Merkle M, Brehm K, Kern P, Manfras B (2003a) Effect of amphotericin B on larval growth of Echinococcus multilocularis. Antimicrob Agents Chemother 47:620–625

Reuter S, Buck A, Grebe O, Nüssle-Kügele K, Kern P, Manfras BJ (2003b) Salvage treatment with amphotericin B in progressive human alveolar echinococcosis. Antimicrob Agents Chemother 47:3586–3591

Reuter S, Manfras B, Merkle M, Härter G, Kern P (2006) In vitro activities of itraconazole, methiazole, and nitazoxanide versus Echinococcus multilocularis larvae. Antimicrob Agents Chemother 50:2966–2970. https://doi.org/10.1128/AAC.00476-06

Ricken FJ, Nell J, Grüner B, Schmidberger J, Kaltenbach T, Kratzer W, Hillenbrand A, Henne-Bruns D, Deplazes P, Möller P, Kern P, Barth TFE (2017) Albendazole increases the inflammatory response and the amount of Em2-positive small particles of *Echinococcus multilocularis* (spems) in human hepatic alveolar echinococcosis lesions. PLoS Negl Trop Dis 11:e0005636. https://doi.org/10.1371/journal.pntd.0005636

Ritler D, Rufener R, Sager H, Bouvier J, Hemphill A, Lundström-Stadelmann B (2017) Development of a movement-based in vitro screening assay for the identification of new anti-cestodal compounds. PLoS Negl Trop Dis 11:e0005618. https://doi.org/10.1371/journal.pntd.0005618

Ritler D, Rufener R, Li JV, Kämpfer U, Müller J, Bühr C, Schürch S, Lundström-Stadelmann B (2019) In vitro metabolomic footprint of the Echinococcus multilocularis metacestode. Sci Rep 9:19438. https://doi.org/10.1038/s41598-019-56073-y

Romig T, Deplazes P, Jenkins D, Giraudoux P, Massolo A, Craig PS, Wassermann M, Takahashi K, de la Rue M (2017) Ecology and life cycle patterns of Echinococcus species. Adv Parasitol 95:213–314. https://doi.org/10.1016/bs.apar.2016.11.002

Rufener R, Ritler D, Zielinski J, Dick L, da Silva ET, da Silva Araujo A, Joekel DE, Czock D, Goepfert C, Moraes AM, de Souza MVN, Müller J, Mevissen M, Hemphill A, Lundström-Stadelmann B (2018a) Activity of mefloquine and mefloquine derivatives against Echinococcus multilocularis. Int J Parasitol Drugs Drug Resist 8:331–340. https://doi.org/10.1016/j.ijpddr.2018.06.004

Rufener R, Dick L, D'Ascoli L, Ritler D, Hizem A, Wells TNC, Hemphill A, Lundström-Stadelmann B (2018b) Repurposing of an old drug: in vitro and in vivo efficacies of buparvaquone against Echinococcus multilocularis. Int J Parasitol Drugs Drug Resistance 8:440–450. https://doi.org/10.1016/j.ijpddr.2018.10.011

Salm LA, Lachenmayer A, Perrodin SF, Candinas D, Beldi G (2019) Surgical treatment strategies for hepatic alveolar echinococcosis. Food Waterb Parasitol 15:e00050. https://doi.org/10.1016/j.fawpar.2019.e00050

Sangster NC, Prichard RK, Lacey E (1985) Tubulin and benzimidazole-resistance in Trichostrongylus colubriformis (Nematoda). J Parasitol 71:645–651

Schmid M, Samonigg H, Stöger H, Auer H, Sternthal MH, Wilders-Truschnig M, Reisinger EC (1995) Use of interferon gamma and mebendazole to stop the progression of alveolar hydatid disease: case report. Clin Infect Dis 20:1543–1546

Schmidberger J, Weimer H, Schlingeloff P, Kratzer W, Grüner B, Echinococcosis Working Group, Ulm (2018) Health-related quality of life in patients with alveolar echinococcosis: a cross-sectional study. Infection 47:67. https://doi.org/10.1007/s15010-018-1219-0

Schmidberger J, Baumann S, Kratzer W, Schlingeloff P, Lachmann R, Stark K, Peters L, Gruener B, Ramharter M, Hoepffner PJ, Strohäker J, Blum A, Müllhaupt B, Lagler H (2021) An update on the epidemiological situation of alveolar echinococcosis: recording and reporting structures in Austria, Switzerland and Germany. Gesundheitswesen (Bundesverband Der Arzte Des Offentlichen Gesundheitsdienstes (Germany)) 83:e51–e56. https://doi.org/10.1055/a-1138-0692

Schubert A, Koziol U, Cailliau K, Vanderstraete M, Dissous C, Brehm K (2014) Targeting Echinococcus multilocularis stem cells by inhibition of the polo-like kinase EmPlk1. PLoS Negl Trop Dis 8:e2870. https://doi.org/10.1371/journal.pntd.0002870

Sen N, Majumder HK (2008) Mitochondrion of protozoan parasite emerges as potent therapeutic target: exciting drugs are on the horizon. Curr Pharm Des 14:839–846. https://doi.org/10.2174/138161208784041024

Siles-Lucas M, Casulli A, Conraths FJ, Müller N (2017) Laboratory diagnosis of Echinococcus spp. in human patients and infected animals. Adv Parasitol 96:159–257. https://doi.org/10.1016/bs.apar.2016.09.003

Siles-Lucas M, Casulli A, Cirilli R, Carmena D (2018) Progress in the pharmacological treatment of human cystic and alveolar echinococcosis: compounds and therapeutic targets. PLoS Negl Trop Dis 12:e0006422. https://doi.org/10.1371/journal.pntd.0006422

Smyth JD (1990) In vitro cultivation of parasitic helminths. CRC Press

Spicher M, Naguleswaran A, Ortega-Mora LM, Müller J, Gottstein B, Hemphill A (2008a) In vitro and in vivo effects of 2-methoxyestradiol, either alone or combined with albendazole, against Echinococcus metacestodes. Exp Parasitol 119:475–482. https://doi.org/10.1016/j.exppara.2008.02.012

Spicher M, Roethlisberger C, Lany C, Stadelmann B, Keiser J, Ortega-Mora LM, Gottstein B, Hemphill A (2008b) In vitro and in vivo treatments of echinococcus protoscoleces and metacestodes with artemisinin and artemisinin derivatives. Antimicrob Agents Chemother 52:3447–3450. https://doi.org/10.1128/AAC.00553-08

Spiliotis M, Tappe D, Sesterhenn L, Brehm K (2004) Long-term in vitro cultivation of Echinococcus multilocularis metacestodes under axenic conditions. Parasitol Res 92:430–432. https://doi.org/10.1007/s00436-003-1046-8

Spiliotis M, Lechner S, Tappe D, Scheller C, Krohne G, Brehm K (2008) Transient transfection of Echinococcus multilocularis primary cells and complete in vitro regeneration of metacestode vesicles. Int J Parasitol 38:1025–1039. https://doi.org/10.1016/j.ijpara.2007.11.002

Stadelmann B, Scholl S, Müller J, Hemphill A (2010) Application of an in vitro drug screening assay based on the release of phosphoglucose isomerase to determine the structure-activity relationship of thiazolides against Echinococcus multilocularis metacestodes. J Antimicrob Chemother 65:512–519. https://doi.org/10.1093/jac/dkp490

Stadelmann B, Küster T, Scholl S, Barna F, Kropf C, Keiser J, Boykin DW, Stephens CE, Hemphill A (2011) In vitro efficacy of dicationic compounds and mefloquine enantiomers against Echinococcus multilocularis metacestodes. Antimicrob Agents Chemother 55:4866–4872. https://doi.org/10.1128/AAC.00478-11

Stadelmann B, Aeschbacher D, Huber C, Spiliotis M, Müller J, Hemphill A (2014) Profound activity of the anti-cancer drug Bortezomib against Echinococcus multilocularis Metacestodes iden-

tifies the proteasome as a novel drug target for Cestodes. PLoS Negl Trop Dis 8:e3352. https://doi.org/10.1371/journal.pntd.0003352
Stadelmann B, Rufener R, Aeschbacher D, Spiliotis M, Gottstein B, Hemphill A (2016) Screening of the open source malaria box reveals an early Lead compound for the treatment of alveolar echinococcosis. PLoS Negl Trop Dis 10:e0004535. https://doi.org/10.1371/journal.pntd.0004535
Stettler M, Fink R, Walker M, Gottstein B, Geary TG, Rossignol JF, Hemphill A (2003) In vitro parasiticidal effect of Nitazoxanide against Echinococcus multilocularis metacestodes. Antimicrob Agents Chemother 47:467–474
Stoll K, Bergmann M, Spiliotis M, Brehm K (2021) A MEKK1 - JNK mitogen activated kinase (MAPK) cascade module is active in Echinococcus multilocularis stem cells. PLoS Negl Trop Dis 15:e0010027. https://doi.org/10.1371/journal.pntd.0010027
Tappe D, Müller A, Frosch M, Stich A (2009) Limitations of amphotericin B and nitazoxanide in the treatment of alveolar echinococcosis. Ann Trop Med Parasitol 103:177–181. https://doi.org/10.1179/136485909X385036
Thompson RCA (2017) Biology and systematics of Echinococcus. Adv Parasitol 95:65–109. https://doi.org/10.1016/bs.apar.2016.07.001
Tielens AG (1994) Energy generation in parasitic helminths. Parasitol Today (Personal Ed) 10:346–352
Tielens AGM, van Hellemond JJ (2006) Unusual aspects of metabolism in flatworm parasites. In: Parasitic flatworms: molecular biology, biochemistry, immunology and physiology, pp 387–407
Timson DJ (2016) Metabolic enzymes of helminth parasites: potential as drug targets. Curr Protein Pept Sci 17:280–295
Torabi N, Dobakhti F, Faghihzadeh S, Haniloo A (2018) In vitro and in vivo effects of chitosan-praziquantel and chitosan-albendazole nanoparticles on Echinococcus granulosus Metacestodes. Parasitol Res 117:2015–2023. https://doi.org/10.1007/s00436-018-5849-z
Torgerson PR (2013) The emergence of echinococcosis in Central Asia. Parasitology 140:1667–1673. https://doi.org/10.1017/S0031182013000516
Torgerson PR, Keller K, Magnotta M, Ragland N (2010) The global burden of alveolar echinococcosis. PLoS Negl Trop Dis 4:e722. https://doi.org/10.1371/journal.pntd.0000722
Torgerson PR, Devleesschauwer B, Praet N, Speybroeck N, Willingham AL, Kasuga F, Rokni MB, Zhou X-N, Fèvre EM, Sripa B, Gargouri N, Fürst T, Budke CM, Carabin H, Kirk MD, Angulo FJ, Havelaar A, de Silva N (2015) World Health Organization estimates of the global and regional disease burden of 11 foodborne parasitic diseases, 2010: a data synthesis. PLoS Med 12:e1001920. https://doi.org/10.1371/journal.pmed.1001920
Trotz-Williams LA, Mercer NJ, Walters JM, Wallace D, Gottstein B, Osterman-Lind E, Boggild AK, Peregrine AS (2017) Public health follow-up of suspected exposure to Echinococcus multilocularis in southwestern Ontario. Zoonoses Public Health 64:460–467. https://doi.org/10.1111/zph.12326
Tu Y (2016) Artemisinin-a gift from traditional Chinese medicine to the world (Nobel lecture). Angew Chem Int Ed Engl 55:10210–10226. https://doi.org/10.1002/anie.201601967
Umhang G, Possenti A, Colamesta V, d'Aguanno S, La Torre G, Boué F, Casulli A (2019) A systematic review and meta-analysis on anthelmintic control programs for Echinococcus multilocularis in wild and domestic carnivores. Food Waterb Parasitol 15:e00042. https://doi.org/10.1016/j.fawpar.2019.e00042
van der Giessen J, Deksne G, Gómez-Morales MA, Troell K, Gomes J, Sotiraki S, Rozycki M, Kucsera I, Djurković-Djaković O, Robertson LJ (2021) Surveillance of foodborne parasitic diseases in Europe in a one health approach. Parasite Epidemiol Control 13:e00205. https://doi.org/10.1016/j.parepi.2021.e00205
Veit P, Bilger B, Schad V, Schäfer J, Frank W, Lucius R (1995) Influence of environmental factors on the infectivity of Echinococcus multilocularis eggs. Parasitology 110(Pt 1):79–86

Walker M, Rossignol JF, Torgerson P, Hemphill A (2004) In vitro effects of nitazoxanide on Echinococcus granulosus protoscoleces and metacestodes. J Antimicrob Chemother 54:609–616. https://doi.org/10.1093/jac/dkh386

Wang J, Alexander P, Wu L, Hammer R, Cleaver O, McKnight SL (2009) Dependence of mouse embryonic stem cells on threonine catabolism. Science (New York, N.Y.) 325:435–439. https://doi.org/10.1126/science.1173288

Wang J, Cardoso R, Marreros N, Müller N, Lundström-Stadelmann B, Siffert M, Vuitton DA, Boué F, Lin R, Wen H, Gottstein B (2018a) Foxp3+ T regulatory cells as a potential target for immunotherapy against primary infection with Echinococcus multilocularis eggs. Infect Immun 86. https://doi.org/10.1128/IAI.00542-18

Wang J, Jebbawi F, Bellanger A-P, Beldi G, Millon L, Gottstein B (2018b) Immunotherapy of alveolar echinococcosis via PD-1/PD-L1 immune checkpoint blockade in mice. Parasite Immunol 40:e12596. https://doi.org/10.1111/pim.12596

Wang J, Marreros N, Rufener R, Hemphill A, Gottstein B, Lundström-Stadelmann B (2020) Short communication: efficacy of albendazole in Echinococcus multilocularis-infected mice depends on the functional immunity of the host. Exp Parasitol 219:108013. https://doi.org/10.1016/j.exppara.2020.108013

Wang S, Ma Y, Wang W, Dai Y, Sun H, Li J, Wang S, Li F (2022) Status and prospect of novel treatment options toward alveolar and cystic echinococcosis. Acta Trop 226:106252. https://doi.org/10.1016/j.actatropica.2021.106252

Wenker C, Hoby S, Wyss F, Mengiardi B, Vögtli R, Posthaus H, Deplazes P, Gottstein B (2019) Alveolar echinococcosis in western lowland gorillas (gorilla gorilla gorilla): ALBENDAZOLE was not able to stop progression of the disease. J Zoo Wildlife Med 50:243–253. https://doi.org/10.1638/2018-0064

Xiao SH, Yang YQ, Guo HF, Zhang CW, Jiao PY, You JQ, Jiao W (1990) [effect of mebendazole, albendazole and albendazole sulfoxide on glycogen contents of Echinococcus granulosus cysts in infected mice]. Zhongguo Yao Li Xue Bao = Acta Pharmacol Sin 11:546–549

Xiao SH, You JQ, Guo HF, Feng JJ, Sun HL, Jiao PY, Yao MY, Chai JJ (1992) [effect of mebendazole on glucose uptake of Echinococcus granulosus cysts]. Zhongguo Yao Li Xue Bao = Acta Pharmacol Sin 13:473–477

Xiao SH, Feng JJ, Guo HF, Jiao PY, Yao MY, Jiao W (1993) Effects of mebendazole, albendazole, and praziquantel on succinate dehydrogenase, fumarate reductase, and malate dehydrogenase in Echinococcus granulosus cysts harbored in mice. Zhongguo Yao Li Xue Bao = Acta Pharmacol Sin 14:151–154

Xiao SH, Feng JJ, Guo HF, Jiao PY, Yao MY, Jiao W (1994) Effects of mebendazole, albendazole, and praziquantel on fumarate hydratase, pyruvate kinase, and phosphoenolpyruvate carboxykinase of Echinococcus granulosus cyst wall harbored in mice. Zhongguo Yao Li Xue Bao = Acta Pharmacol Sin 15:69–72

Xiao S, Feng J, Guo H, Yao M (1995) Effects of mebendazole, albendazole and praziquantel on alanine aminotransferase and aspartate aminotransferase of Echinococcus granulosus cyst wall harbored in mice. Chin J Parasitol Parasitic Dis 13:107–110

Xin Q, Yuan M, Li H, Lu J, Song X, Jing T (2019) In vitro efficacy of ampelopsin against Echinococcus granulosus and Echinococcus multilocularis. J Vet Med Sci 81:1853–1858. https://doi.org/10.1292/jvms.19-0347

Xin Q, Lv W, Xu Y, Luo Y, Zhao C, Wang B, Yuan M, Li H, Song X, Jing T (2022) 2-deoxy-D-glucose and combined 2-deoxy-D-glucose/albendazole exhibit therapeutic efficacy against Echinococcus granulosus protoscoleces and experimental alveolar echinococcosis. PLoS Negl Trop Dis 16:e0010618. https://doi.org/10.1371/journal.pntd.0010618

Yamashita J, Ohbayashi M, Konno S (1957) Studies on echinococcosis VI. Secondary Echinococcus multilocularis in mice. Jpn J Vet Res 197–202

Yao MY, Xiao SH, Feng JJ, Xue CL, Shimada M (1994) Effect of mebendazole on free amino acid composition of cyst wall and cyst fluid of Echinococcus granulosus harbored in mice. Zhongguo Yao Li Xue Bao = Acta Pharmacol Sin 15:521–524

You JQ, Xiao SH, Guo HF, Jiao PY, Mei JY, Yao MY (1997) Effect of mebendazole and praziquantel on glucosephosphate isomerase and glyceraldehydephosphate dehydrogenase in Echinococcus granulosus cyst wall harbored in mice. Zhongguo Yao Li Xue Bao = Acta Pharmacol Sin 18:75–77

Yuan M, Luo Y, Xin Q, Gao H, Zhang G, Jing T (2016) Efficacy of osthole for Echinococcus granulosus in vitro and Echinococcus multilocularis in vivo. Vet Parasitol 226:38–43. https://doi.org/10.1016/j.vetpar.2016.05.016

Zhang W, You H, Zhang Z, Turson G, Hasyet A, McManus DP (2001) Further studies on an intermediate host murine model showing that a primary Echinococcus granulosus infection is protective against subsequent oncospheral challenge. Parasitol Int 50:279–283. https://doi.org/10.1016/S1383-5769(01)00086-1

# Cross- and Multidisciplinary Interventions Used for Echinococcosis Control

14

Haining Fan

**Abstract**

Echinococcosis is a complex disease, and the diagnosis and treatment involve imaging, surgery, pathology and drugs. Multiple diagnostic and therapeutic modalities have emerged in the progress of research on the treatment of echinococcosis. Multidisciplinary treatment (MDT) plays an important role in the diagnosis and treatment. MDT can avoid the limitations of single-disciplinary treatment, provide one-stop full medical services for patients, promote interdisciplinary communication, and actively establish and improve treatment principles and clinical practice guidelines based on multidisciplinary expert consensus. Diagnosis and treatment cannot be limited to a simple superimposition or sequential combination of various techniques.

**Keywords**

Cystic echinococcosis · Alvealar echinococcosis · Multidisciplinary treatment

## 14.1 Introduction

The onset, development, evolution and regression of echinococcosis involve intricate mechanisms. Therefore, multiple diagnostic and therapeutic modalities have emerged in the progress of research on the treatment of echinococcosis, which has been described in detail in the previous sections. Implementing specific protocols

H. Fan (✉)
Affiliated hospital of Qinghai University, Xining, China
e-mail: fanhaining@medmail.com.cn





J. Li et al. (eds.), *Echinococcus: Control and Elimination of Echinococcosis with a Focus on China and Europe*, Parasitology Research Monographs 19,
https://doi.org/10.1007/978-3-031-54479-8_14

requires cross- and multidisciplinary interventions to be discussed and decided, simply called Multidisciplinary Treatment (MDT).

MDT can avoid the limitations of single-disciplinary treatment, provide one-stop full medical services for patients, promote interdisciplinary communication, and actively establish and improve treatment principles and clinical practice guidelines based on multidisciplinary expert consensus. The selection of rational treatments and drugs requires high-level evidence-based medical evidence, as well as consideration of individual patient differences and regional health economics. Through an effective MDT model, patients can better benefit from individualized, clinical evidence-based clinical treatment decisions based on standardization.

Combining various effective treatment modalities for echinococcosis into one and fully exploiting the advantages of the MDT integrated treatment model to maximize the diagnosis and prognosis of patients with echinococcosis.

## 14.2 Cross- and Multidisciplinary Interventions for Diagnosis

Imaging techniques are key to the diagnosis of echinococcosis, and serological tests for parasite-circulating cell-free DNA and echinococcosis-specific antibodies can both prove the diagnosis of echinococcosis to some extent. Different application scenarios require the cooperation of different clinical departments.

The applications of ultrasound, CT, and MRI imaging techniques have been described in detail in the previous sections, and different imaging techniques can be selected for different needs. In primary care hospitals, ultrasound can help in rapid mass screening for echinococcosis and is important in the initial screening of echinococcosis. The dynamic observation of ultrasound can see the continuity of the evolution of the lesion, which has some delineation and significance for each stage of development. CT scan is an important tool to confirm the diagnosis of hepatic echinococcosis, which can accurately diagnose qualitatively, show the relationship between the lesion and the intrahepatic vascular structures, and help surgeons perform territorial surgery. However, the lack of imaging physicians in primary hospitals limits its application to a certain extent. By building a tele-imaging consultation platform, we can help primary hospitals to make correct typing and staging and combine MDT to develop the best treatment plan.

On the other hand echinococcosis-specific antibody serology plays an important role. The AE serology is more reliable than CE (The sensitivity and specificity of echinococcosis-specific antibodies in CE patients are lower). A commercialized Em2plus enzyme-linked immunosorbent assay (ELISA) has been used extensively for clinical diagnosis of AE (Knapp et al. 2014), with sensitivity and specificity exceeding 90% (Cisak et al. 2015). Nevertheless, serology cannot be used as a first-intent diagnostic tool for mass screening in areas of endemicity, where a proportion of the human population exhibits positive serology without AE lesions (Bartholomot et al. 2002). Serology has also been shown to be frequently negative in immunosuppressed individuals with AE. Thus, it should not be used on its own as an argument against a diagnosis in such patients (Chauchet et al. 2014). For both CE and AE, serology is thus now used only to confirm imaging results.

The sensitivity and specificity of Parasite Circulating Cell-free DNA is much higher than that of echinococcosis-specific antibodies. Still, this technique is not recommended as the method of choice due to cost constraints (Fan et al. 2021). In summary, no single diagnostic modality can be applied perfectly in all settings. These diagnostic modalities are complementary, and the diagnosis of echinococcosis requires the multidisciplinary team to combine multiple findings (FIGX).

## 14.3 Cross- and Multidisciplinary Interventions for Treatment

Depending on the imaging presentation AE and CE have different therapeutic measures that can be used alone or in combination with various options to treat CE and AE, including (1) surgery, (2) non-surgical interventions, (3) anti-infective benzimidazole medication, and (4) a "watch and wait" approach (FIGX). Current recommendations for the management of encapsulated disease take into account the cancer-like nature of the disease, which is prone to recurrence. Therefore, cancer care management models that promote a multidisciplinary approach are used, with treatment decisions made through interdisciplinary team consultation.

The core of integrated multidisciplinary management of echinococcosis is patient-centered. Diagnosis and treatment cannot be limited to a simple superimposition or sequential combination of various techniques. Consideration should be given to combining different mechanisms of various treatment techniques and integrating the advantageous specialties of each discipline in treating liver cancer to achieve a mutually enhancing and complementary treatment model.

## References

Knapp J, Sako Y, Grenouillet F et al (2014) Comparison of the serological tests ICT and ELISA for the diagnosis of alveolar echinococcosis in France. Parasite 21:34

Cisak E, Sroka J, Wójcik-Fatla A et al (2015) Evaluation of reactivity to Echinococcus spp. among rural inhabitants in Poland. Acta Parasitol 60(3):525–529

Bartholomot G, Vuitton DA, Harraga S et al (2002) Combined ultrasound and serologic screening for hepatic alveolar echinococcosis in Central China. Am J Trop Med Hyg 66(1):23–29

Chauchet A, Grenouillet F, Knapp J et al (2014) Increased incidence and characteristics of alveolar echinococcosis in patients with immunosuppression-associated conditions. Clin Infect Dis 59(8):1095–1104

Fan H, Gai W, Zhang L et al (2021) Parasite circulating cell-free DNA in the blood of alveolar echinococcosis patients as a diagnostic and treatment-status indicator. Clin Infect Dis 73(1):e246–e251

# Vaccines for Echinococcosis: Where Are We Now?

Heinz Mehlhorn

Among a broad spectrum of other species (see Table 9.1, Chap. 9) the species *Echinococcus granulosus* and *E. multilocularis* are the most common and most important ones with respect to human infections (see Chap. 9). Chemotherapy by albendazole and mebendazole helps when used at an early stage of infection.

On the other side, vaccination would be most useful as protection. The present short look presents the papers, which have given deep insights into the present stage of some vaccine trials.



J. Li et al. (eds.), *Echinococcus: Control and Elimination of Echinococcosis with a Focus on China and Europe*, Parasitology Research Monographs 19,
https://doi.org/10.1007/978-3-031-54479-8

# References

Abdybekova AM, Zhang Z, Sultanov AA, Abdibayeva AA, Zhaksylykova AA, Junisbayeva SM, Aubakirov MZ, Akhmetova GD, Torgerson PR (2019) Genotypes of *Echinococcus* isolated from domestic livestock in Kazakhstan. J Helminthol 94:e69

Anvari D, Rezaei F, Ashouri A, Rezaei S, Majidiani H, Pagheh AS, Rezaei F, Shariatzadeh SA, Fotovati A, Siyadatpanah A, Gholami S, Ahmadpour E (2021) Current situation and future prospects of *Echinococcus granulosus* vaccine candidates: a systematic review. Transbound Emerg Dis 68(3):1080–1096

Avcioglu H, Guven E, Balkaya I, Kirman R, Akyuz M, Bia MM, Gulbeyen H, Yaya S (2021) *Echinococcus multilocularis* in Red Foxes in Turkey: Increasing risk in urban. Acta Trop 216:105826

Azizi H, Kazemi B, Bandehpour M, Mohebali M, Khamesipour A, Aryaeipour M, Yaghoobi H, Rokni MB (2016) Modulation of the immune response to DNA vaccine encoding gene of 8-kDa subunit of *Echinococcus granulosus* antigen B using murine interleukin-12 plasmid in BALB/c mice. Iran J Parasitol 11(4):480–489

Gan W, Zhao G, Xu H, Wu W, Du W, Huang J, Yu X, Hu X (2010) Reverse vaccinology approach identify an *Echinococcus granulosus* tegumental membrane protein enolase as vaccine candidate. Parasitol Res 106(4):873–882

Hassan H, Al Hadithi TS, Al Sakee HM (2016) Experimental trial with a heat-shocked protoscolex extract as a vaccine candidate for protection against hydatid disease. Turkiye Parazitol Derg 40(1):1–8

Kandil OM, Abdelrahman KA, Mahmoud MS, Mahdy OA, Ata EB, Aloufi AS, Al-Olayan E (2020) Cystic echinococcosis: development of an intermediate host rabbit model for using in vaccination studies. Exp Parasitol 208:107800

Larrieu E, Gavidia CM, Lightowlers MW (2019) Control of cystic echinococcosis: background and prospects. Zoonoses Public Health 66(8):889–899

Mehlhorn H (2016) Animal parasites: diagnosis, treatment, prevention. Springer, Cham

Mehlhorn H (2023) Human parasites: diagnosis, treatment, prevention, 2nd edn. Springer, Cham

Mehlhorn H, Eckert J, Thompson RC (1983) Proliferation and metastases formation of larval *Echinococcus multilocularis*. II. Ultrastructural investigations. Parasitol Res 69(6):749–763

Pourseif MM, Moghaddam G, Saeedi N, Barzegari A, Dehghani J, Omidi Y (2018) Current status and future prospective of vaccine development against *Echinococcus granulosus*. Biologicals 51:1–11

Pourseif MM, Yousefpour M, Aminianfar M, Moghaddam G, Nematollahi A (2019) A multi-method and structure-based in silico vaccine designing against *Echinococcus granulosus* through investigating enolase protein. Bioimpacts 9(3):131–144

Tamarozzi F, Deplazes P, Casulli A (2020) Reinventing the wheel of *Echinococcus granulosus* sensu lato transmission to humans. Trends Parasitol 36(5):427–434



J. Li et al. (eds.), *Echinococcus: Control and Elimination of Echinococcosis with a Focus on China and Europe*, Parasitology Research Monographs 19,
https://doi.org/10.1007/978-3-031-54479-8

Toews E, Musiani M, Checkley S, Visscher D, Massolo A (2021) A global assessment of *Echinococcus multilocularis* infections in domestic dogs: proposing a framework to overcome past methodological heterogeneity. Int J Parasitol 51(5):379–392

Wen H, Vuitton L, Tuxun T, Li J, Vuitton DA, Zhang W, McManus DP (2019) Echinococcosis: advances in the 21st Century. Clin Microbiol Rev 32(2):e00075–e00018

Woolsey ID, Miller AL (2021) *Echinococcus granulosus* sensu lato and *Echinococcus multilocularis*: A review. Res Vet Sci 135:517–522

Yu M, Zhu Y, Li Y, Chen Z, Sha T, Li Z, Zhang F, Ding J (2021) Design of a novel multi-epitope vaccine against *Echinococcus granulosus* in immunoinformatics. Front Immunol 12:668492

# Index





J. Li et al. (eds.), *Echinococcus: Control and Elimination of Echinococcosis with a Focus on China and Europe*, Parasitology Research Monographs 19,
https://doi.org/10.1007/978-3-031-54479-8

SPRINGER NATURE

# GPSR Compliance

*The European Union's (EU) General Product Safety Regulation (GPSR) is a set of rules that requires consumer products to be safe and our obligations to ensure this.*

*If you have any concerns about our products, you can contact us on ProductSafety@springernature.com*

In case Publisher is established outside the EU, the EU authorized representative is:

Springer Nature Customer Service Center GmbH
Europaplatz 3
69115 Heidelberg, Germany

The manufacturer's authorised representative in the EU is Springer Nature Customer Service Centre GmbH, Europaplatz 3, 69115 Heidelberg, Germany. If you have any concerns regarding our products, please contact ProductSafety@springernature.com

Printed and bound by CPI Group (UK) Ltd, Croydon, CR0 4YY
01/10/2026
02225899-0005